A Simple Guide to Technology and Analytics

A Simple Guide to Technology and Analytics

Brian J. Evans

CRC Press is an imprint of the
Taylor & Francis Group, an **informa** business
A CHAPMAN & HALL BOOK

First edition published 2022
by CRC Press
6000 Broken Sound Parkway NW, Suite 300, Boca Raton, FL 33487-2742

and by CRC Press
2 Park Square, Milton Park, Abingdon, Oxon, OX14 4RN

CRC Press is an imprint of Taylor & Francis Group, LLC

Library of Congress Cataloging-in-Publication Data

Library of Congress Cataloging-in-Publication Data
Names: Evans, Brian J., author.
Title: A simple guide to technology and analytics / Brian J. Evans.
Description: First edition. | Boca Raton, FL : CRC Press, 2022. | Includes bibliographical references and index.
Identifiers: LCCN 2021019341 | ISBN 9780367608613 (hardback) | ISBN 9780367634766 (paperback) | ISBN 9781003108443 (ebook)
Subjects: LCSH: Automation. | Electronic apparatus and appliances. | Computer science.
Classification: LCC T59.5 .E88 2022 | DDC 670.42/7—dc23
LC record available at https://lccn.loc.gov/2021019341

ISBN: 978-0-367-60861-3 (hbk)
ISBN: 978-0-367-63476-6 (pbk)
ISBN: 978-1-003-10844-3 (ebk)

DOI: 10.1201/9781003108443

Typeset in Minion Pro
by KnowledgeWorks Global Ltd.

Contents

List of Figures

List of Tables

List of Equations

Acknowledgements

During the writing of this book, I tried to make my explanations of technology simple and understandable. My wife Margaret read every word which was very tiring for her, especially since my command of English grammar (as an engineer) was not always ideal. So I really thank her for all the corrections she made and in turn, I am sure she will thank me for finally finishing it.

This book was intended to explain technology and analytics in a simple manner, but the reality is that, for the reader to maximize their knowledge, some basics of this complexity must be understood. The most complicated chapter in the book takes the reader from an understanding of simple control, to designing fully automated control systems. I would like to thank Professor Ahmed Barifcani for reviewing the automation section, and bringing my attention to the errors I had made. Ahmed is an outstanding engineer and I really appreciate his input and help with getting the details right.

I also acknowledge my boyhood friend – marine surveyor Mike Wall. Mike is a globally-known marine consultant and has written many books about marine issues. It was through his tips about making the book readable, that I hope I finally got there.

Acknowledgements

About the Author

Before I studied at university, I completed an apprenticeship as an electro-mechanical draftsman with an electrical switchgear company in Liverpool UK, leading to my being able to repair all manner of electrical devices. Then during my studies, I worked part-time as a student engineer for a local power company and graduated in 1969 as an electrical control design engineer. My first job was as a graduate automation control engineer and I thought all the world was automated. It wasn't.

I worked for GEC Control Systems in Bromsgrove UK which involved designing automated control circuits to run paper mills and newspapers (an automation engineer was not a trendy job then like it is these days). I was part of a young team using the latest electrical and electronic systems of the 1970s to maintain a smooth automated processing flow. I then transferred to GEC Automation in London with the task to automate the London traffic lights. I was part of the team that tested new 'digital computer' equipment to make London traffic lights automated, so that there would be no traffic jams.

I worked with the new *MODEM*s- that is '**mod**ulator-**dem**odulators'. A MODEM consists of a complex circuit which changes the electrical output as cars travelled over wires at traffic lights into pulses fed into a counter which when filled by a set number of cars, caused a circuit to change the lights according to traffic flow – to keep traffic moving and not jam.

We tested this at a T-junction on London Heathrow airport's Ring Road which worked reasonably well and we linked it with a large message display gantry so that each convoy of vehicles would enter the lights at a defined speed. This way traffic flow in London would <u>never</u> be jammed. Then we installed the first overhead gantry on London motorway M1, and it worked well for a few hours with traffic changing speed when we changed the speed limits. After a few days of tests, we handed this to the local police station to set future commands and speed limits. Much to our surprise 24 hours later, the M1 had its first-ever traffic jam just around the spot where we had raised the gantry. The police apparently had fun constantly changing the speed limit. I stayed with GEC for only a short time after this, because I didn't think I would have a great future.

In 1970, automation i.e. electrical control systems, was rarely used in anything other than process control, where the process was steady and continuous (such as filling beer bottles). Computer systems were large and bulky – the 32K core-store system I operated consisted of a copper wire matrix of 32,000 junctions of wires crossing each other in the x, y and z directions with a toroidal magnet holding them at their junction. One day I came to work and after pouring a coffee I turned to pick the cup up and I nudged the computer cabinet with my elbow. I watched as the line of flashing lights above each switch turned off. I had vibrated the cabinet so much that the cores had touched the crossed wires, and had shorted out losing their electromagnetic fields.

Oh catastrophe for me, my job was on the line! I then started re-coding the machine for the next two hours pushing the switches up for a '1' and down for a '0' followed by the last switch which represented an 'enter' button. So I got to know how to treat a computer system with care and how to panic re-program. I decided to look for a job since I was clearly not cut-out for a computing career.

North Sea oil and gas production was expanding and money could be made if I got onto those rigs. I got a job in Paris and then Dallas, subsequently enjoying life as a rig geologist/mud logger and later as an offshore geophysical operations engineer, travelling the world trouble-shooting geophysical equipment failures, and had some of the most fascinating times.

I set up company in Australia as an oil and gas consultant based in the seismic industry which to this day still uses the most complex computer and data recording/processing systems.

Eventually I took Masters and PhD degrees in Geophysics at Curtin University, which involved building recording systems and using different data analytical methods to evaluate the recorded data. So during my career, I kept abreast of the technologies involved in everyday data recording/processing and analysing. From then on I lectured data acquisition, processing and analysis. In the 1990s, I became a Professor of Geophysics and Head of Petroleum Engineering, retiring in 2019, so new *technology* and *data analytics* has been in my blood since I was in my 20s.

During 2019, it appeared to me that many of the *smart* new technologies were deliberately confusing. All things can be explained with basic physics and you don't need to be smart to understand them. Fellow educators referred to *STEM* (Science, Technology, Engineering and Mathematics) as being the way of the future- but they didn't explain why. These educators were more often administrators who had little knowledge of the practical technologies. It is said that in 2012, Exxon automated one-third of their offshore Gulf of Mexico production platforms in order to operate them from onshore, and within three years the cost of this automation was fully recovered. To me, this was automation in practice and the idea of *smart* technology was only their computer control (like the London traffic lights).

I attended seminars on STEM but they were focused on sales of data systems and computer software. They did not embody my understanding of STEM, which was *the application of knowledge to control a process.* Consequently, I started putting together a book explaining in simple terms, the meaning of *smart* and technological devices. All devices use simple physics and hence, the real science of *smart* and how *analytics* is part of that smart, is really very easy to understand.

It took me a year to put in words, simple explanations of technology in everyday words. But I do hope it doesn't take you a year to read it! After reading, please use it as a reference book because that is how it is probably best used, and the Glossary should help as a simple ready-reference dictionary.

Preface

Dear reader, I hope this book is beneficial to you. The book is written in simple English, to help explain everyday technologies and the application of analytics with illustrations, so anyone can understand how they work. At the end of most chapters, I have inserted an exercise or two which are practical examples of technology discussed in the chapter. I have also put some ideas for further reading after the exercises if you are keen, making sure that most of them are on the web so that you don't have to work hard to get them (just copy and paste from an e-book). At the back of the book, there is a Glossary so that if there is a word in technology that seems trendy, complicated or weird, I have tried to explain it in simple everyday words. The back of the book also has the worked answers to the questions as well as the Index.

After reading this book, and maybe doing the exercises, you will understand more about the technologies you are typically using, and subsequently make better choices with future technologies you may buy. In the new computer age of *quantum computing* (introduction around 2030), you will be able to do all manner of things we can't do now- so an understanding of technology will help you and your family use technology better in the future.

But to understand where we are going on our technology road, you have to understand the basics. So at the start of this book, we have the basic background of technology in Chapter 1. This is a short history of computers and networks, and their relative speed because it is computer network speed increase that is becoming the driver of technology and automation. Technology uses this speed change in practice, and the industry which is adopting the new technologies fastest is 'sport'- which is now using *data analytics* of players to get ahead and win trophies.

To watch sport we mainly use TV monitors. The TV broadcast companies have realised that sport action (and therefore paying customer numbers) can be improved by showing more game analysis. On the way, we learn about how TV monitors are being improved to give a better viewing experience. So the difference between types of TV, phones, towers and radiation issues all form part of that explanation.

The book tries to steer clear of equations because this book is for the average person who isn't involved with working out mathematical details- except maybe when we are driving our car and figuring out if we are exceeding the speed limit. So to understand speed, we must understand other basics such as what frequency means and how it works (the faster we go, the faster the wheels turn). Showing sports on the TV is enhanced by viewing the action from different angles. So the book explains how we use multiple camera positions to see the action in greater detail.

Chapter 2 then takes up the theme of how these multiple cameras are used, and how we calculate simple statistics like ball speed and trajectory- a necessity when we are using *Hawk-Eye* to predict if a tennis ball lands on or outside a line. In today's sports, we want to feel as if we are in the sports stadium. Cameras are now giving multiple shot locations so we can re-run a play in a game using simulated 3D. In the future with *quantum computing*, this will evolve into a full 3D *immersive experience* for the viewer (where you can be either in the coach's box or on the terraces).

Tracking players and the ball allows us to work out relative player positions, but how do we do that? Technology of today allows us to work out the statistics (aka *stats*) to do this. And then there is the speed gun, so Chapter 2 looks at how we can track not just a ball but any object in 3D, if we have sufficient camera positions. This also involves satellite tracking so there is a brief explanation of how satellites work.

Having images of action on a field is further enhanced if we can somehow track them across the screen. The question of how we can do this is answered through *pattern recognition* which is explained in Chapter 3 so that the reader can understand how computers are able to recognise patterns very quickly. *Pattern recognition* is used in everyday life (such as at the Immigration Gate) so the book simply explains without equations, how the computer automatically recognises people and how we can do our own simple pattern recognition- it's not hard.

The *pattern recognition* method needs its input data to be analysed first, so methods to analyse and then predict future events (such as where a ball will travel) is explained in Chapter 4. This is one of the most technical chapters, explaining in simple terms how we perform *predictive analytics*.

Since we now know how to make simple predictions, the prediction of what happens next with the ball on the sports field is applied in Chapter 5 along with some of the new evolving technologies such as *Reality Technologies* and *3D visualisations*. This leads then into the most common sensors we have, with natural, active and passive sensors being explained in Chapter 6 from first principles, as well as how we transmit, process and display their data. The incredible future of automation will be made possible by *quantum computing*, so the chapter finishes by explaining the concept of *quantum computing* and how it may be used to revolutionise the speed of everything.

Automation and *simulations* are then reviewed in Chapter 7, which takes the reader from the basic process control system towards how to automate a typical system which is presently manually controlled. This is probably the equal hardest chapter (alongside Chapter 4) for the non-technical person, which then finishes with introductions to the *digital twin*, *blockchain* and issues involving *cybercrime* which are part of our expected future – all explained in the simplest manner.

Chapter 8 is devoted to explaining how household appliances work. As is often said, there is more computer technology in the everyday washing machine controller than there was aboard the moon-landing Apollo 11 vehicle and its pod. So it is good to know how typical kitchen appliances work, so that when the time comes to replace them, a good choice can be made of their replacement based on knowledge. This is the point of this chapter that may be the most useful of all.

Finally, the last chapter discusses the future of living in the 2040s and onwards, after the new quantum technologies are developed by the use of the new *quantum computer.* There will be new forms of communication through *holograms* and *haptic technologies*, and some of the concepts displayed in the Star Wars movie, are in fact possible in the future.

There will be new games to play in which *3D visualisation* will be simply remarkable. The *smart* new technologies of the 2040s and onwards will make today look pedestrian. I hope to live to see my dream come true. After reading this book, you will be abreast of all of the present technologies and ready to forge your own in the future, so please enjoy the book like I have done so writing it.

Learning Outcomes

This book was designed as a simple guide for the person who has some basic understanding of everyday technology, but wants to get a better understanding of the new *smart* devices. It does not go into great depth in each technology, and has just a few well known basic equations (such as *Pythagoras*) to help the reader understand the next step. It is well illustrated with over 200 graphs, photos and exercises all compiled by the author to explain how *analytics* works with *technology* to understand what has happened and then *predict* the future. This is a requirement of *technology* and *automation*.

After reading this book, you will:

- Have a better understanding of the most common *smart* technologies that use simple physics and analytics. There is even an easy-read chapter on kitchen appliances.
- Be aware of the changes in data transmission networks that have caused *smart* devices to go from *3G*, to *4G*, to *5G*, *Bluetooth*, and *wifi* and then be able to assess the future transmission technologies as they are developed.
- Appreciate how the frequency spectrum is used in new tools to employ *infrared*, *ultraviolet*, *laser* and potentially *haptic* technologies.
- Understand how *pattern recognition* works, which is used for example: to pick-out faces in a crowd, in security services for passport ID checks, in finger-print recognition and in automatically analysing results from surveys that had been previously labour intensive.
- Be able to work out the basics of the science behind predicting future trends, with applications not only in sports but also in industry and the financial sector.
- Become familiar with the physics of over 100 common sensors that are used by industry, and how they are used to automate processes.
- Understand how robots will fit in our future, ranging from a sensible discussion with *Siri* to physically playing 3D *haptic* football or cricket with the greatest players of past and present.
- Know what is meant by the *STEM* buzz words, including *machine learning*, *recursive filtering*, *artificial intelligence*, *deep learning*, and *quantum computing*.

- Be familiar with how a process can be automated, including the parameters of equipment selection and remote operation.
- Have the tools to consider the future of an automated world, possible processes yet to be invented, and how that could affect your daily life.
- Be appliance-literate and able to make choices about which home kitchenware you want, based on simple everyday logic and knowledge.
- Have an immediate reference Index which can be referred to when someone uses a new 'trendy' technology word you don't understand.
- Have access to further reading, mainly from the internet so that an e-book will link with the web site for ease of additional reading when needed.

CHAPTER 1

Background to Technology

1.1 OVERVIEW OF BASIC TECHNOLOGY AND WHY THE RAPID INCREASE IN NETWORK SPEED IS IMPORTANT

1.1.1 Introduction to Basic Technology

Over time, new materials are being discovered or developed, with new properties which can be used or adapted to some new form of improved technology. For example, over the last couple of decades, the US *National Aeronautical and Space Administration* (NASA) *Apollo* space program has applied *Teflon* (Polytetrafluoroethylene – PTFE) coating to the heat shield of its earth re-entry vehicles because *Teflon* takes a high level of heat before it melts as well as an ability to have objects bounce off it. In similar manner, a spacecraft has a metal surface and on re-entering the atmosphere passes through the sky and clouds so fast that the friction on its surface can get very hot (Figure 1.1) and it needs a surface which causes particles coming past it to bounce off.

Teflon was actually discovered in 1938 by chance in a DuPont Company [(1)] lab as a heat-resistive coating for metals. Its everyday use was not really recognised until it was adapted for re-entry vehicle heat shield use by NASA, and thereafter was used in common everyday life as the surface coating of non-stick household frying pans.

It takes typically 20–30 years for these sorts of technologies to develop from initial material discovery to broad acceptance and technology application in the community. A new technology often has spin-off effects leading to the development of other technologies. Technologies can sometimes add on, adapt, and improve some aspect of existing technologies over time – computers being an example. It has been well accepted that the first electromechanical computer was built in Germany in 1938 by Konrad Zuse called the Z1 but didn't progress much further. The first electric programmable computer (Figure 1.2) was called *The Colossus* and was developed in 1943 by Alan Turing (an English mathematician who broke the German secret code produced by the German *Enigma* machine – made famous in 2015 by the movie *The Imitation Game).*

Up to the early 1950s, computers were mainly developed to solve mathematical problems, and were built for official government work programs. Many other machines were developed over the years until the first industrially commercial computer using transistors

DOI: 10.1201/9781003108443-1

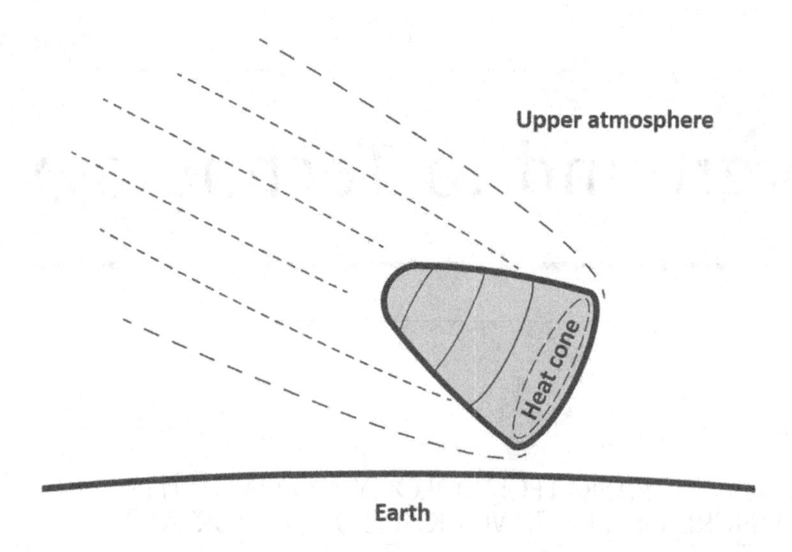

FIGURE 1.1 Space capsule returning to earth. (Courtesy: C. McCartney.)

was demonstrated in 1956 by *Massachusetts Institute of Technology* (MIT). This was called the *TX-0* (Transistorized Experimental Computer). The use of small transistors replaced the use of large electric control vacuum tubes or valves, so had the effect of reducing the size of the computer while making it faster. The new technology in this case was the transistor, which allowed much smaller desktop computers to be developed, the first by *Olivetti* in 1964 called the *Programma 101*, followed by the *Hewlett Packard HP9100A* in 1968, which was the first mass-produced and marketed desktop computer.

Because a lot of labour was involved with soldering the transistors to printed circuit board metal tracks, companies started developing very small transistors embedded in insulated

FIGURE 1.2 Colossus. (Courtesy Geoff Robinson, photography/Shutterstock.)

materials which were then called an *integrated circuit* (IC) or *chip* so that mass production line printed circuit boards could be made, allowing a major increase in the number of transistors on the same board, producing what is now known as the *micro-processor* IC. By 1975, portable calculators were using these processors, with some calculators such as those made by Texas Instruments being mounted on small printers so that their computation results could be printed soon after calculating. This led to many different types of mobile devices being developed. In that same year, *International Business Machines* (IBM) introduced its first portable computer, the *IBM 5100* series which had a monitor screen and tape drive for reading/recording data (Figure 1.3). IBM continued to lead the world market making large computers for business, before introducing its first *Personal Computer* (PC) in 1981 (*the Acorn*) followed by the first laptop in 1986. IBM computers were considered expensive at the time, and often cheaper PCs retailed for a third of the IBM price. For example, the first *Apple 1* computer retailed at $666.66 in 1976 but was little more than a single printed circuit board which required the purchase of a separate power supply, display, and keyboard. You had to be technically competent to put it together so it wasn't easy to make a PC.

IBM PCs were recognised as very dependable and well-built machines, not liable to early failure as the alternative circuit board type of PC had been. However, they were the most expensive but now the IBM machine set the bar in personal computing – however, its software required the user to have some knowledge of computer languages (thereby limiting its general use). Soon after, *Apple* developed its *Windows* software being the most user-friendly software compared with the IBM software (which used *Microsoft Disk Operating System* [MS-DOS]). By the late 1980s, *Microsoft* used a *Graphical User Interface* (GUI) overlay screen to give it a *Windows* appearance and capture the market with cheaper PCs but using user-friendly *Windows* – this now made cheap user-friendly computing available to the general public and *Apple's* sales started taking a dive. In 1998, a US court ruling over the *Microsoft* copying of the *Apple Windows GUI* made *Microsoft* leader Bill Gates

FIGURE 1.3 IBM 5160, the first IBM Personal Computer. (Courtesy: Shutterstock.)

purchase $150 million of Steve Jobs' *Apple* shares – so in fact, *Microsoft* continues to have a large investment in *Apple*. This was serendipity since although smaller than *Microsoft* at the time, *Apple* later expanded introducing the *smartphone* to become the larger company of the two – as it still is.

The evolution in computing brought increased functions and speed in which financial banks and other data processing centres expanded. At the same time, computers were being tied to each other, so that the output of one computer performing operations could go to another computer performing other operations. Different industries were built out of this computer *networking*, such as payment of shopping bills by automatic payment cards which was easier than paying by cheque or cash when shopping, the payment operator (e.g. *Visa*) would be one company, while the actual payment would go from a bank account of another company.

As computers ran faster, the use of analogue copper-wired networks (like wires to telephone lines) to connect them together was relatively slow (and still is!). The rate of transfer of data from one computer to another and the delays experienced in the computing transaction is called *latency*. This latency between connecting computers needed improved technology to make the data transmission faster, so instead of using an analogue network, the digital network was developed (after all, computers became digital as soon as transistors were developed – this will be discussed in later chapters).

It wasn't long before embedded transistors became much smaller, the concept being that if you can put electronic circuits physically closer to each other, the distance the signal travels becomes shorter and so the IC becomes faster and computer latency reduces. Since 1965, *Moore's Law* (named after founder Gordon Moore of Fairchild Semiconductors) states that "the number of transistors doubles every two years", which results in doubling the speed of computing every two years. Figure 1.4 is a graph showing the number of transistors they

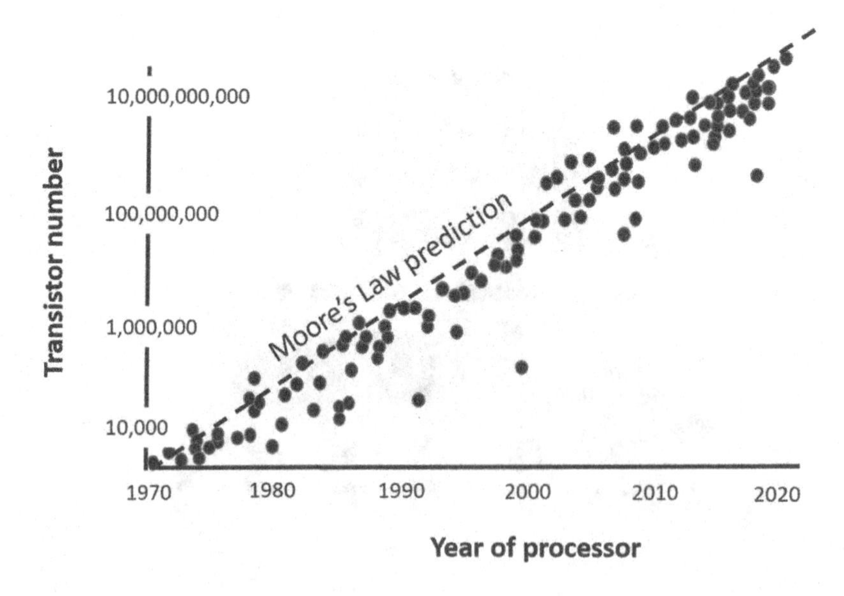

FIGURE 1.4 Computers follow *Moore's Law* in which transistor numbers double every year.

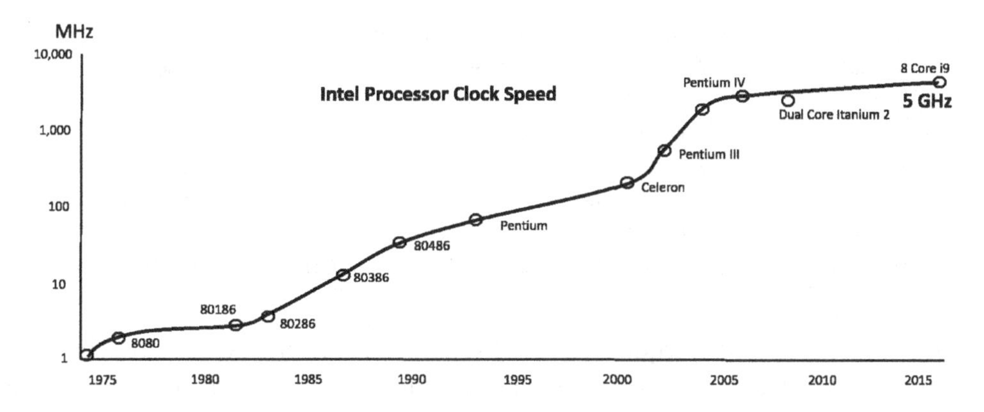

FIGURE 1.5 Intel CPU trends indicate that speed increase has almost flattened.

have been able to fit into the same sized space each year and it can be seen that the slope of *Moore's* Law is pretty linear. However, how long that will continue is questionable.

Over the years, different computers have had faster ICs fitted, and this branded the computer type. For example, as more transistors were packed into miniaturised ICs, the *computer processing unit* (a CPU contains many ICs) became faster and each version was given a name – a first example being the *Intel 286* computer. Then *Intel* named their processors in grand manner, calling them *Pentium* computers, followed by the faster *Pentium 4*. To make them run even faster, the idea was then to have two CPUs together, so that became known as *Dual Core* operations (Figure 1.5). This running of two blocks of ICs instigated the idea of putting a number of computer CPUs together, and modifying the computer code so instead of computer instructions being single line instructions in a *serial* manner for one computer CPU to work with, the code addressed a number of CPUs at the same time (*in parallel*). This then led to the term *parallel computing* or *supercomputer*. As can be seen in Figure 1.5, computer speed has grown steadily since the late 1970s, but the speed of growth has reduced recently with 8-core computing at around 5 GHz speed, the maximum for a single CPU.

1.1.2 What Is Network Speed, 3G, 4G, 5G, *Wifi*, and *Bluetooth*?

With so many computers talking to each other becoming faster and faster, it became clear that the blockage on faster computing was a result of the speed of computers actually communicating with each other. The first generation of computer data transmission was via the copper-wired telephone lines and modulators/demodulators (*MODEMS*) which was very slow. It could take a half minute or longer to just get a connection, so in order to make sure the user knew something was happening, modems were wired for sound and produced beeping noises to inform the user it was actually working. So far data transmission had used analogue street telephone lines which were inherently slow and were, in fact, the first generation of data transmission (now recognised as *1G*).

So, in 1979, the first digital network transmission towers were erected in the United States, and it took over a decade for them to be used around the world. The 1G analogue

TABLE 1.1 Speeds for Each Transmission Generation

Generation	Bit Rate	Type	Year	Transmission
1G	2.4 kbps	Analogue	1979	
2G	64 kbps	Digital	1993	25 MHz
3G	144 kbps–40 Mbps	Digital	2001	25 MHz
4G	100 Mbps–1 Gbps	Digital	2009	100 MHz
5G	10 Gbps	Digital	2016	30–300 GHz
Bluetooth	25–250 Mbps	Digital	2001	2.4 GHz

networks had sent data at an equivalent of 2400 bytes per second, i.e. 2.4 Kb/s (bits/bytes and computer data words will be explained later). But the second generation (*2G*), which was the introduction of the digital network in 1993, transmitted data via telecommunication satellites and towers in digital form at a bit rate of 64 kbps – over 20 times faster than the analogue network. Table 1.1 shows how the improvement in the use of ever faster satellite and transmission equipment increased the speed of operations around every eight to ten years. The higher frequencies of transmission resulted in reduced latency so that computers transmitting data to each other could do so at ever faster rates. This reduced the time for a credit transaction at a supermarket checkout counter to be reduced from 15 seconds to 1 or 2 seconds. While the computers were doubling in speed every two years, the transmission network lines were not.

In Figure 1.6, it is clear that there has been a major change in speed between 3G, 4G, and 5G networks. This is simply because the 5G network uses higher transmission frequencies than 4G, and has many more towers (aka *antennae*) that are closer to each other than 4G had. It is of interest to note that one of the world's major suppliers of 5G – *Huawei* (a company owned by the government of China) holds no patents of the technology (two owned by US companies and *Samsung* of Korea owns the other), yet they have the full technology – this is one reason why a number of nations do not want *Huawei*-installed 5G networks (because they haven't paid their dues). The other reason is that in every 5G contract with *Huawei*,

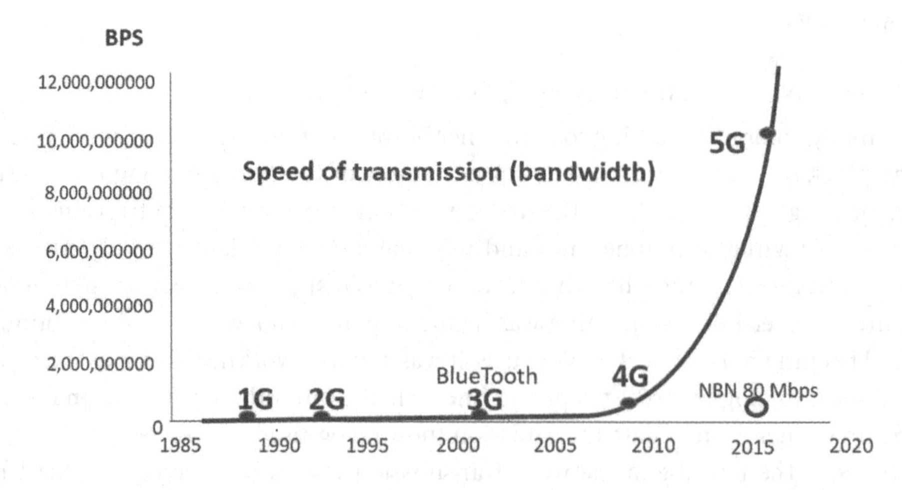

FIGURE 1.6 Network Speeds graphed.

the hardware agreement with *Huawei* is that it can be switched off at any time at the behest of the company owner (the Chinese Communist Party). Hence, *Huawei*-installed networks can be terminated as required by the Chinese government, which could be an important action at the time of any conflict.

It may also be noted that some other country networks (labelled here as NBN – *National Broadband Network* – acronym for the network in Australia) that use landline fibre-optic cables rather than satellite are much slower than the 5G networks. A major problem with nationally introduced landline fibre-optic systems is the cost of upgrading the controlling electronics along the line. This is both more expensive and slower than the launch of a new satellite with new faster electronics to replace old satellites with old electronics. As such, satellite network systems will always be faster than landlines and as computer speed increases, the fibre-optic lines will always lag behind and have greater latency.

The main controlling factor in data transmission is *operation frequency* (2). This will be discussed later, but it is enough to say that, as transmission frequency increases, more data bits can be transmitted in the same period of time and so the network transmits data faster. The first invention to make a computer talk to a remote computer in the same room without a wired connection was patented by John O'Sullivan in 1992, a research engineer with the *Australian Commonwealth Scientific Industrial Research Organisation* (CSIRO). The technique allowed one computer to connect to other computers within the CSIRO laboratories and was named *Wireless Fidelity* (Wifi). It also connected to its wired equivalent known as the *Ethernet* (introduced in 1983), which eventually connected directly to the internet.

To allow mobile phones to talk to the *local area network* (LAN) at both transmission and receiver ends, short-range (less than 10 m) *Bluetooth* was developed by *Ericsson* and first used commercially around 2001, in the *IBM ThinkPad* laptop computer. It has a speed similar to 3G, having been introduced around that time with those electronics, and, so with upgrades, would soon be expected to be around the 4G–5G rates. This slow transmission from your computer to your computer network may be blamed for slow speeds rather than the network itself. One way to check *wifi* speed is to do a *Speed Test* on a mobile phone. Ideally, a local speed transmission test has an *up-load* and *down-load* speed. If we run a speed test on the mobile phone using the same *wifi* placed next to the computer, it will appear as shown in Figure 1.7 where the speed is a very reasonable 46.51 Mbps. My own transmission speed is

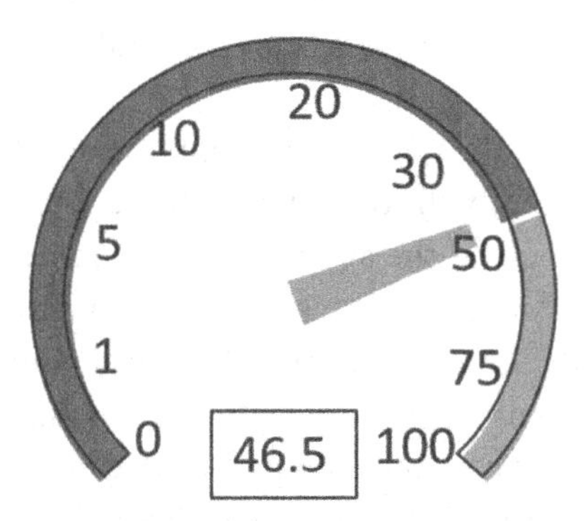

FIGURE 1.7 Data transmission speed dial.

around this number if I stand next to the *Bluetooth* and *wifi* transmitter, but in reality, my computer operates in an office with two brick walls between it and the transmitter – causing the speed to drop to less than 10 Mbps. At times, this transmission speed can reduce to under 7 Mbps when the internet freezes and transmission *drops out* (it can't get through the walls). The only way to solve this is to have a transmission booster which amplifies the signal from the transmitter to the computer. Otherwise, it is best to move the *wifi* transmitter in the line-of-sight with the computer or vice versa if possible.

1.1.3 Advantages of Higher Network Speed

A rapidly developing computer speed requires a fast network so that computers can talk with each other equally fast. If the network speed is relatively slow compared with the computer, then the complete computer-to-computer data transfer rate is reduced by that speed. The old maxim that during World War II, 'the slowest ship in any convoy across the Atlantic slowed the whole convoy of ships to the same slow speed', still works today. If any part of a network is slow, all of the network slows to its speed.

A slow landline or *wifi* will retard a computer's overall speed, despite the increasing speed of MODEM electronics. So, the whole system needs to be upgraded, often in practice piece by piece rather than at the same time. When 5G was introduced by switching frequencies higher and putting additional transmission/receiver towers in the ground or on top of buildings, the whole network became much faster to the point that existing supercomputers were now becoming more useful than ever.

By increasing speed of TV screens and monitors, more information can be provided in real time. By having faster communications, more data can be analysed and more mechanisms can be controlled remotely. Now, for the first time, we could lecture remotely to students in real time with a low level of latency – seminars became *webinars* with multiple students sitting through lectures at home especially during the time of the COVID-19 virus. Activities could be monitored remotely – such as bringing seminars via the internet as *webinars* to the mobile phone, and special remote monitoring allowing videos for the first time to be broadcast to a *smartphone* of people passing or knocking on your front door. The ability to monitor 24/7 was not lost on the security guards of London City, where all of the streets of the downtown area were security monitored 24/7 for 365 days of the year by 2012 This helps not just security but also could help resolve traffic problems. So a very fast network speed is essential and has been the catalyst for a lot of the new technologies that are being introduced almost daily.

However, the introduction of the *Quantum computer* over the next 20 years, which will have an astonishingly fast speed, will force the continued need to have ever faster transmission lines. The end point will be the application of Quantum electronics to fibre-optic transmission in their MODEMS and satellites. That is when, finally, the fibre-optic landline will equal the satellite speeds. More about the Quantum computer later. But the benefit of having such high-speed communications between computers (whether they are used in a TV, monitor, mobile phone or electronic pad) is that multiple pictures and images of events will soon be able to be displayed in real time along with statistical analysis of what is happening at that time. There will be real-time vision updates of events happening on the other side of the world, as they happen. This is what the industry calls *connectivity*, and so it is important to have the best and optimum information available, so that we can use that data for our own benefit. This calls for a high level of understanding of technology and its applications in data analysis.

1.1.4 Why STEM and Analytics Have Become Important

The miniaturisation of CPUs with lots of *chips* causing the ever-increasing speed of computer data transfer has resulted in computing speeds becoming faster and faster. When computers talk with each other, they need to be equally fast. However, it is only recently that computer transmission networks have started to catch up, and the increased transmission speed of the 5G network has resulted in the increased ability for computers to transmit to each other very fast. Having very fast transmission of data allows other processes to develop as a by-product. In the past, local area *Bluetooth* has connected the mobile phone or computer with the *wifi* server, but as *Bluetooth* increased in speed with the faster 5G network, remote monitoring and recording of local industrial operations that were not possible before 5G became more possible. The more automated a process becomes, the more remote control and *data analytics* play a part.

As a result of the need to understand and implement automated processes, during 2019 and 2020, there was a push to have more *Science, Technology, Engineering, and Mathematics* (STEM) graduates from universities. However, what was really needed was more good quality *Mechatronics* graduates, that is, technical workers and management who understand the mechanics of the problem and then use mechanics with electronics to make a process automated (mathematics, engineering, and technology are already taught as part of the mechatronics courses, as well as the computing to implement it). Lectures on the benefits of automation have entered the management schools but one wonders why this is not more widespread when it is management who have to decide on providing the finances to automate.

In order to effectively automate, we need to understand the various forms of technologies that are available to be automated. So the next section (Section 1.2) will introduce some of the technologies required for remote data transmission and the basic tools of data analysis which are commonly applied in today's technological world.

1.2 APPLICATION OF NEW TECHNOLOGIES IN SPORT AND INDUSTRY

1.2.1 Introduction to New Technology Analytics

Faster computer and data transmission speeds result in the ability to record and display data in real time, and also to analyse that data – known as *data analytics*. In baseball, if we know how fast and far off the direct track a baseball pitcher can throw a ball, then we know how easily the pitcher can fool the batter so that the batter misses the ball. This fact was not lost on the first data analysts in sport – that was baseball in the late 1890s – when they were trying to find out who was their best pitcher. In those times, it was enough to find a pitcher who could curve a ball in-flight, but today we also need to know how many *Walks plus Hits per Innings Pitched* (WHIP), which is how many base runners the pitcher allows on both hits and walks – this is a proven method of pitcher efficiency since the pitcher who knows how the batter will respond to a ball is ahead in the game. By 1910, teams did better if they had the best pitcher, since analysis of pitcher actions could win games but in today's world more data is needed to provide that edge.

It is similar in terms of industrial practices, where automation can increase the speed of manufacturing goods, and consequently boost profit. After World War II, *Ford Motor Company* found that increasing the speed of the production line using robots to weld steel car parts together (replacing human welders) made production faster because robotic welders were fast, efficient, did not need a lunch break and did not have sick-days off. Even when robotic machines appeared to falter in operation, checks on their operations from a central monitoring room beforehand could use *analytics* to predict when a machine might fail, and so a time could be made on the production line for preventative maintenance of any failing robot. So the concept of *Predictive Analytics* was born, which resulted in robot monitoring equipment being added to the production line, and the ability to optimise a continuous process through *analytics* became accepted practice [3].

"After World War II, *Ford Motor Company* found that increasing the speed of the production line using robots to weld steel car parts together (replacing human welders) made production faster because robotic welders were fast, efficient, did not need a lunch break and did not have sick-days off. Even when robotic machines appeared to falter in operation, checks on their operations from a central monitoring room beforehand could use *analytics* to predict when a machine might fail, and so a time could be made on the production line for preventative maintenance of any failing robot. So the concept of *Predictive Analytics* was born, which resulted in robot monitoring equipment being added to the production line, and the ability to optimise a continuous process through *analytics* became accepted practice."

Today, monitoring any form of action whether in sports or the workplace is normal, but the difference is that now with fast networking, the monitoring of equipment performance can be done at the local level, with major issues being passed on to the monitoring centre for action and potentially human intervention if necessary. This approach makes the whole workplace efficient, allowing human involvement only when non-standard issues arise.

1.2.2 Application to Sports

The application of technology in sports was discussed earlier of how analytics of on-field actions in baseball help the team management to improve overall team performance. When considering the requirements for analytics today, the focus is on being able to see real-time events analysed and repeated to ensure any failure is fully assessed, and resolved. This involves mass gathering of statistics, such as those of the baseball player, the pitcher and the base runners. But at the same time, there are other players on the field, whose job can be equally complicated.

Consider a soccer field where there are 11 players in each team. The front three players are generally chosen for their running speed and their ability to kick the ball while running around (*dribbling past*) a defensive player, all the while controlling the ball. The speed of this player could be observed if the player ran from one point to another in a straight line by the use of a timer, but when having to dodge around another, the timing becomes inaccurate because the player is no longer running in a straight line and so will speed up at times and slow down when dribbling around the opposition. Also, the ability to dribble around another player needs to be assessed – if the ball is passed from one foot to another, the player rounded, and then picked up by the other foot to continue the run – is hard to assess unless you count how many times the ball hits both feet. However, if we put a data transmitting sensor on the player, which can read how many times the player kicks a ball and how fast the player moves, then a statistical database can be constructed of that player's dribbling performance. (Somewhere in there you will need to program the statistics computation software with how many kicks and movements of the body are needed and in which direction.)

This is the same situation for the defender, who can wear a sensor to allow the number of times the player kicks the ball to be recorded and displayed. Whether a defensive manoeuvre is successful or not still requires someone to annotate that action, so even with the technologies we have today, we still require human intervention to ensure the data is correct for the correct result. However, human intervention will not be required in the future as computer networks improve in speed, which will be explained in later chapters.

1.2.3 Application in Industry

New technologies are being developed to optimise the performance and output of continuous industrial operations. For example, in 2012, a major international company operating around 400 large and small offshore platforms in the Gulf of Mexico decided to automate the majority of its offshore operations. Before that time, each offshore platform or vessel would be producing oil and gas from a reservoir deep beneath the seabed. The produced hydrocarbons would have pressure, temperature, flow rate and therefore volume of each platform checked daily by workers on the rig.

However, having workers on the rig required them to work often in extreme weather such as when a hurricane passed through the area, blowing strong gale-force winds, and rain against the rig, apart from the potential to develop sea-storm conditions. The supply of food and other necessities was also an ongoing operation. Under potentially extreme weather circumstances, the production wells would be shut-in by the workers on the rigs, and then the workers would need to evacuate by helicopter or standby vessel to safety. So production would be shut down before the storm's arrival, but what if a storm changed course and did not arrive? The decision process could cause unnecessary loss of income to the production company, apart from the fact that it takes far longer to restart a shut-in well than it takes to shut it down.

The solution was to automate as much of the oil and gas production monitoring and activity process as possible (70% was eventually automated). They started by installing monitoring equipment which could transmit data via satellite to shore-based monitoring control rooms, thereby removing as many personnel as possible. They also included a great number of local monitoring networks on the platforms, so that conventional adjustments of valves to optimise the processes could be done by observers in the comfort of the onshore control centres.

This development was welcomed by the petroleum production industry, since it removed uncertainty in operations and, because production could be maintained longer by the production platforms, the cost of the automation process paid for itself after three years of operation. In addition, many of the platforms or vessels could be continuously checked and using predictive analytics, future problems could be resolved by a skeleton maintenance crew visiting the platforms once a month if needed, using the power of predictive analytics to guide their preventative maintenance scheduling to maximise operations and therefore optimise profits. *Predictive analytics* are the way of such offshore operations in the future to guide maintenance programs and these could not be developed further without the development of new technologies.

1.3 BASIC PHYSICS OF EVERYDAY TECHNOLOGY INNOVATIONS

1.3.1 Basic History of Technology Development Leading to Data Analytics

When new technologies are introduced, they are accepted because there are other supporting technologies which are commonly used at the time. There is no point in having a very smart and fast operating TV camera broadcast lots of high-resolution colourful images of sport when the audience is watching an old black and white analogue TV. So while computers and networks were increasing in speed, similar electronic developments were happening in the TV marketplace. Screens went from *analogue* operations to *digital*, phosphorus-lit screen to *light-emitting diode* (LED) screen. Computing itself went from the *IBM Acorn* screen which was housed within the computer body to individual monitors – from heavy, bulky TV-type screens to the very light flat-screen of today; from large computer mainframe machines to small laptops having similar power. It was mentioned that today's washing machine has more computer power than was housed in the 1969 *Apollo 11* moon landing module. Of course, other parallel technologies advanced at the same time, such as the landline telephone being replaced by the digital wireless mobile cell phone, and this meant that eventually pictures of sporting contests and games could be watched (and in the case of computer games – played) while the viewer was mobile.

Much of the technology was developed during the 1980s for the video gaming industry – that is, the ability to play fast computer games between different people at first on a single computer console, and later in the 2000s between players each having a mobile phone. The games were mainly animations or word-type games, and some gave statistics of player responses to make the game more challenging.

So running statistics during live sports matches, with faster transmission speeds, became of interest to the sports industry, and this was aided by the introduction of the sports performance tracking vest. This will be discussed later in more detail, but it is enough to know that the tracking unit which is worn by each player or athlete on their back or arm monitors their physical condition and movement, and broadcasts this data using local *wifi* so the team coach can use a computer to monitor and record statistics while watching the player move.

For many years before 3G, it had been a slow process to review an action on the sports field by rewinding the recording and re-displaying it, because data was stored on a magnetic tape and rewinding a tape to the correct position was time-consuming. With the use of *compact discs* (CDs) and *digital video discs* (DVDs), it became possible to very quickly jump across the tracks on the disc and have rapid access without rewinding the disc very far. Now with the faster computers/transmission speeds/DVDs, it became possible for a TV camera to observe an action, record it, and rapidly review TV data of interest to the viewer. Since more than one TV camera had been used for decades to view at the same player from many angles, the developing computer and network technologies started using other methods for complementing viewing – starting with animations (and simulations of games were already happening), after which player tracking and player analysis became a major part of the broadcast and professional sports – at first for the coach to improve the team's performance, but later this could be broadcast for the TV audiences to enjoy the subsequent analysis by experts.

An important sports statistic is always speed of action or player reaction. During 1991, IBM used car speed radar to determine tennis serve speed, and from that point onwards, two radars were used on tennis courts to track the speed of the ball once it was hit. Later in 2001, software developed by Paul Hawkins was produced as *Hawk-Eye* to animate a ball hitting a line (since we knew its speed), and so ball tracking became a common tool to use when disputes over whether a tennis or cricket ball or batter was *in* or *out*. The major use initially was in televised cricket matches, but today there are multiple cameras to get all the angles needed to make an analysis of the action as well as a simulation of alternative actions – more in-depth analysis will be discussed later.

So during the 2010 decade, analysing on-screen actions and subsequently providing in-depth data analysis was becoming more widely accepted as a statistical tool to help both coaches and audiences understand both successes and failures of sports athletes. It was noticeable how universities absorbed and further developed player data analysis as part of their sports degrees, and how coaches started developing sports analysis departments since it gave an edge to their team performances (the winner of the English Football League in 2019–2020 season was Liverpool, considered by many to have improved because head-coach Jürgen Klopp was a sports analytics graduate, who set up a department and thereafter used the data analytics to the club's advantage).

However, real-time on-screen data analysis was not possible until recently, because computer networking was still too slow. Software was developed to rapidly view the game from different angles, using different cameras, to give a sense of 3D action. TV even started using data analysis more frequently, with commentators in sports, the financial world and even during the COVID-19 crisis using data analytics to show the weak and strong performances of actions in their respective industries.

This sudden upsurge in the use of data analysis to improve performance was matched by industry. The increase in speed of networks allowed data from industrial processes to be recorded and analysed with rapid display of results. Since the 1970s, it had been possible to use electronics to monitor and adjust continuous processes only (such as the continuous manufacture of paper), but, during the late 2010s, the faster speed of operations introduced by faster CPUs and networks meant that it was becoming possible to control

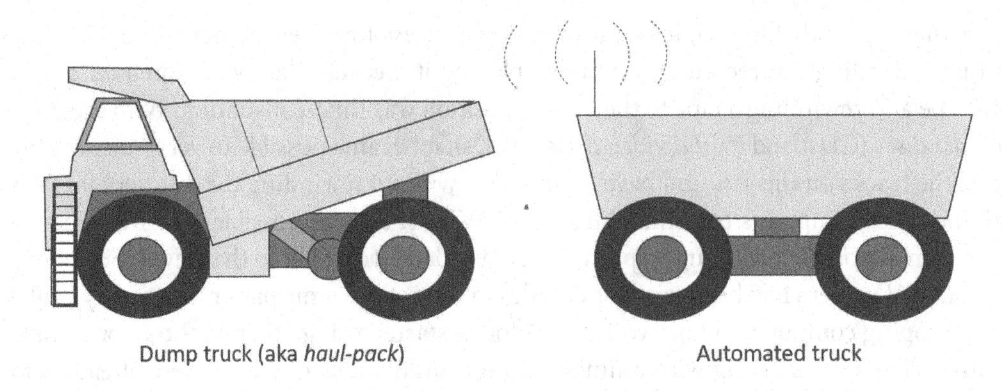

FIGURE 1.8 Autonomous driverless mining trucks are rapidly replacing driver dump trucks.

semi-continuous operations remotely for the first time, and this led to a plethora of companies developing remote operations – true *automation* was finally arriving.

Now we have company departments that implement the remote control of semi-automatic processes, so that remote data analytics is continuously used to optimise the local process, particularly in the mining industry where dump trucks (Figure 1.8) manually driven around the mine site have been replaced by automated driverless, wireless trucks transmitting data and receiving commands from a remote control centre where their operations are analysed and controlled [(4)].

The mining industry argued that, apart from having a driverless cab (having a human driver was a safety issue), mining ore productivity would increase 15%–20%, fuel consumption decrease 10%–15%, tyre wear decrease of 5%–15%, maintenance decrease by 8%, because all of these operational conditions could be optimised. For example, fuel consumption could be monitored so that the distance to be travelled with remaining fuel could be computed to within metres before automatic refuelling. This optimisation would be ongoing 24/7 potentially saving both lives and costs.

New technologies are entering the sports and industry at a greater pace than at any other previous time, and it is becoming clear that if the public is aware of the new technologies and how they work, they too can play their role. We have experienced a rapid increase in technological products over the last 10 years and this is set to continue over the next 30 years to a point when 2000 will really be considered the tipping point of automation and robotics of the future [(5)]. The aim now is to make these technologies understood, so that we can see how they can be further developed to improve our everyday living in the future.

"We have experienced a rapid increase in technological products over the last 10 years and this is set to continue over the next 30 years to a point when 2000 will really be considered the tipping point of automation and robotics of the future. The aim now is to make these technologies understood, so that we can see how they can be further developed to improve our everyday living in the future."

1.3.2 The Physics from Analogue TV to Smart Viewing – LED, QLED, LCD, OLED

With faster and faster electronics, different technologies were adopted to improve the performance and speed of communications in both TVs and telephones. The original analogue TV developed from the wartime use of *RAdio Detection And Ranging* (RADAR) equipment, in which an *electromagnetic* (EM) pulse was transmitted out to incoming aircraft, and the pulse was reflected back to a receiver, which then triggered an electron gun to fire electrons towards a phosphorescent screen. Because this phosphorescent dot moving across the screen allowed the observation of the incoming plane's movement, they could figure out the distance of the plane to the radar point and also its speed (Figure 1.9).

In Figure 1.9, a radar is positioned on an island, and its transmitter/receiver dish rotates around (with the dotted line showing the transmitted signal path) so that it gets reflections from all objects around it. In this figure, the radar shows the shape of the island with the waterside rocks having reflected its transmitted waves back to the dish at the centre. The circular lines show distance from the centre of the radar dish to any objects, with one square object appearing about 4½ miles out to sea, and a few smaller objects around it, which could possibly be rain clouds which have reflected the transmitted pulses.

In a TV, the pulse of electrons can be moved in-flight by a deflecting coil to hit a phosphorescent screen and on impact change energy from electrons to photons (a dot of light). This deflecting coil was developed so that we could control electrons passing through its centre. To understand this better, consider that the earth's globe has a magnetic field with poles at the north and south points and the field lines travel from South Pole to North Pole as indicated in Figure 1.10. (If we arrange a set of tiny iron filings on a piece of paper above a magnetic bar, the lines of magnetic force would be shown as the filings line up in their own north-south direction as shown on the right side of the figure.)

If we were to pass an electrical wire through the bar magnet's force field, an electrical current would flow in the wire, and in turn, the wire would generate its own separate magnetic field around it. So if we think about this the other way around, we could have a

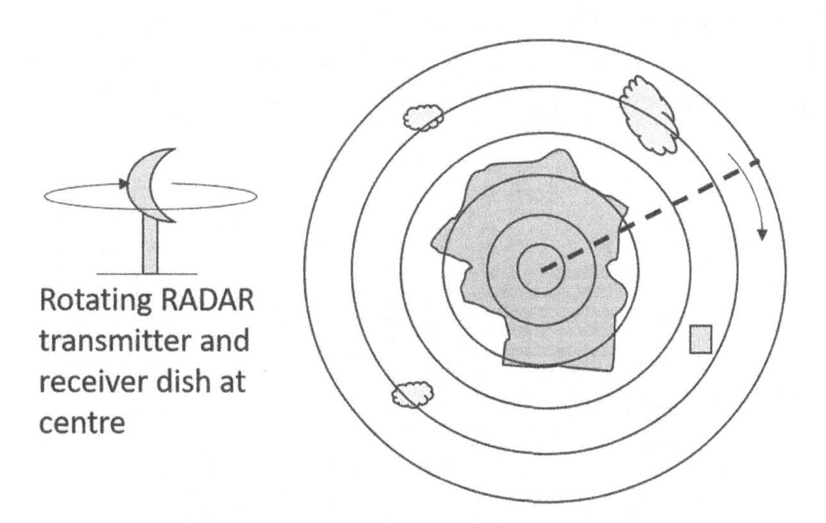

FIGURE 1.9 Rotating radar dish can track objects reflected back to it.

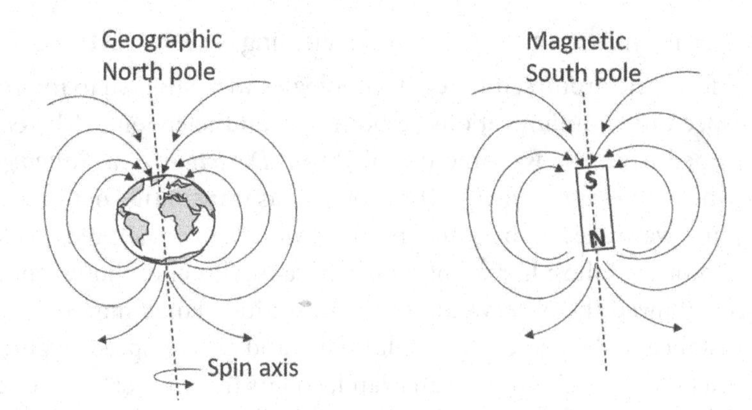

FIGURE 1.10 Magnetic fields for earth and a bar magnet – the geographic north is the magnetic south pole.

ferromagnetic ring like a doughnut, and wrap this wire around it. Then if we pass current through, the amount of current would cause a change in its magnetic field through the middle and this would change the flight path of any electron stream that would be passing through the doughnut. If we have two separate wires wrapped around, we could then manipulate the magnetic field by changing the currents in the two different wires, thus moving the electron stream in any direction we want it to go. In Figure 1.11, we have a doughnut-shaped ring with two separate wires wrapped around it (upper and lower coils). When we pass equal current through the coils, the electron stream coming through the centre carries on through in a straight line, but if we put more current through one wire coil than the other, this causes the magnetic field to change and it can cause the electron stream to bend away from its path. So by adjusting the two coil's magnetic field strength, we can control the direction of electrons passing through the coils. The shape of this doughnut is a toroid shape, so this type of electromagnet field is known as being made by a *toroidal coil.*

This toroidal coil when applied in radar would move or deflect the electron stream around so that its position of arrival on the screen would be similar to the distance the plane was from the radar receiver, so that wherever the dot appeared on the screen, it could be tracked as it moved towards or away from the centre of the screen.

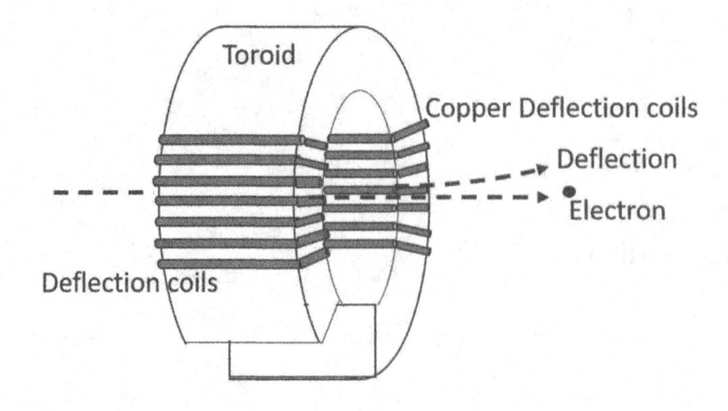

FIGURE 1.11 Electromagnetic field developed by a toroid deflects electron stream.

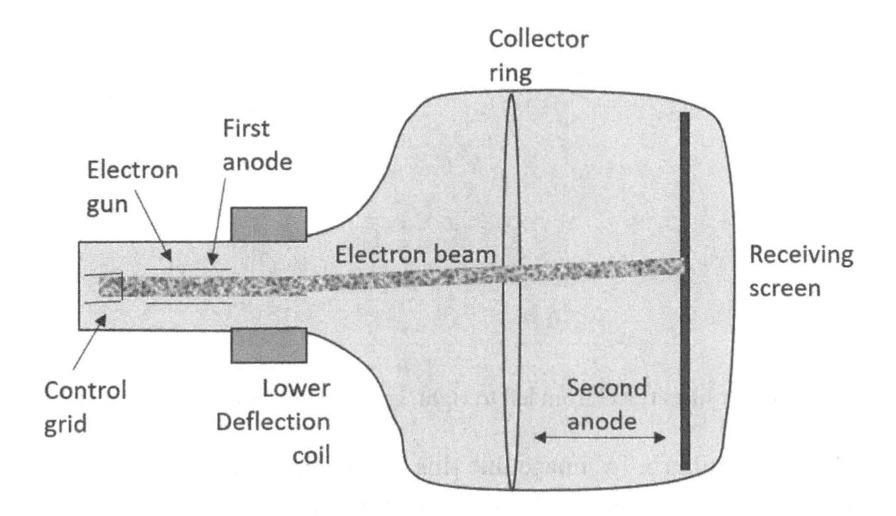

FIGURE 1.12 Cathode ray tube aka CRT oscilloscope.

TVs were manufactured using the same principle that you could guide any stream of electrons onto a screen using a controlling EM coil. Figure 1.12 shows one of the first schematics of how the analogue TV worked. In this sense, an analogue TV received signals as a continuous stream compared with digital TV which receives signal as separate blocks of data (i.e. *digitised*).

Just like electrons travel in a car battery from negative to positively charged plates – we refer to the positive plate as the *anode* and the negative plate as the *cathode* (we say the current goes from positive to negative only for simplicity) – so in a *cathode ray tube*, electrons were fired by a negatively charged *electron gun* towards a positive anode, through the deflecting coils and then through a positive collector ring. The electron stream finally passed through a positively charged sheet and onto a phosphorescent-coated screen. Different coatings of plate were used to improve the appearance of the arriving burst of light (the original colour being green), and later more colours were added by changing some frequencies within the transmission *frequency spectrum*. That's a bit complicated to explain without an understanding of frequencies, so we will just discuss for now the basics of the TV.

In the radar, the dot on the screen was a short pulse of light, but because the dot was repeated very quickly, it looked like a steady dot, moving slowly on the screen. When applied to a TV, it could be guided from the left side of the TV screen to the right side to form a line, and then jump across to start the next line down as shown in Figure 1.13. On arriving at the end of the last line at the right side, the dot would jump back to the top left to start again. As the fluorescent screen would retain some of the light after the electron hit it, we have a matrix of dots but the dots jump so quickly that our eye sees a continuous picture of different shades of light as if the dots of light were constantly shining. This was referred to as an *analogue* TV, so called because it was a *continuous* feed of electrons to produce a display of data representing something (compared with *digital* displays comprising of individual pixels representing something). The word *analogue* derives from 'analogy', meaning a transfer of one piece of information or meaning to another.

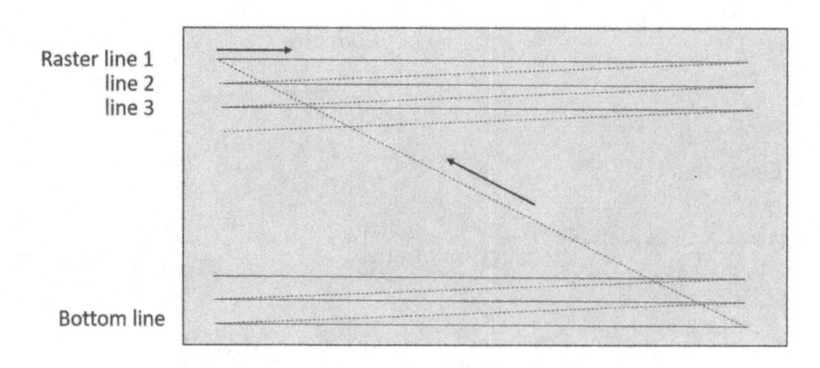

FIGURE 1.13 TV raster lines travel from left to right, going to bottom, then return to top.

The method of forming a TV image like this is sometimes scientifically referred to as *Raster* graphics, but more commonly today we refer to them as a *scan*. The location of each analogue dot is referred to as a *vector* point, meaning that each location is the result of having been steered by the EM field to go in the two directions of X (across) and Y (down) the TV screen. With the original radar, the screen was circular because the transmitter (gun) fired the pulses from a central point while the transmitter/receiver rotated to produce a picture that had an origin at the centre, but now with raster technology, we could produce rectangular TV screens. This then led onto the development of different types of TV, to improve the viewing ability, increase speed, and reduce the cost of manufacture.

LEDs had been commercial since the 1960s but they were discovered way back in 1907 by Mr. Round, an engineer with Marconi UK who stumbled on their lighting effects when developing radios [(6)]. At the time, he noted that 'on applying a potential of 10 volts between two points on a crystal of carborundum, the crystal gave out a yellowish light'. He had no idea of the innovations that this would lead to in later decades. Basically, their structure has not changed over the years and is shown in Figure 1.14. A diode is a semiconductor material (a mix of some different materials) that allows current to flow in one direction only. Commercial LEDs were developed by *General Electric Company* (GEC) from 1962 and pass the current through a p-n junction diode, to emit photons of light as the current's

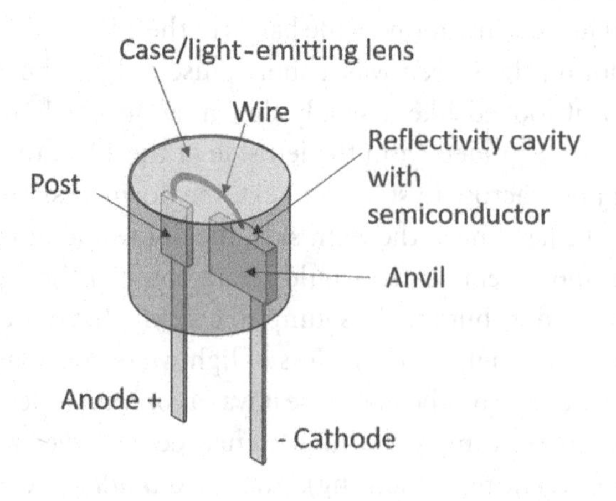

FIGURE 1.14 Light-emitting diode (LED) schematic.

electrons meet resistance (the same as electrons produce light on the fluorescent screen) which is then amplified by its casing. The p-n material can determine the type of light emitted and operates best at about 5 Volts DC (direct current). They can also be used with alternating current.

In the figure, current flows from the anode to the cathode (opposite to the direction of electron flow). The current passes up a frame and along a thin wire embedded in semiconductor material. This glows with light, which is transmitted to the wall of the LED casing lighting it (like the TV electrons lit the fluorescent screen). So a small electrical pulse from anode to cathode had the effect of lighting the end of the LED – and with low voltage means that they are relatively cheap to use compared with other forms of lighting.

LEDs have many applications today, typically in stop lights of road vehicles, signs, computer monitors, TVs, and house lighting. Because they only need a low level of power, they are the cheapest form of TV to operate. Note also that LED must not be confused with *liquid crystal display* (LCD).

LCDs were the forerunner to LED displays being used in early calculators and often used in small watches. To some degree, they were inferior to LED displays because they needed more power and were slower when running. A light is provided to the rear of the screen and is polarised to get the right colour by a film. Polarised light is demonstrated when we take polarising sunglasses and look at a distant object – as we rotate the sunglasses (which are polarised in one direction versus the other to remove unwanted glary sunrays), we can see how the light passes through the lens giving clarity by removing glare, as shown in Figure 1.15. In liquid crystal screens, the polarised light consists of red/green/blue individual crystals. The backlight may be always turned on, which makes this form of display the most expensive in terms of power consumption.

The colour requires a light to pass through a liquid crystal film which was placed between two electrodes and two orthogonal (right angled) polarising filters (like the sunglass lenses). An electric field is applied to the two electrodes, and as the two polarisers are turned on and off, different amounts of light are allowed to pass through (Figure 1.16 left side) so that each LCD emits variable light – useful for small displays and digital watches (Figure 1.16 right photo) but often poor when used as a parking meter display

FIGURE 1.15 Using polarising sunglasses and LCD crystals. (Courtesy: Shutterstock.)

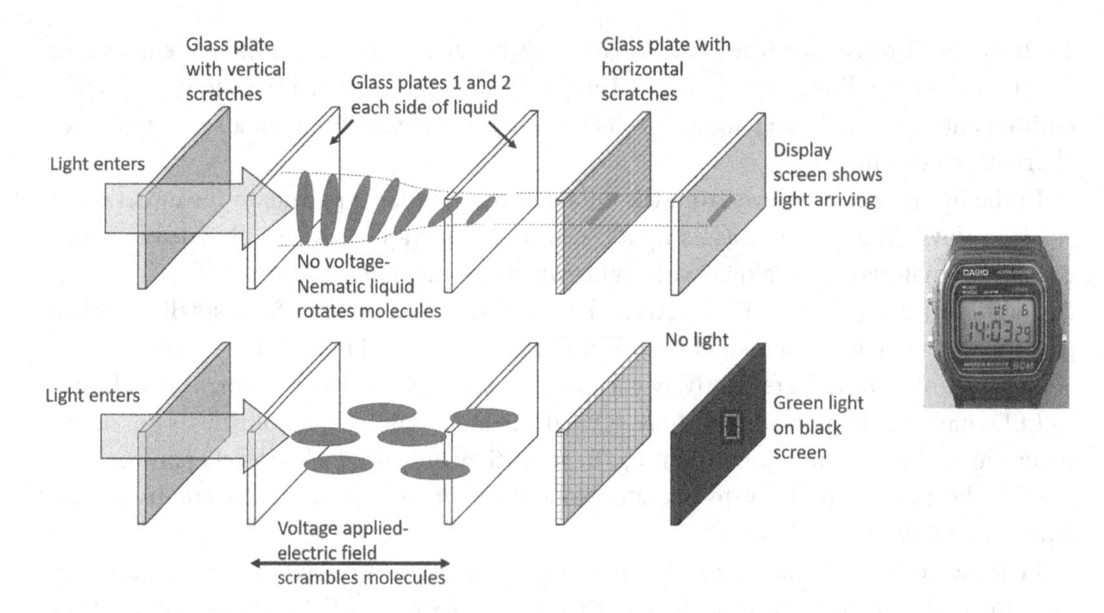

FIGURE 1.16 LCD results from light being polarised before display and the Casio digital watch.

where sunlight shining on the front reduces this contrast (so not advisable for use in cheap parking meters where there is no backlight).

Quantum dot LEDs **(QLEDs)** are similar to normal LED displays, except they use tiny nanocrystals of semiconductor material which fluoresce when light is shone on them. The different colours depend on the size of the nanocrystals, since the size determines the output frequency they give (frequency being related to colour – see later explanations). The nanocrystals are said to be smaller than the COVID-19 virus. Figure 1.17 illustrates these nanocrystals mounted on a film (*quantum dot enhancement film* – QDEF), sandwiched between a display's backlight screen and a liquid crystal matrix (LCM) layer. If nanocrystals of red and green only are used, they can be mixed together and a blue backlight screen can provide a white light output. Filters may now be put to use, as is used in the LCD screen.

Generally, the LED displays are preferred over the LCD because LEDs consume less power and can be controlled faster than LCDs. While the LED displays have a limited

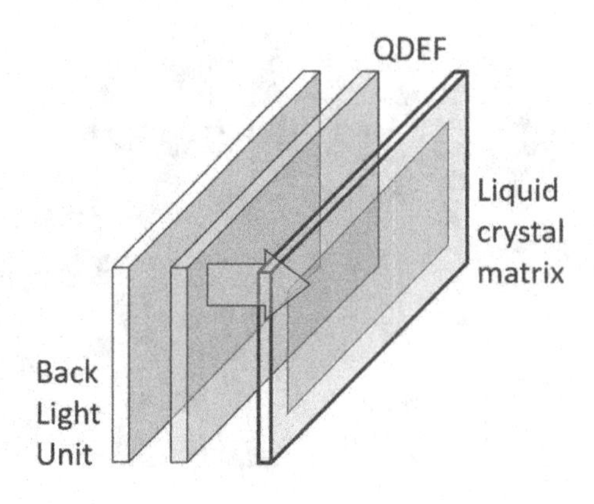

FIGURE 1.17 QLED sheet.

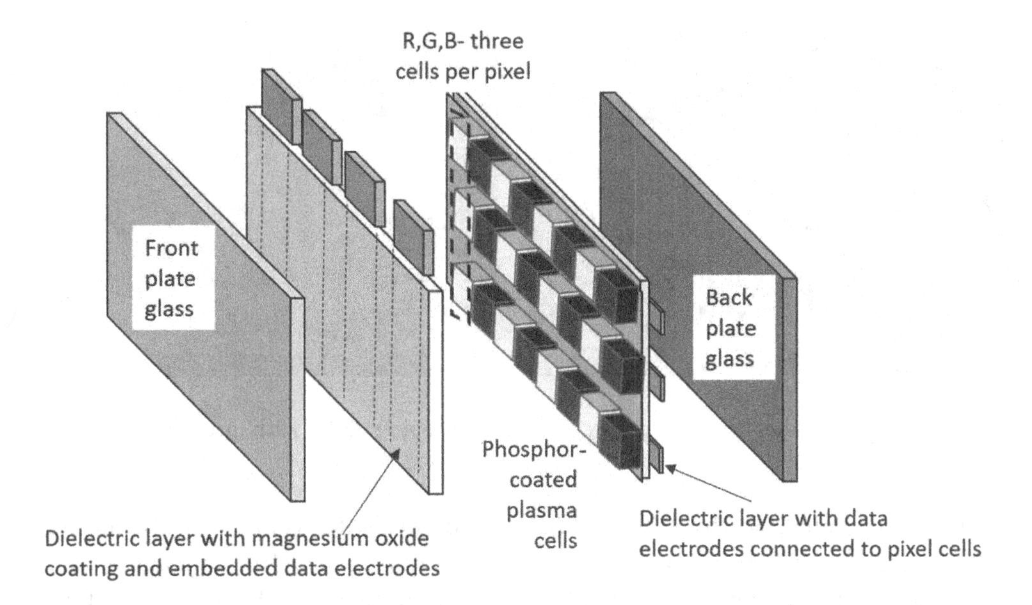

FIGURE 1.18 Plasma screen and how it displays light.

viewing angle, the LCD has broader angle but, because it needs a glass front, the LCD screen is normally heavier than the LED screen.

Plasma has been an alternative form of display screen for many years since its application in 1964 as a first step to replace the *cathode ray tube* (CRT) TV. In fact, the word plasma refers to clouds large or small of inert gas which can emit invisible ultraviolet (UV) light by passing electrons through it, and are found in large quantities in space.

If we inject a tiny pocket of neon or xenon gas within a glass layer, we can then electrically pulse it to make the gas produce UV light, which is then converted into a visible light by the phosphor coating on its screen, as illustrated in Figure 1.18. This is how household neon tubes and neon street signs work. The display has additional coatings so that the light appears as red, green, or blue dots developed in the cell. This is now considered to be a display *picture element*, which has been shortened to the word *pixel* (since 'going to the pix' has been a slang term for movie pictures since the 1930s, while 'el' is short for element).

The plasma screen involves a matrix of tiny glass or plastic cells each containing bubbles of gas with a luminous coating. This sandwich of materials has electrodes that provide the electrical control signal. The dielectric plate sandwich comprises of two plates holding the pixel cells together. A plate of insulating glass holds the whole assembly together at front and rear. The dielectric layer which contains the electrodes has a magnesium oxide coating to disperse the heat generated from the process of the electric signal through the cells generating heat. The electrodes on each side of the cells are orthogonal, so that signals polarised in different directions like the LCD can cause the change in colour of each pixel. These pixel cells are actually very small measuring about 0.01 inches (about ¼ mm).

***Organic LED* (OLED)** is an organic material type of screen that has been recently introduced, to provide greater colour contrast than has been possible – and due to the increased cost of the plastic screen, they tend to only be offered in the larger screens (>55 inch

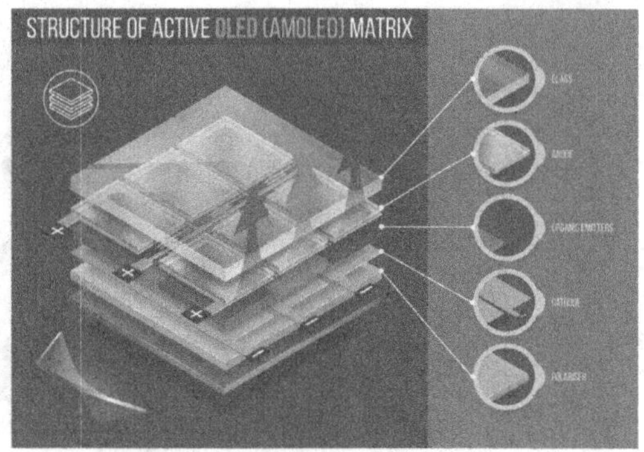

FIGURE 1.19 OLED display uses plastic screen developing light with greater colour range. (Courtesy: Shutterstock.)

measured diagonally). So this type of LED screen is called *organic*, because the plastic conductor plate (rather than semiconductors of the LED) is made from a polymer and the electron passing through producing light is therefore considered to be organically formed. This type of screen consumes less power and is lighter weight than the other forms of display screen – and more expensive to manufacture. By attaching anode and cathode power points across the plastic screen, changes in current produce changes in light produced between them, see Figure 1.19. In the figure, the display to the left indicates how each pixel consists of individual coloured cells while the active matrix layers are on the right.

OLED has greater colour contrast than the other types of display, and it is generally accepted that OLED screens are the brightest in colour and, with a *4K screen*, have better definition and power. Definition or screen *resolution* means the level of detail that can be observed in a picture or on-screen using pixels. This means that the more pixels we have on a screen in a specific area, the more detail we can see.

If we look at a modern car's red stop lights, they appear as one bright red light. As we look closer, however, they are actually individual small LED lights which from a distance appear to join together to form a single light. So screen resolution depends on the number of raster lines and pixels there are per line (which are normally measured in *pixels per square inch* – PPI), and depends on the amount of data on the screen. The greater number of raster lines there are, the greater the level of detail. The more pixels there are along the raster lines of a screen (Figure 1.13), the higher the density of PPI.

A high definition (HD) *screen* or monitor typically having 720 scan lines may have 1280 pixels (whereas ultra HD = 3840 × 2160 pixels) per line as shown in Figure 1.20. The important issue is how many pixels can be squeezed into the same area, and how many raster lines can be used.

HD is an almost square screen when it has 480 lines × 640 pixels across being 30% wider than its height. If we double the pixel number per line making it more oblong, the screen is still considered to be HD at 720 lines × 1280 pixels.

Full high definition (FHD) is the next size up at 1080 lines × 1920 pixels. This is the most popular display size since most computer monitors, laptops, and TVs come in 1080-line resolution.

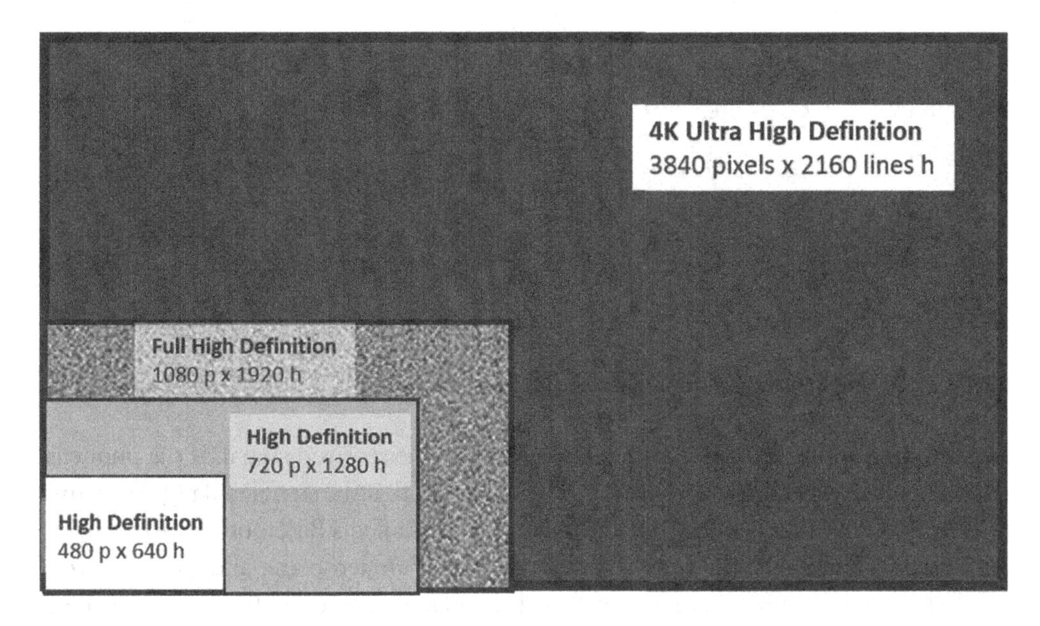

FIGURE 1.20 Screen pixel definition.

Ultra-high definition (UHD) is 3840 lines × 2160 pixels or 4096 lines × 2160 pixels. Because the number of lines are around 4000, these are referred to as 4K resolution screens, and are reserved for the larger screens of home theatres.

Because these screens have built-in computing abilities, they are known as *smart TVs*, since they not only display TV channels, but also can be used to display other information such as the stock market values, weather around the world, and any other form of data which they are programmed to display and link to via the internet.

1.3.3 Smartphones, Towers, and Radiation

When we display data on a screen, that data is simply signals that turn the different LEDs or crystals on or off to display the data as a colour. This data control results from a computer program that has been written to control signals flowing through the various CPUs as explained earlier. If we now reduce the size of our screen to fit into the hand, and use the telephone network to connect the screen with other phones using similar screens, we have the *smartphone* or mobile. In the United States, this is referred to as a *cell* phone, whereas in most of the world outside the United States, this is just a *mobile* phone.

Because these phones are a reduced-size programmed computer, they have software which allows them to be far more versatile than TVs and used for many different applications (or *apps*), from watching live TV channels and recordings, to observing the bus/train/airport arrivals/departures, to determining if the user's blood pressure is high, and so on. These phones transmit and receive data to/from local towers, which send and receive the data transmissions which are line-of-sight. If you pass a building while talking on a mobile phone, you may reduce transmission amplitude strength (the sound volume reduces) as a result of the building masking the line-of-sight to the transmitter, or sometimes the transmission amplitude increases (the sound volume gets greater) because the phone is receiving reflections

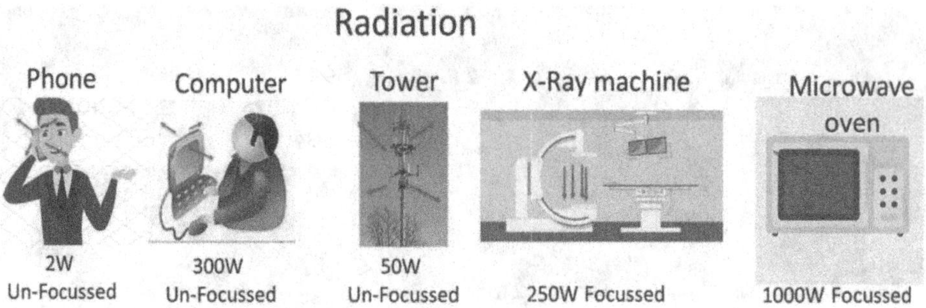

FIGURE 1.21 Comparison of radiation from different sources.

from a building across the street and this is adding to the received signal. If the phone is in an old brick house with thick brick walls, the chances are that the signal will reduce in amplitude as thick brick can block the transmitted signal. This affects Bluetooth also.

And do you remember the time when you boarded an aeroplane and they told you to turn off your mobile phone because it affected the plane's instruments? This was generally not the case – the issue was that when you got airborne, your call to someone from the sky could arrive at many local telephone transmission towers below you (if you are at 30,000 feet above a city, you could be in the line-of-sight of 500 towers at any time), which would cause confusion in the tower receiver electronics, because when many towers are receiving your telephone system, they get confused over which tower is going to handle your call and so many calls can jam-up the system (and it could freeze due to call overload). They have since resolved this issue (whereby now your phone connects with the local phone in the aeroplane cockpit and a single tower is selected) so that now you can call from a plane and the local tower alone (nearest you) picks up your transmission – Figure 1.21.

And what about the *radiation* from telephone towers – can it affect us? Transmission of high-frequency radio energy has not been proven to be an issue for human health, other than the tower spoiling the pleasant looks and ambience of a countryside. The amount of energy used to cause heat in a body over time – such as the energy needed to heat up a cup of coffee in a microwave oven – is in the order of 1000 Watts (1 Kw) focussed at an object over one minute. Compare this with a smartphone's output power of about 2 Watts, and a tower's output of up to 50 Watts, neither of which is focussed (a typical computer uses about 300 Watts which is unfocussed). And it is known that a microwave oven can interfere with the local *wifi* transmission, when they operate at near frequencies (so how much radiation are you getting in your body from a microwave oven?). These points suggest it is more likely that your laptop is more damaging to you than the local telephone tower.

Does a mobile phone when pressed against the head cause medical issues? It is again argued that holding a 2-Watt transmitter against the skull will cause some high-frequency power transmission through the brain, but when this is compared with 250 Watts used by X-ray machines focussed on imaging the body without harm to the body, the answer is that really, it shouldn't disturb the brain cells (and is yet to be proven otherwise) [7]. We rarely use our phone constantly, and perhaps the transmission frequency is the issue rather than power considerations – let's look at transmission frequency.

1.4 FREQUENCY AND THE BASICS OF SAMPLING SPEED

When we consider the pace of technology, there are a lot of basic changes in physics that have been made to make the technology operate. The recent increase in network speed, and the discussion around faster computers, has resulted from changes in operating frequency.

1.4.1 Use of the Frequency Spectrum – More Than Just Colours

Frequency is a fundamental occurrence in nature, and is a pre-requisite part of physics which should be understood before going further. Simple questions like 'why do twinkling stars change their colour?' 'Why do we see a rainbow and how do I hear different sounds?' The simple answer is that these natural observations are caused by changes in frequency, so we need to understand this fundamental characteristic of nature [8].

Let's consider how we see objects around us – when we look at the sky, we see our sun shining on a blue sky if there is no cloud, and a white cloud which is a group of tiny raindrops, and a black cloud when the whole cloud volume is filled with larger raindrops. The sun is shining a 'white' light, which is then arriving at the white cloud, but when there are a lot of raindrops in that cloud, the drops disperse the light coming through and act as a filter so that if very little light passes, the output light is a low level of light appearing to be dark gray or black. This dispersing is what we know as *refracting* the light, causing the light to bounce away off the drops, so the drops stop light passing through the cloud towards us. When the sunlight meets the edge of a small cloud of rain, the light refracts into its different frequencies and we see all the colours of the rainbow, which have a curve the same as the shape of the earth – it is just a projection of the earth's round curve onto a cloud's raindrops (where the light is being refracted) and we see the colours because they are different frequencies of light that we can see. Imagine how colourful our world would be if we could see all frequencies! It would be quite chaotic because we would never have a dark night and so find it hard to sleep.

Sunlight, sound, and EM energy travel, depending on the medium, at different relative velocities in the form of waves, with light and EM travelling at just under 300 million m/s while sound travels at a much slower speed around 1500 m/s in water and 340 m/s in the air depending on which direction the wind is blowing.

White light travels with all frequencies of the light spectrum (the same as the different frequencies we have in our radio and TV stations). It is actually a lot of very excited particles we call *photons*, and contains all frequencies which can be separated into various *bandwidths* as shown in Figure 1.22. If you imagine a particle of dust on the sea surface and a wave comes by, the particle may rise to the crest of the wave, and then fall to its trough – this total distance of rise and fall is *amplitude* of the wave – when it falls from one peak to a trough and back to a peak, this is a full cycle which we call a *Hertz* (Hz). The distance between the wave peaks is its wavelength (measured in metres), and the number of times a single wave cycle passes a fixed point in a fixed time, is called the wave's frequency (measured in cycles per second or *Hz*). The higher frequency a wave has (the more the wave compresses and is shorter in length), the more chance it is filtered out or *attenuated* because the tiny photons hit those clouds and bounce inside them. Consequently, there

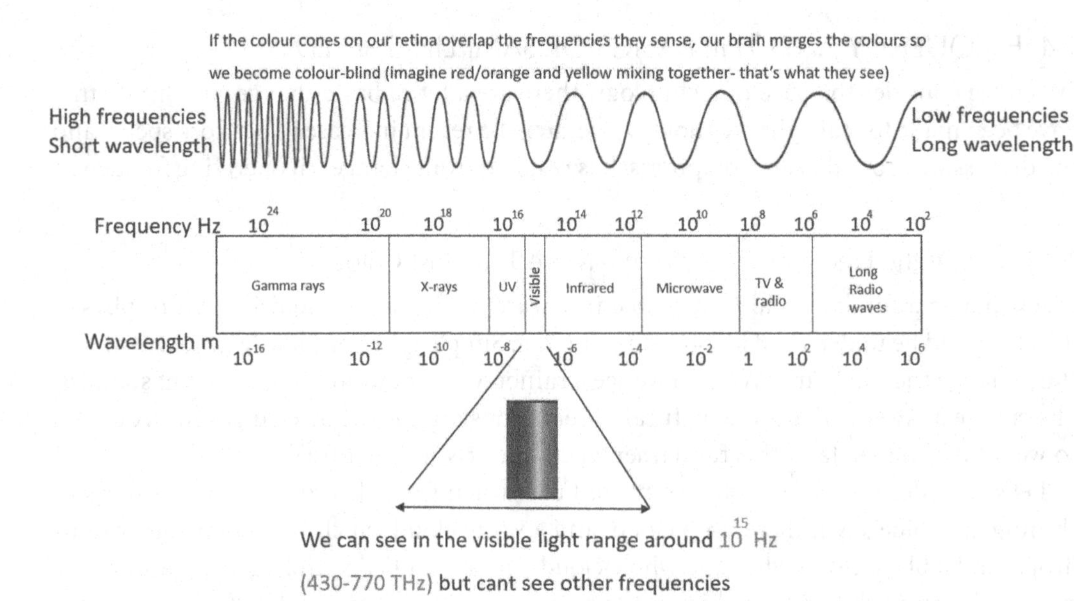

FIGURE 1.22 Frequencies and wavelengths (we can/can't see).

are many high-frequency waves that don't make it to the earth through the clouds, and we often just see the lower frequencies.

1.4.2 What You Can See (and Hear) – Is Not What You Always Get

When we look at white light, it has a small *bandwidth* compared with the full bandwidth of the EM frequency spectrum of around 10^1 Hz (which means about 10 Hz) to around 10^{26} Hz (which is 10 with 25 zeros behind it) – very high. The frequency content of sunlight is relatively narrow at about 410–790 trillion Hz (THz) in the 10^{15} Hz range, and is in the middle of the full frequency range. What this means in a practical sense is that, what we see in the light bandwidth is small compared with what is out there and we don't see.

We have made use of these other frequencies as follows:

Gamma rays – Used for medical scanning and to kill cancer, to detect defects in metal products, and to preserve food by removing bacteria.

X-rays – Used for medical scanning in checking out human body bone fractures and larger cracks in metal.

UV rays – Medical treatment of skin, killing bacteria in ponds and pools, in some fluorescent lighting and to see very hot objects in space.

Infrared (IR) *rays* – To see hot objects (such as looking at someone who has high temperature or fever) and tracking planes or missiles in-flight, and also using the glow from heat to be able to see in the dark (aka *night vision*).

Microwave rays – Used for cooking because they vibrate molecules in water, TV transmission, radar, telecommunications such as are used in mobile phones and *wifi*.

Radio and TV waves – Radios and TV in which these radio waves are converted into sound so you can hear a radio program or someone talking on radio or TV.

Long waves – Used in radio transmission, because their longer wavelength allows them to travel longer distances (over hills and around the earth's curvature) than normal radio waves. Over the last few decades, these have been reduced in use because satellite transmission (at between 0.2 and 32 Giga Hz, i.e. 0.2–32 GHz) allows signals to be sent to a satellite in space and back, so that longer range transmission can be done over very large distances.

While we have long waves, we hear reference sometimes made to short-wave transmission, which is the normal visible light bandwidth. We also hear people referring to *AM* and *FM* radio channels, which basically refer to amplitude modulation or variation (AM is continuous frequency analogue transmission but the amplitude peaks and trough heights of the waves are translated as sound) and frequency modulation or variation (FM is digital transmission where the different radio stations transmit using a constant amplitude of waves, but the sound is caused by compression and expansion of the frequencies). AM has a better range or distance of transmission, whereas digital FM only transmits essentially over line-of-sight distances (and so is used in cities where transmission antennae are mounted on office block roofs), but it has better sound quality.

What we have learnt then is that there is a small visible bandwidth that we can see. It is even smaller if we are to consider hearing frequencies, with the human ear being able to hear frequencies from about 20 Hz to just under 20,000 Hz (2 kHz or 2×10^4 Hz), depending on the amplitude which we call the 'volume'. When we walk in the street, we can often hear the buzz of a power line – this is because the power line operates (dependent on where you are) at 50 or 60 Hz, which is well within the human hearing bandwidth. By comparison, a dog can hear up to about 40 kHz which is why the higher frequency of dog whistles can only be heard by dogs and not humans.

One of the things about sound that we don't often understand is that it travels well in a room or street where there is no wind. But as soon as wind occurs, it can blow against the direction of sound travel so we hear less or no sound, but when the wind blows the other way, we can hear the sound well because it travels with the wind [9]. And when you hear a plane coming towards you, the jet engine sound has higher frequencies (just over 2 kHz) which continues to get higher in frequency and also amplitude as the jet comes closer. Then the sound goes lower as the jet moves past and away, reducing frequency to about 1.5 kHz.

In total, the change in frequency bandwidth is about 500 Hz. This frequency change is known as the *Doppler shift* or *Effect* and is caused by the jet's sound being compressed at the front of the jet as it comes towards you (becoming higher frequency to the ear), and sound reduces as the jet flies away from you (reducing frequency). The same effect can be heard when ambulances with their sirens blaring out pass you by – they are much higher frequency when the ambulance comes towards you, and much lower when they are driving away – see Figure 1.23. If you drive behind or in front of a bus, the sound is just the same

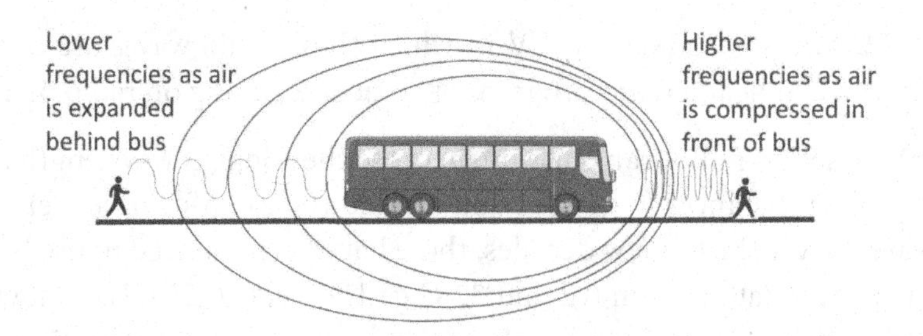

FIGURE 1.23 Doppler shift occurs when bus comes towards and then away from observer, going from high-frequency noise in front of bus to low frequencies as bus drives away.

because it is not being compressed or expanded – the effect only happens when we are stationary observing the moving object.

The *Doppler Effect* is very important to our everyday life (and hard to believe it was first explained by Christian Doppler in 1842). It is used in our mobile phones (connecting with satellites which are moving) to figure out our location on earth. It is used by police radar to compute speed of a travelling vehicle; in astronomy, it is used to figure out the speed of planets travelling in space with respect to the earth; and it is used in the sports arena to work out the speed of a ball travelling towards and away from players. Of course, this latter use requires multiple camera positions, which can be used to introduce how we can use technologies like *Doppler* to improve our performance in sports using data analytics.

1.4.3 Basics of Clock Rate (aka Timing Frequency or Sample Speed)

Most technologies operate at different frequencies, and we should have an awareness of how we can detect them in order to use them. When a wave is received in its analogue form, it arrives as a stream of pulsing energy – be it sunlight or wind on the skin, or other radio waves on our smartphone or TV screen. Sometimes we can watch action on TV such as car chases or helicopter flights where the wheels (or blades) go around in normal manner and then they appear to slow down and go backwards. Old western movies often show wagon wheels going forwards and then backwards, while the wagon is clearly moving forwards on the road (see Figure 1.24).

If we are looking at a wagon wheel rolling forwards, we can see the spokes roll around clockwise, but if we start blinking our eyes quickly (assuming the vision is clear enough to see the spokes), then we will naturally see that the speed of our observing (blinking the eyelid) determines if a mobile image of the spokes appears to roll forwards or backwards. This is referred to as *aliasing* an image – that is, when we see the wheel roll forwards, it is the correct image, but when it appears to roll backwards, we see a representation or an *alias* of the image, which is not real (it is just that our mind thinks it is rolling backwards). When applied to viewing on-screen activities, this is seen when digitising and reshowing an action scene on TV – its scanning or rasterisation across the screen has to be faster than the actual scene otherwise things become a mess. So if a ball travels across a screen, the scanning and speed of the display must be fast enough to allow the eye to track its movement from one side of the screen to the other.

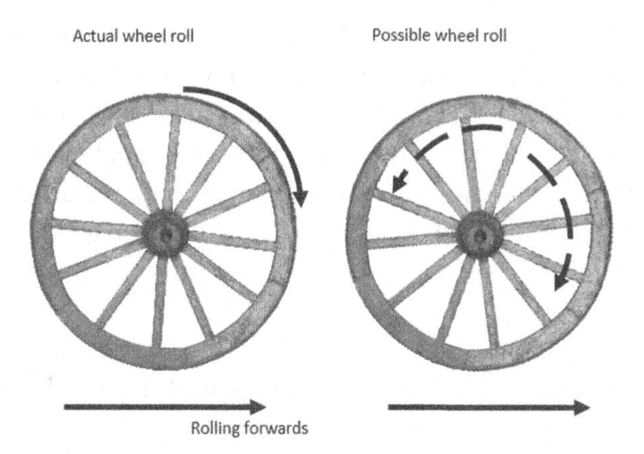

FIGURE 1.24 Wagon wheel effect.

TV transmissions are in the frequency bandwidth of radio waves [10], so we refer to TV data as the same as EM radio waves, which travel at the speed of light. The TV data arrives at the speed of light, and then goes through a series of electronic connections through the MODEM which slows the data down. It is then converted to light in all the screen types explained previously, and displayed.

The normal human eye reacts no faster than ¼ second, but the screen scan moves across lighting one pixel and then to another (Figure 1.25). The pixel raster scan is faster than the eye reaction and so the eye doesn't see the movement from one pixel to the next [11]. For example, if a 31-inch screen has 120 pixels per inch (0.02 mm between pixels) and the pixels have a colour refreshment rate of 200 Hz, then it will take 0.00002/200 = 0.1 μsec to appear to travel from one pixel to the next. This is faster than the eye can track.

So computers may operate at different frequencies (aka *clock rate*),

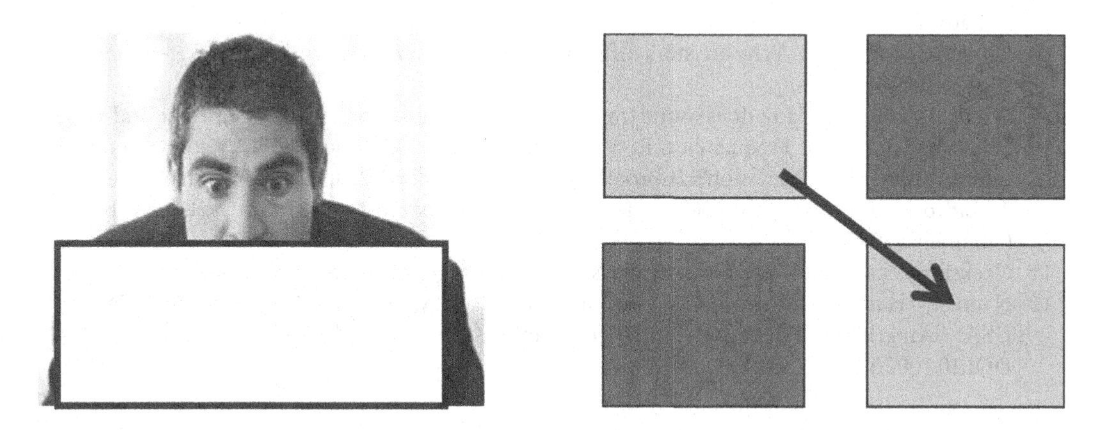

FIGURE 1.25 Data producing light can move very fast between pixels, and is not easily seen to be moving.

Even if we look closely at the screen, the eyes won't see the image move because at 0.1 μsec speed between pixels, the image move is 2½ million times faster than what we can see (250,000 μsec/0.1 μsec). Clearly at this speed, we have no problems with aliasing (our problems would have started if the pixel refresh rate were about every 60,000 μsec which is known as the *Nyquist Criterion* – problems occur when the relative speeds approach a quarter of the number of completed cycles per second of the image being displayed [(12)]).

1.5 EXERCISE

What do we mean by clock rate/timing frequency/sample rate and aliasing?

Given a computer operates at a frequency of 3 GHz (three gigahertz which is three billion cycles per second), what is the sample rate for this frequency, and what is the least sample rate we can use before the picture is aliased?

FURTHER READING

1. Plunkett, R.J. and Evans, W.L, 1938. The Mechanism of Carbohydrate Oxidation: The Action of aldehydo-d-Glucose and of aldehydo-d-Galactose in Alkaline Solutions. *Journal of the American Chemical Society*; 60, 2847–2852: DOI:10.1021/ja01279a007
2. York University, 2011. Data Transmission Chapter 3: https://www.eecs.yorku.ca/course_archive/2011-12/F/3213/Notes/chapter_3.pdf
3. Nyce, C., 2007. Predictive analytics white paper: https://www.the-digital-insurer.com/wp-content/uploads/2013/12/78-Predictive-Modeling-White-Paper.pdf
4. MSkyfi Labs, 2021. Mechatronics v robotics – what is the difference: https://www.skyfilabs.com/blog/mechatronics-vs-robotics-what-is-the-difference
5. Costa, I.F. and Mocellin, A., 2007. Noise Doppler-shift measurement of airplane speed. *The Physics Teacher*; 45, 356–358: DOI:10.1119/1.2768692
6. Round, H.J, 1907. Letter to the editor: a note on carborundum, Electrical World, Retrieved 10 January 2021: https://www.electronics-notes.com/articles/history/light-emitting-diode-led/led-history.php#:~:text=%20LED%20History%3A%20story%20of%20the%20light%20emitting,the%20late%2Fmid%201960s.%20These%20LEDs%20early...%20More%20
7. ANSTO, 2021. What is radiation? https://www.ansto.gov.au/education/nuclear-facts/what-is-radiation
8. Universe Today, 2021. Why are stars different colors? https://www.universetoday.com/130870/stars-different-colors/
9. Sciencestruck, 2021. How does sound travel? https://sciencestruck.com/how-does-sound-travel
10. Television Broadcast Frequencies, Retrieved 23 February 2021: http://otadtv.com/frequency/index.html#:~:text=Television%20Broadcast%20Frequencies%20A%20television%20station%20Radio%20Frequency,into%20VHF-Lo%20(channels%202-6),%20and%20VHF-Hi%20(channels%207-13)
11. Geeksforgeeks, 2021. Raster-scan displays: https://www.geeksforgeeks.org/raster-scan-displays/
12. Nyquist, Harry, January 1932. Regeneration theory. *Bell System Technical Journal*. USA: American Telephone and Telegraph Company (AT&T); 11 (1), 126–147: DOI:10.1002/j.1538-7305.1932.tb02344.x

CHAPTER 2

Tracking and Triangulation

It's Simple

In televising a sports game, we need to consider how we can calculate real-time position and transmit that to the viewer. Much of the work is done by very fast computers mounted in the control centre of a truck, which is recording the event. This computation is synchronised with the camera pictures, while the control centre that is monitoring many events is probably not at the local game but recording the data at some other locations many kilometres away (Figure 2.1). In some instances, like the English Premier League football matches, they use the *Video Assisted Referee* (VAR) system, in which officials view the video performance of a match, so that important incidents during a game can be independently viewed to advise the on-field referee of any errors of judgement or correct any mistakes.

The remote control centres can have many such remote referees who can monitor and discuss incidents with (in the order of 10) on-field match referees at the same time. With on average three VAR assistants per match, a full remote VAR centre has a lot going on. At the present time, the statistics (*stats*) and computing of positions and velocities are not in synchronisation with the monitoring, simply because the speed of the networks and computers handling the data are not sufficiently fast or have the software to allow this to happen. So, often, the VAR system fails because it does not account for all on-field issues that generate data.

2.1 TRACKING REAL-TIME POSITION

2.1.1 Animations of a Moving Object Using Multiple Fixed Cameras

When we refer to tracking something, we are monitoring an object moving across a screen, which is translated to the screen as moving the appearance of an object across a group of pixels on a screen. A ball is a group of similarly coloured pixels having a combined typical shape in a known colour, Figure 2.2. The benefit of using TV technology is in its ability to display the real-time image of an object and record it in preparation to replay the scene as well as the ability to freeze the object at a point of interest to the onlooker. For example, when a tennis ball hits a line, any part of the ball that touches the line is considered to be *In*

DOI: 10.1201/9781003108443-2

FIGURE 2.1 Video Assisted Referee (VAR) uses remotely located independent referees. (Courtesy: Hawk-Eye Innovations.)

the field of play, whereas if it were fully outside of the line, it may be considered to be *Out* of the field of play. Seeing this scene in real time would be very difficult to determine if all of the ball had bounced across the line from right to left, and so a screen freeze helps the viewer decide on whether this happened.

To take this picture, the camera position would have to be almost along the line of play. This is not possible in any sport, since collisions with players can result in harm to the players as well as the potential for damage to the expensive camera equipment. Instead, this

FIGURE 2.2 Ball on line. (Courtesy: Shutterstock.)

picture is an animation of the prediction of where the ball would bounce if a camera had been close to the line and would have been constructed from a number of TV cameras that were recording the events placed around the tennis court.

This technique was invented by a British scientist Paul Hawkins when he worked for Siemens subsidiary Roke Manor Research, calling the method *Shot Spot,* with the technology being patented in 2001, and sold to Sony for use in cricket.

For decades, TV cameras have been employed to record and transmit the action on the sports field. However, with the ability to use radar and simple trigonometry to determine the vector speed of a ball (in a direct line), it was now possible, using *Hawk-Eye,* to bring the two technologies together and improve not just the statistics of ball travel but also the graphics that the viewer could observe, including statistics, animations of the action, and study of player performance from different angles in 3D.

2.1.2 Calculating Speed (Velocity) of a Moving Object Using Multiple Fixed Cameras

Each pixel on a screen is located at an X- and Y-value. We use Pythagoras geometry to determine the distances between any objects on the screen but the view is from some camera distance away from the objects so something more than a simple computation is required. Also we can determine an object's apparent moving speed as seen from that location, as it crosses the screen from one pixel location to the next. The use of just two coordinates of X and Y assumes a flat surface, but the apparent speed is not the true velocity, which needs the third dimension of depth (Z) input data, Figure 2.3.

Imagine, in this figure, that you are looking at a TV screen that has a graph on it with the horizontal axis in the X-direction (along the pixel scan lines) and vertical axis in the Y-direction. We use Pythagoras triangulation to determine the distance from an object at say 25 pixels along X and 35 pixels up along Y (call this point X_{25} by Y_{35}) to a chosen point at X_{40}, Y_{15}. If the object moves across the screen, we can now determine its apparent velocity on the 2D screen (which is not true velocity because we don't know how far into or out of the screen it moves).

We can use simple Pythagoras right-angled triangulation methods to work out distances and speeds, as shown in Figure 2.4. In this figure, the two sides of X and Y lengths are at

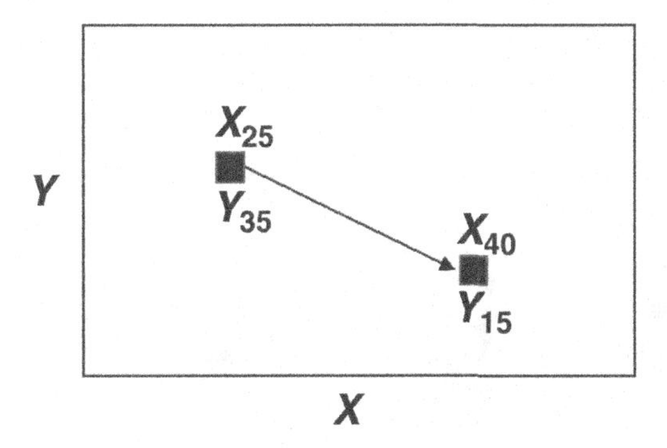

FIGURE 2.3 Without depth dimension Z, objects appear flat and move about on a screen in X and Y dimensions only.

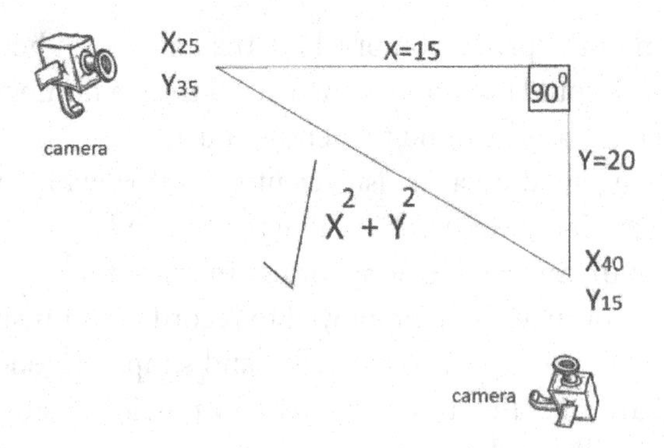

FIGURE 2.4 Pythagoras triangulation can be used to determine distance and speed that objects travel.

90° to each other with the third side called the 'hypotenuse'. The length of the hypotenuse is equal to the two side lengths squared, and summed together ($X^2 + Y^2$) and then square rooted. So it is easy to find the length of the third side of a triangle when you have lengths of X and Y. In Figure 2.4, if X = 15 m and Y = 20 m, the hypotenuse length will be $\sqrt{X^2 + Y^2} = 25$ m.

If we now have two cameras of known location in X and Y but at right angles to each other on a tennis court providing ball speed values in the X and Y direction, we can then work out the apparent speed the ball is travelling from the equation, Speed = distance (m)/time (sec). To get true speed, we have to project this onto the field of play, which will be discussed – and we will need another camera at a different position.

An alternative approach to tracking true speed is to use the radar and *Doppler Effect* discussed earlier and used when police are checking the speed of a car (Figure 2.5). If we set up a small *radar gun* in a fixed position on a tennis court, approximately in line-of-sight to the direction a tennis ball could be served, the ball would travel towards the radar gun reflecting the *electromagnetic* (EM) energy to it (Figure 2.5 in which *d* is distance to the

FIGURE 2.5 Radar guns track objects like tennis balls moving towards/away from them. (Courtesy: Shutterstock.)

ball). Because EM travels at the speed of light (just under 300 million m/s – or 300,000 Km/s), it is almost instantaneously received [(1)]. So we can track the ball from one position to its next position in real time and quickly compute its travel distance *d* because we know the time when we fired the pulse and when it was reflected back to the gun, and therefore it's speed using *speed* = *distance/time.*

If it takes a millisecond (1/1000th second) for a pulse to be fired from a radar gun, travel to an object and be reflected back, how far away is the object?

The answer is 1/1000 sec × 300,000 Km/sec = 300 Km.

Of course at sea level, we can't see further than about 3 miles or 5 Km in line-of-sight transmission because objects further away are over the horizon. But, if we get higher (like a radar dish on the top of a ship), we can see much further and maybe as far as 200 Km. This demonstrates that a radar gun working on a sports field or police radar checking car speed transmits and receives the pulse faster than the finger can click and return the fire button (which normally takes about 100 msec).

There has been some controversy over which type of radar speed gun is more accurate – an EM gun or a *laser* (Light Amplitude-Stimulated Emission Radiation) gun. Basically, an EM radar gun transmits a broad beam of EM frequencies, whereas a laser gun – sometimes referred to as a *LIDAR* gun (Light Detection and Ranging) – transmits a very narrow pencil-width light-pulsing beam, as shown in Figure 2.6. The difference is that if there were vehicles travelling in a group towards the EM gun, no individual vehicle could be picked out by the radar as having a different speed from the rest of the group, whereas with the laser gun, an individual vehicle can be picked out or 'spotted' by the gun [(2)]. The difficulty of use of the laser gun over the EM gun is that the laser gun requires a highly accurate pointing of the laser at a specific object, and it is easier to do this from an overhead bridge or standing out in front of the traffic. This is not easy to do, and hence the laser radar gun is not the police speed-checking gun of choice.

The speed-checking radar gun uses the *Doppler Effect* to obtain the car speed, using the shift in reflected frequencies to determine a car's first position, and then shortly afterwards the car's next position. The speed is only approximate, since for personal safety reasons, we cannot be directly in front of the oncoming traffic. But, if we have a second radar gun,

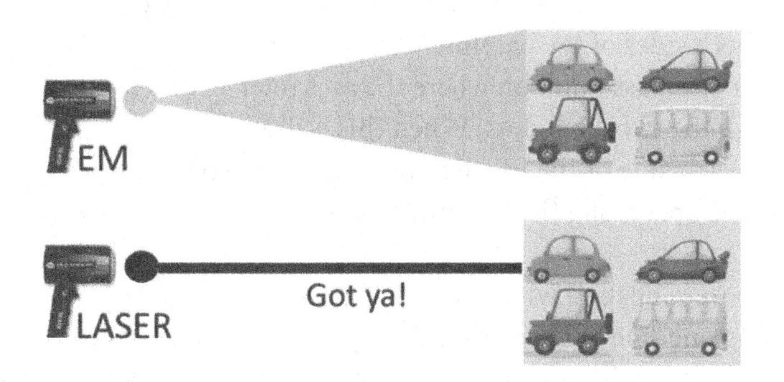

FIGURE 2.6 EM and laser radar guns are used in different applications. (Courtesy: Freepick.com.)

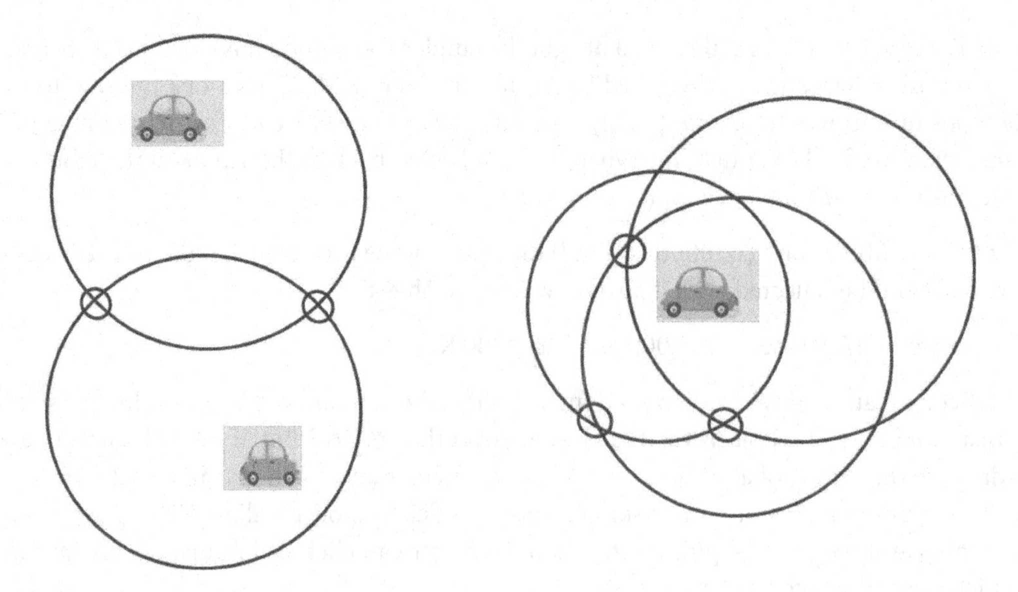

FIGURE 2.7 Location of an object from two radar guns is less accurate than using three radar guns.

with their locations precisely known, we can then make an assessment of locations of the car by triangulation.

Consider two fixed position radar guns tracking a moving car using *Doppler Effect* to detect its movement past them. If we just have two guns tracking the car, we can determine by triangulation if the car is one side of them or on the other side, as seen in the left side of Figure 2.7, where the open dots are the radar guns and the circles show the area in which the triangular computation can be performed. To do this, we need to be able to understand the basic geometry to get these locations.

2.1.3 Calculating Multiple Body Tracking

To draw the location of these circles, we use *Euclidean geometry* (which is the normal geometry taught in school). First, we draw a line (called the *baseline*) between the two open dots. Then we have to bisect the baseline, by drawing a line at right angles (the *normal*) to this baseline and to do this, we must put a geometry compass point on one open dot and draw the arcs on both sides of the baseline – see Figure 2.8.

Then we put our compass point on where the arcs intersect each other and draw a circle that will go through the two open dots. When this is drawn, the circles look like Figure 2.9. The circles drawn each side of the baseline between the two radar guns is called the *good angle area*, while the area along the baseline where the two arcs intersect is called the *bad angle area*.

Of course, if we were dealing with two radar guns pointed at a car on a road, we would be only considering one side of the guns, so let's just deal with that side. If we were to draw a line from both radar gun locations to intersect with the circle, the angle made would be 30 degrees as shown in the good angle area of Figure 2.10. But if we draw the lines from the

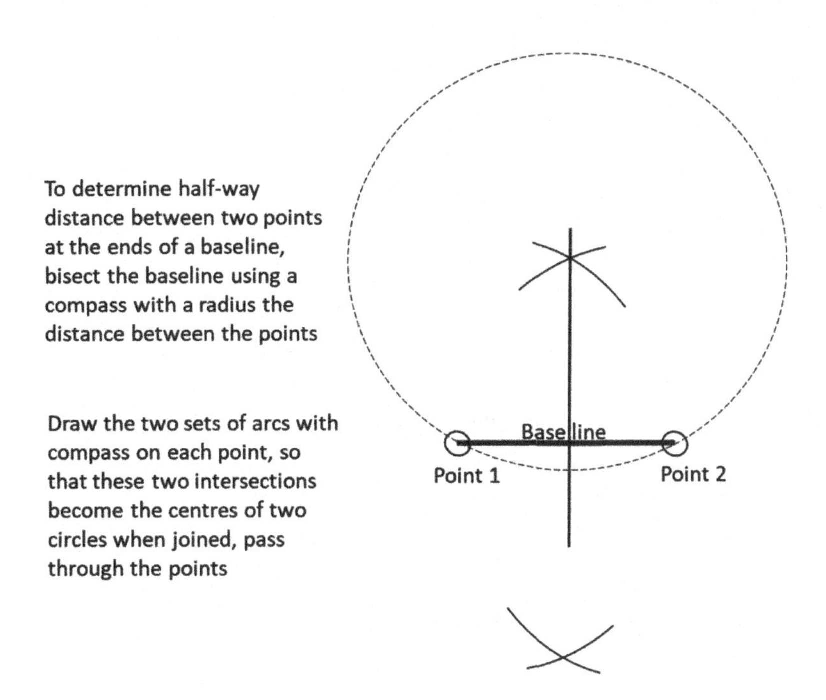

FIGURE 2.8 Euclidean geometry for bisecting a line (here called a *baseline*) in half.

dots to the bad angle area, that angle would be 150 degrees. In order to obtain an accurate location of any object from two observation points (the radar guns), the object must be positioned between 30 and 150 degrees. We call the larger area (within 30 degrees) that allows accurate positioning computations as one of *good angle,* whereas the area where a poor or inaccurate computation is performed (within 150 degrees) as the *bad angle* area.

If we were to try to get an accurate position of any object outside of the good angle area, our accuracy would plummet immediately. To get an absolutely accurate location of any object in any area, we therefore need a third radar gun. This allows us to have the object in a good angle area of any pair of stations or points, as shown in Figure 2.7 right side, where the car is close to bad angle from the top and right side radar guns, but this is compensated by good angle from the bottom two radar guns.

During World War II, bomber aircraft from the United Kingdom and Germany started to use this form of radar station guidance at night to drop their bombs using just two stations operating at 25 MHz (10^6 Hz). However, three stations at around 2.8 GHz (10^9 Hz) were used by the United States to devastating effect in the bombing of Vietnam during that war. This gave the United States high precision in their pinpoint bombing (whereas during World War II a single station failure, due to lost signal at night, made bombing very erratic so bombing during day time when you could see the target was preferred). Accuracy was therefore increased when we have three radar stations operating at higher frequencies, rather than two (in the knowledge that one of the three radar stations could be airborne in aircraft known as *Airborne Warning and Control System* – AWACS).

When used to track a car or a ball, three-station radars using the *Doppler Effect* can now compute position to within millimetres and speed to within micrometres per second. On a

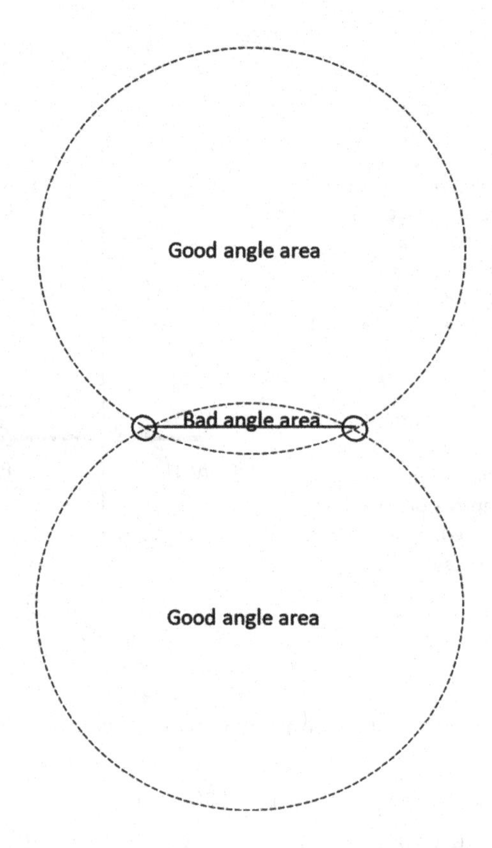

FIGURE 2.9 Good angle and bad angle locations.

tennis court, they are not so interested in the absolute accuracy of the ball's speed so they only use one radar gun at a time [3].

This laser gun is the same laser that is used by military forces, in conjunction with infrared night light (see later) for seeing in the dark, which is used for both shooting missiles at houses or firing gunshots at other military objects. For instance, when we see a 'SWAT' team go into a drug-running house, we can clearly see their laser dots being pointed at people or doors. The technology involved in the missile being able to *home-in* on a target is simple - the airborne jet fires the missile generally towards a target (usually a house) it wants to destroy, and then the jet flies over the house pointing a laser gun at the target house [4]. The missile now in transit in a general direction towards the house, sees the laser EM reflections from the building (like the conventional radar reflections), and then steers a course towards them - Figure 2.11).

If the laser is turned off or moved from that target before the missile hits it, the missile is likely to miss, so the laser gun must be continuously pointed at it with the missile getting reflections from the target until it hits.

In the counter-espionage wars of Iraq and Afghanistan, secret agents would walk around buildings at night and point the laser gun at roofs of houses that the agent wanted

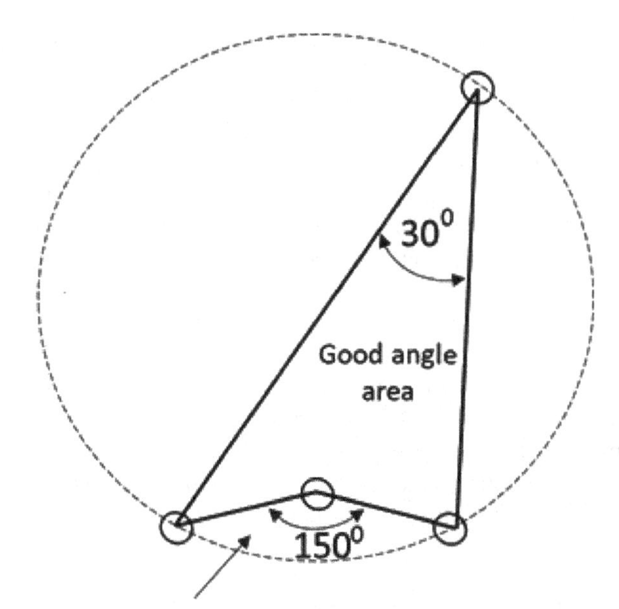

FIGURE 2.10 The most accurate location is within 30 degrees.

to be destroyed. The jets flying above the city would know what time the agent on the ground would turn on the laser (Figure 2.11), so the jets simply aimed their missiles in the approximate area of the target and the missile would do the rest. Because the laser gun spot on the roof was tiny, no one could see the spot from the street. However, this could prove difficult for the agent because, if there is a cloud of dust between the gun and the roof, anyone passing by would see the laser beam pointed at the roof. It was better for agents to take a vantage point from another building's roof to point the laser. Of course,

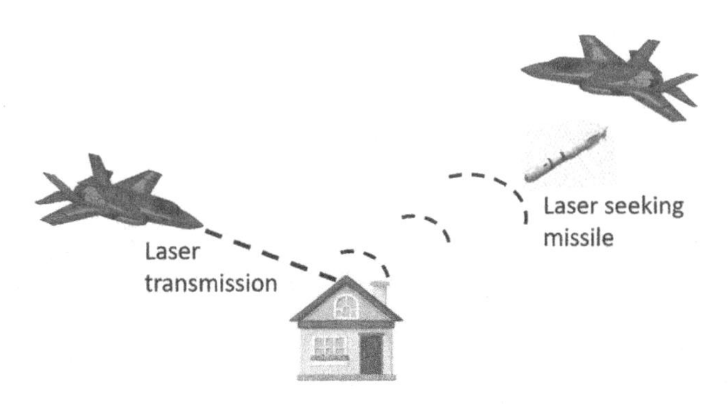

FIGURE 2.11 Using laser scatter, a missile is directed to the roof of a house.

this could then put the agent in the firing line if the agent was disturbed during the laser gun shooting at the roof!

> "The technology involved in the missile being able to *home-in* on a target is simple- the airborne jet fires the missile generally towards a target (usually a house) it wants to destroy, and then the jet flies over the house pointing a laser gun at the target house. The missile now in transit in a general direction towards the house, sees the laser EM reflections from the building (like the conventional radar reflections), and then steers a course towards them. If the laser is turned off or moved from that target before the missile hits it, the missile is likely to miss, so the laser gun must be continuously pointed at it with the missile getting reflections from the target until it hits."

2.2 CALCULATING POSITION FROM THE TV SCREEN VIEW

When we look at an object, one eye alone can only see a simple image that the brain processes as flat on a screen. Two eyes closely placed, the slightly different images allow the brain to process it as a 3D image having depth (Figure 2.12). So equally we can use a camera to view and record images from two close locations and convert them later into a movie with a 3D depth view (this has been used to great effect by subsea cameras [5], where a single camera could be towed past a shipwreck on the ocean's bottom, and later a 3D movie has been made allowing the viewer to see the wreck in full 3D).

2.2.1 Tracking in 2D on the Sports Field

A 2D location on a screen can be referenced to a field position but it needs a second view to get the third depth dimension. A minimum of two cameras/screens are needed

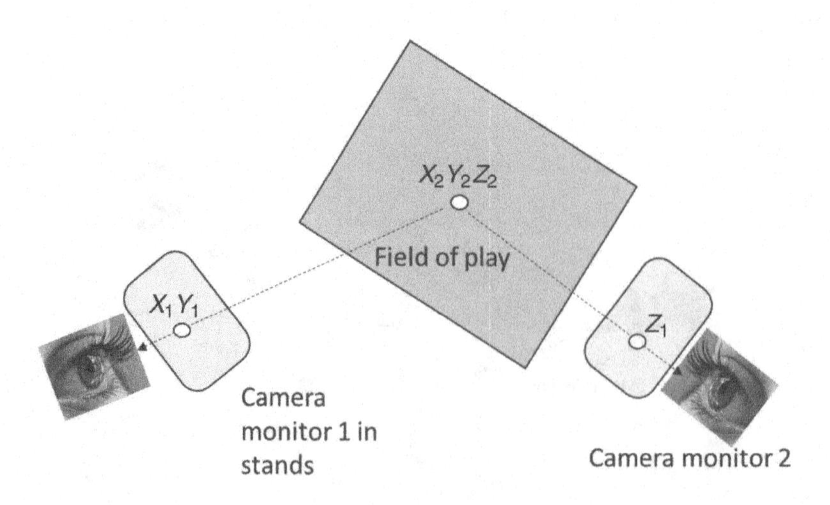

FIGURE 2.12 Getting a 3D location using two camera positions.

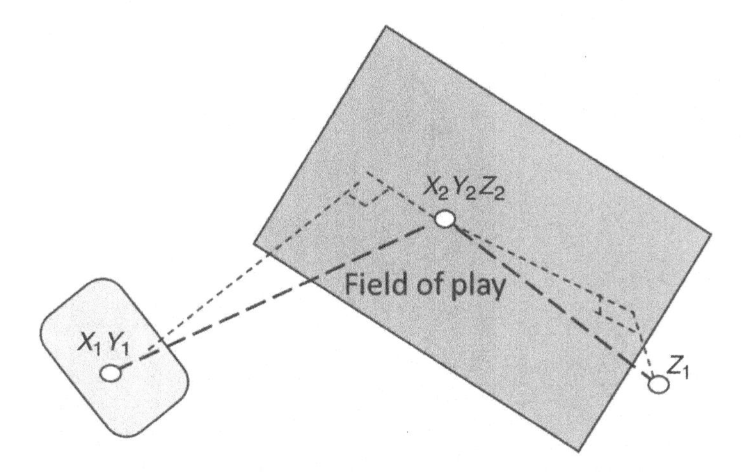

FIGURE 2.13 Position on the field can be translated to pixels on the screen.

for 3D location. What you see on the monitor screen is what the camera sees. A position for a single camera of X_1 and Y_1 can translate to field position X_2 and Y_2, but it needs the second camera monitor 2 to get the depth position Z_1 (which translates to Z_2 on-field).

In 3D therefore, to get the distance on the field to relate to pixels on a screen, we need to:

- Know the location of each object on the field
- Know where this sits in pixel form on the screen
- Compute the distance from the two cameras to the field, using Pythagoras and the positions from two laser guns
- Result is to have a pixel on each camera screen that translates to a field location

This gives us the screen location for each field position in Figure 2.13, from which we can then compute not just location, but speed of travel of a ball or a player.

The speed of a ball can be determined using the screen images – we can see on the field an object has moved from one point to another, see Figure 2.14. To get this distance, we know the 3D location of both points, so we can compute the 3D distance and the speed between the two using Pythagoras. We know the distance between one point and the other on the field, so we can determine the speed since this is distance/time.

Working out locations (*tracking*) for more than one body (of players) is a little more complex. It is the same concept, except, in 3D, the two bodies may collide both in real life on the pitch and also on the screen, in which case we have to track them both at the same time but separately, and so screens on their own are inadequate. Instead, we attach a tracking device to each player, which gives an output signal to either a local short-range network (*wifi*) or a *satellite network*. The navigation satellite network, which is a series of satellites orbiting the earth at a height of between 6,000 and 12,000 miles (10,000 – 20,000 Km), is

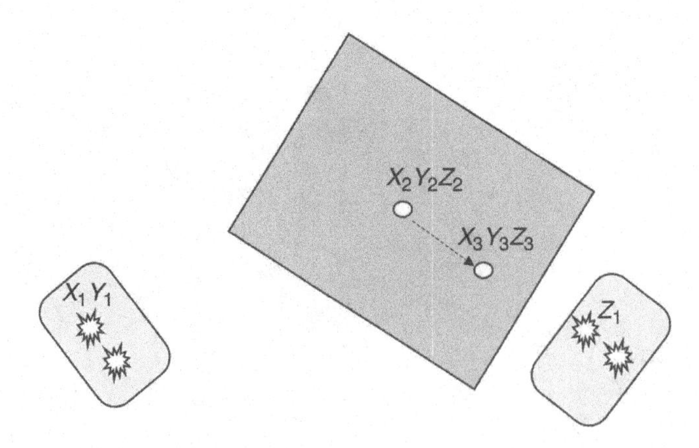

FIGURE 2.14 Speed of a ball or player can be determined using screen images. (Truck courtesy: Shutterstock.)

often referred to as *sat-nav*, but better known as the *Global Positioning System* (GPS) that is now installed in new cars and mobile phones [6].

> "The navigation satellite network, which is a series of satellites orbiting the earth at a height of between 6,000 and 12,000 miles (10,000 – 20,000 Km), is often referred to as "sat-nav", but better known as the *Global Positioning System* (GPS) that is now installed in new cars and mobile phones."

As with a mobile phone, the tracking device is constantly transmitting its position up to the satellite system along with other parameters such as player speed, acceleration, the body vibrations as the player kicks the ball as well as medical information. The data is retransmitted back to any receiver that connects with it asking for this information. This would normally be the control centre receiver on the ground (Figure 2.15). There may in fact be satellite tracking for the general TV audience as well as the local short-range tracking by the coach and staff of the home team.

There are usually two systems for tracking players – the most accurate being the local short-range *wifi* network, which is configured around the playing field, compared with the less accurate satellite network.

The short-range network can operate with many transmitters and cameras around the ground (from about 3 up to about 25 for important games like the Premier League in the United Kingdom or American Football), with positioning accuracy in centimetres (Figure 2.16 box camera). The benefit of the short-range network is that player data and pictures can be rapidly recorded at a control centre, with the *wifi* providing positioning accuracy, to go with the satellite pictures that compute with little accuracy (no better than 10 m). When used in conjunction with the local *wifi* network at the ground (Figure 2.16 large cameras), it allows extension of *Hawk-Eye* abilities to compute whether a player was offside at the time a ball was kicked towards another player – the result of which could be worth millions of dollars to either team. However, at times, the pictures are still not always absolute proof of position.

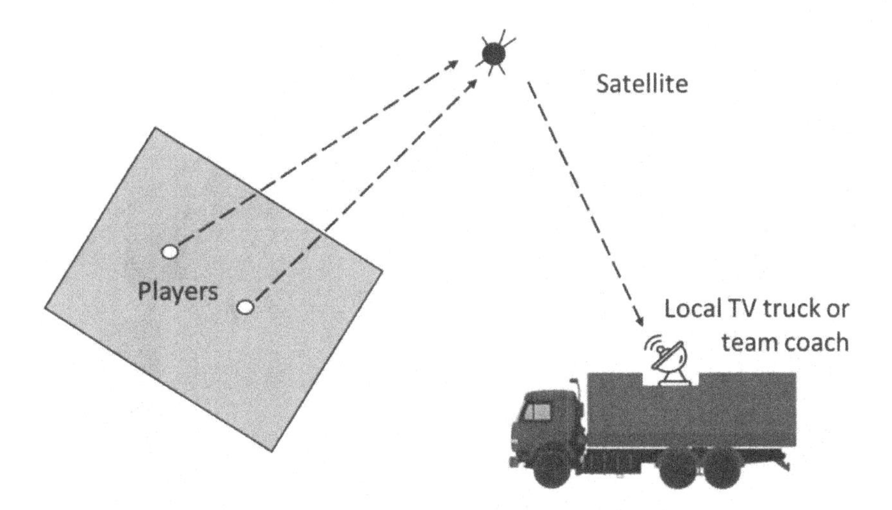

FIGURE 2.15 Tracking players using satellite capabilities.

2.2.2 Tracking in 3D from Multiple Observation Points

Having player stats flash up on the viewer's screen during controversial events still takes time, because that would require much faster data transmission and computation – which is not there yet. But what is there at present is the ability to predict where a ball will travel, provided data is stored of its past say 20 positions in 3D. This is known as *predictive analytics* [7].

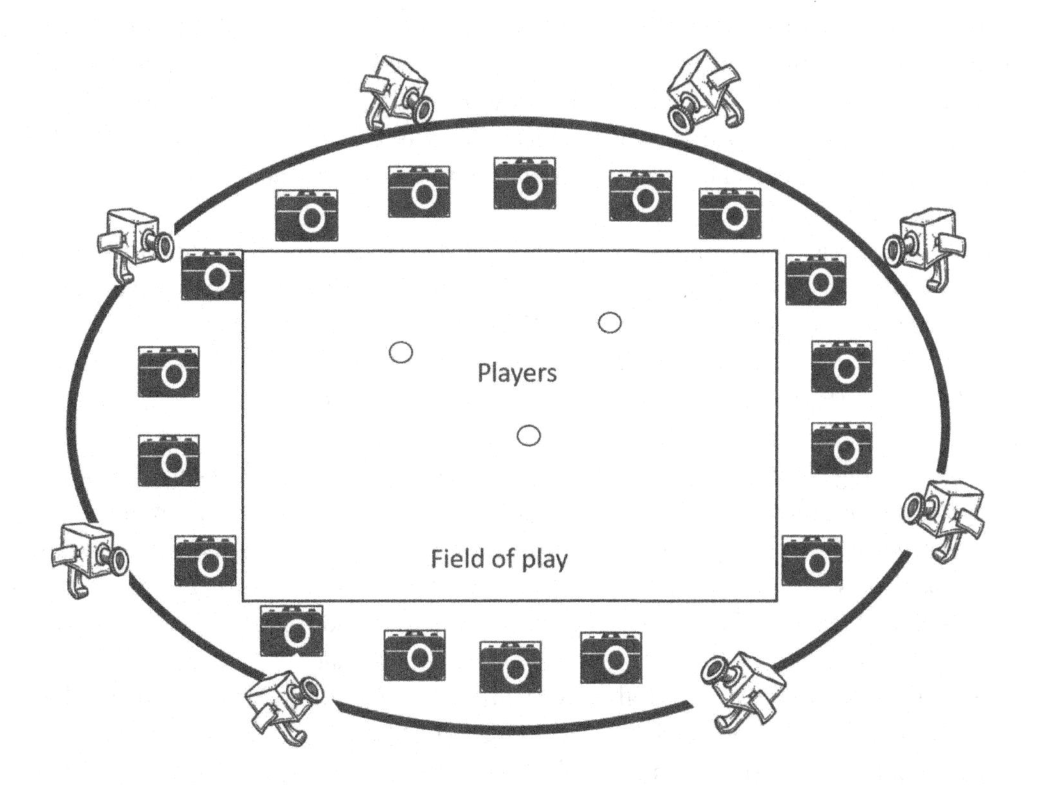

FIGURE 2.16 Data transmitters and TV cameras (any adjacent camera pair provide 3D images).

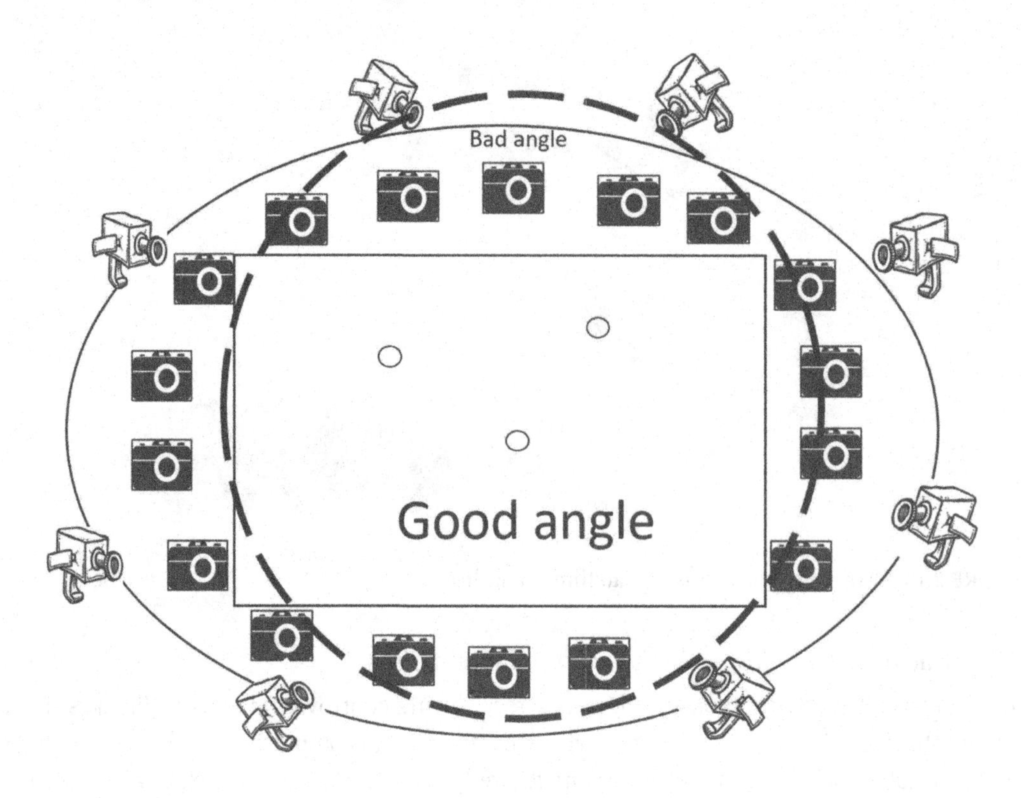

FIGURE 2.17 With two cameras, most of the ground has good angle TV coverage.

Put simply, if we can obtain 2D data of ball position in the *X*- and *Y*-directions by one camera, and can do the same with another camera in the *Y*- and *Z*-directions as discussed for Figure 2.4 (and using say the top two TV cameras in Figure 2.16), then, providing the ball and players are in good angle with respect to them both, we can start predicting where the ball might track next since we can join their outputs as a very accurate vector position in 3D (Figure 2.17).

Today each professional sports ground has the short-range tracking monitors around the ground, the positions of which (in *X*, *Y*, and *Z*) are precisely known to a surveying first order (which is sub-millimetre accuracy). These monitors receive data from each player who has a tracking monitor worn between the shoulder blades, tucked into the shirt or separately worn (Figure 2.18). A special player tracking vest is available, which was pioneered by an Australian company called *Catapult Sports*, which is used in professional football including *National Football League* (NFL), cricket, rugby, and baseball. Ice hockey is a very fast sport and it is tough to keep a close track on their players. German company *Kinexon* produces a tracking product that uses web-based computing and display, mainly for indoor sports with a major focus on software indicators for training control, injury prevention, and tactics. With Irish company *STATSports* products, they connect to other wearables (e.g. watches, tablets, and iPods). *Polar USA* is a major competitor for *Catapult*. *GPEXE* a subsidiary of *Exelio* of Italy has additional metabolic equations and are cheap, while *Catapult*-owned *PlayerTek* is the cheapest available and has easy to use software.

Each player monitor transmits continuous data from its own location to at least one local station in the ground – using a GPS receiver for position, an accelerometer sensor that

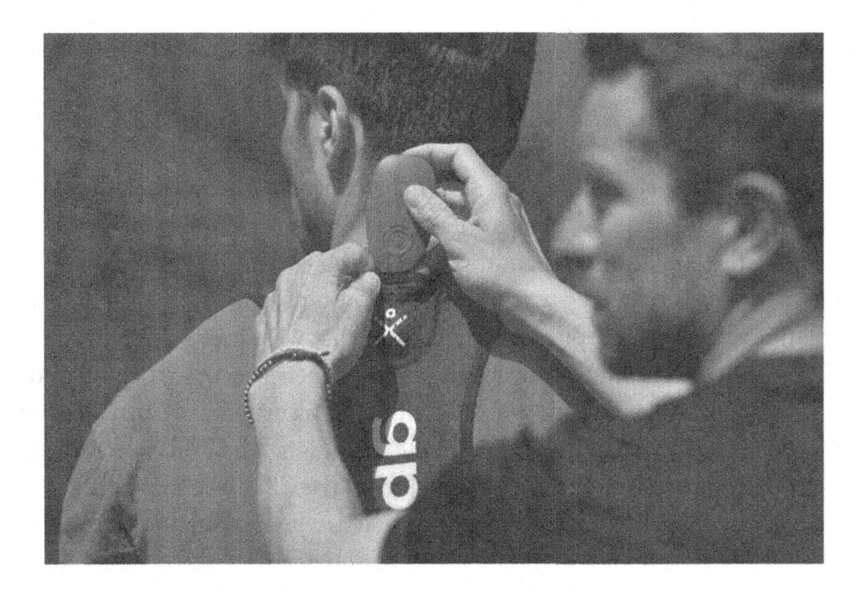

FIGURE 2.18 Tracking sensor *Pro2* on a sports person's back-vest (Courtesy: GPEXE)

measures changes in force using strain gauges, a gyro for position orientation, and a magnetometer for compass information. Sometimes foot sensors are used to monitor the load on a player's foot – providing *biometric* data to the coaches along with the *interactive motion unit* (IMU) data. These IMU sensors are linked to a small *computer processing unit* (CPU) in the package that computes the different data needed by the coach on individuals or the full team (Figure 2.19).

The accuracy of player and ball position depends on the game to be played. For example, in professional soccer, many times the important data is player position not just ball position – a player can be caught *offside* by a screenshot while another screenshot freezes the moment a player releases the ball. However, in cricket, the ball position with respect to

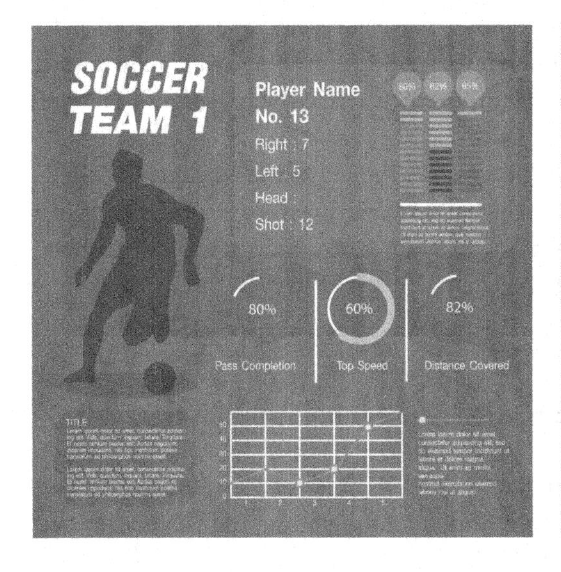

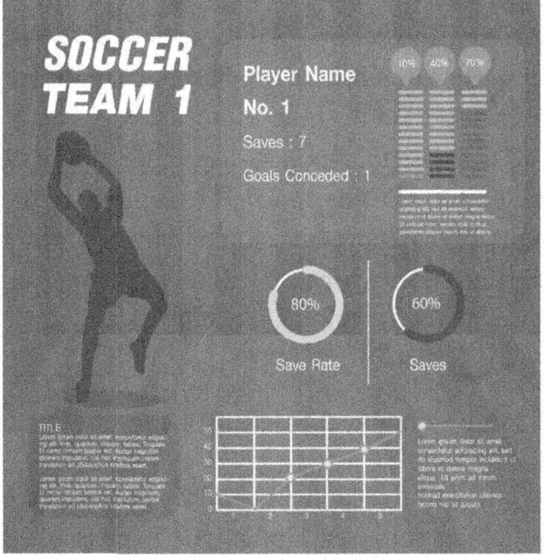

FIGURE 2.19 Team data shows the coach, player and team statistics (Courtesy: Shutterstock).

the wicket is needed to determine if a batsman is to be declared leg before wicket (and therefore *out*).

2.2.3 Doppler Positioning by Satellite

Originally, the positioning satellites were launched into space by the United States and Soviet armies in the 1960s and 1970s as military positioning satellites, but when the Soviets managed to crack the equations and coding used by the US military in the late 1960s/early 1970s, the United States opened up all their positioning satellites for public use (they of course later launched smarter replacement satellites that were more difficult to crack).

While GPS satellite positioning is relatively poor (of metre accuracy) compared with the local network (which would be sub-metre), by contrast, screenshots are considered as being of great accuracy when needed by the referee to make a decision. All of this positioning computing is done through equations in the software coding in the CPU. The trend is to refer to them as *algorithms*, which makes them seem very complicated, when in fact, they are just equations.

The GPS positioning is done using its own set of equations, based on satellite frequency shifts involving the *Doppler Effect*. Satellites are positioned in orbit around the world and they transmit their position on a number of frequencies. Anyone with a smartphone that has a GPS receiver can receive the different frequencies and position information of the satellite as it transits past our GPS receiver. The satellite signal is a frequency that is higher as it moves towards the receiver and lower as it moves away, which as explained earlier is what we call the *Doppler Shift* or *Effect*. Knowing this change in frequency from one known position to another allows us to insert this information into a Doppler equation [(8)], and this then allows us to compute where our GPS receiver is located. Just like we worked out the *good* and *bad angles* in the Section 2.1.3 on positioning, we use two or more satellites to work out our GPS position, with some Doppler equation solutions not being accepted when the satellite gives a result from a *bad angle* – this can happen when a satellite is low crossing the horizon or when it is immediately above our location.

We have seen that equations are used to compute position using GPS, and then more equations are used in sports to compute and predict player abilities, body responses to activity, and actions. All these data help us understand the effort and load on a player and, in some cases, where the ball tracking is important, predict where the ball would go over a short period in time.

Let's therefore look at data prediction in 2D, in which each data set has a value in time – called a *number* or *time series*. Computing where the ball goes requires a basic understanding of graphical representation of data and numbers – first, we need to understand the number or time series which we will do in the next chapter.

2.3 EXERCISE

Exercise on object tracking a car – better to have one camera in the line of travel.

We have three police cameras positioned in the *X*- and *Y*-directions around a freeway watching cars pass along. One is positioned on a bridge looking along the road, with the other two on either side of the road.

- Position of camera 1 $(x, y) = 0, 100$ m.
- Position camera 2 = 0, 0 (on the bridge).
- Position camera 3 = 120, 0 m.
- The car travels from Car position 1 at 100, 100 m to Car position 2 at 50, 50 m in 1 second.
- Determine the speed seen by each camera.

If the speed limit was 110 Km/hr (70 miles/hr), has the car exceeded the speed limit?

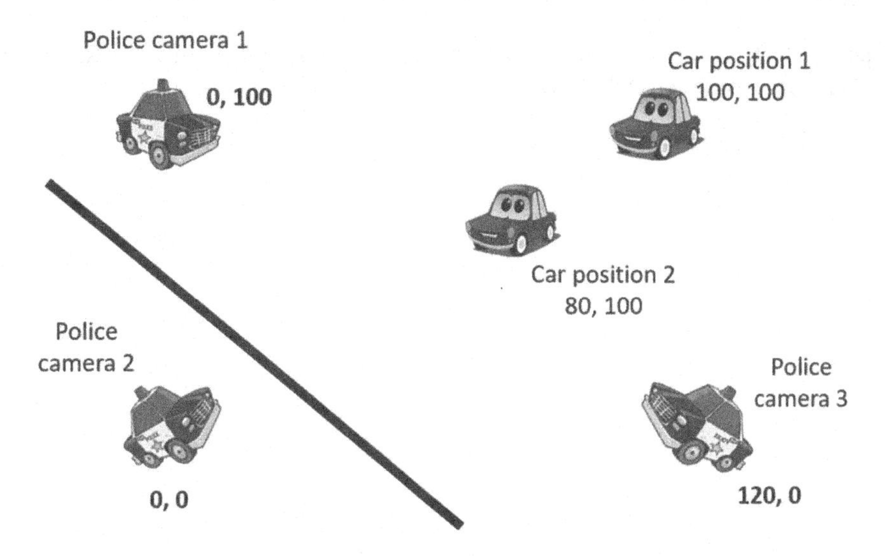

On graph paper, mark the scaled positions and then work out using Pythagoras, the distances first to Car position 1, then Car position 2. The difference in distance between Car position 1 and Car position 2 divided by 1 second gives speed in m/s.

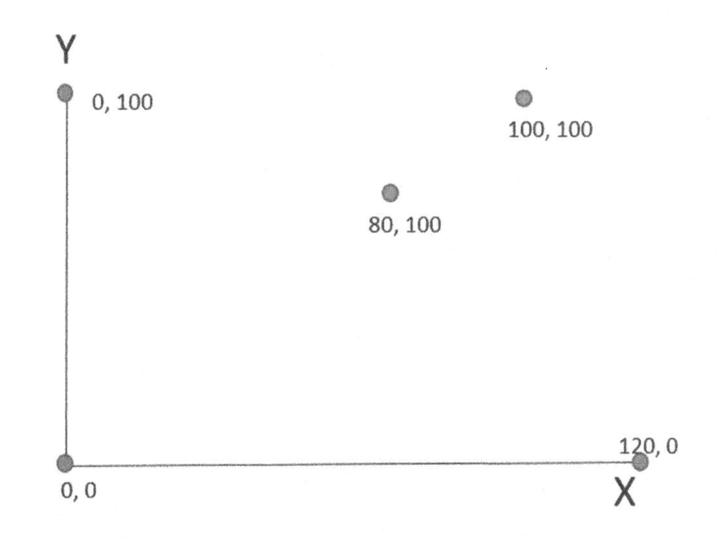

FURTHER READING

1. Radar speed: https://en.wikipedia.org/wiki/Radar_speed_gun#Limitations
2. The Lidar gun: https://en.wikipedia.org/wiki/LIDAR_traffic_enforcement
3. Tennis radar gun on YouTube: https://www.youtube.com/watch?v=DQA0DaArUHE
4. How laser guided missiles are made: http://www.madehow.com/Volume-1/Laser-Guided-Missile.html
5. The science of 3D movies: https://www.scienceabc.com/humans/science-of-3-d-movies-how-things-on-a-flat-screen-appear-to-emerge-depth.html
6. How does a GPS system work? https://www.elprocus.com/how-gps-system-works/#:~:text=The%20global%20positioning%20system%20consists%20of%20satellite,%20control,on%20some%20incidents%20in%20several%20ways,%20such%20as
7. The application of predictive analytics: http://www.ijsrp.org/research-paper-0517/ijsrp-p6564.pdf
8. Doppler Effect equations for sound: https://www.school-for-champions.com/science/sound_doppler_effect_equations.htm#.YD8uBk7iu70

CHAPTER 3

Pattern Recognition and Its Applications

3.1 INTRODUCTION TO NUMBER (DATA) REPRESENTATION

3.1.1 Basics of Data Analysis – the Time Series

The horizontal *X*-axis line in Figure 3.1 represents increasing time, while the vertical *Y*-axis is value of what we are measuring at each equal point in time. If we join the data value points together, it can represent the continuous curvilinear line we see on the graph below the data points. However, if we displayed the data in the form of blocks (in *histogram* form), we would produce the graph of Figure 3.2. This shows that we can change the appearance of the same data set, and so we must be careful of how data is represented so that we easily understand what it means (having data in a histogram block form tends to suggest the data is the same across the peaks, and so tends to emphasise the peaks of the data rather than the average data).

This number series (which has data at values of 1, 1.5, 0.75, −1, −2, −0.8, 0.5, −1.2, 0, 1, 2, 1.5, 1, 1.5, 0.5, −1, 0.5, −1.2, −2, −1, 0, 1, and 0) has all of the sample data points recorded at exactly the same time interval. If we take this original number series and record only every second sample or data point instead, this number series will not represent the original series, because it is less accurate than the original sample. We must sample the data at a time rate (known as time or *temporal sampling*) sufficient to record and reproduce the data properly. Remember the earlier comment about the *Nyquist Criterion* and needing to sample at least half the minimum sample rate to avoid aliasing. Figure 3.3 shows the original graph from Figure 3.1 at the top, with an aliased data time series below it, which clearly no longer looks like the correct input time series.

3.1.2 Number Systems Using BITS and BYTES

The time series of numbers can provide a graph of perhaps a player's movement sideways, and results from the accelerometer data transmitted from the sensor. The number series may be the outputs of the sensors in volts of 5, 0.7, −1, −2, which may represent a jump to

DOI: 10.1201/9781003108443-3

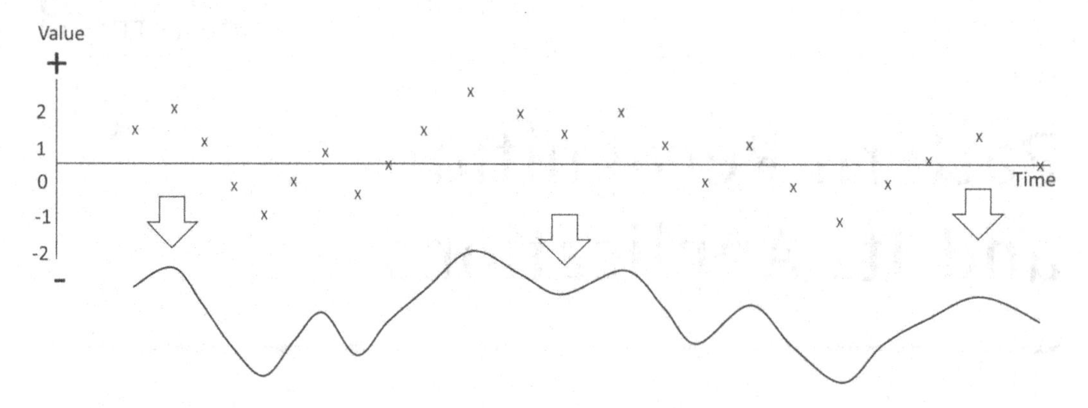

FIGURE 3.1 Graphical representation of data over time known as a *time series* of data.

the left (5 V) followed by a slower movement to the right (0.7, –1, –2 V) as the player runs around another player. This number series is digitised and then transmitted in a code form that covers a wider range of values. We can't transmit a value of 2 V when the actual *computer processing unit* (CPU) electronics package has a low power and uses a 1 V battery, so instead we transmit a code that represents that number.

Binary integers or digits (BITS) is instead used as a code to represent the value of voltage. If our smallest voltage output from a sensor is 0.1 V, then we can say that no volts can be represented by '0' while 0.1 V is represented by '1'. In binary code notation where each number of the code is two times the code next to it, then 0.1 V would be represented by 01, and 0.2 V represented by 10, and 0.3 V becomes 11. If we want to represent 0.4 V that is 100 (where 1 is 0.4, the following 0 means 0 × 0.2, and the final 0 means 0 × 0.1).

So for 0.5 V, we have 1, 0, 1 – which is 1 × 0.4, plus 0 × 0.2, plus 1 × 0.1 = 0.5 V.

For 1 V, we have the code 1, 0, 1, 0, which would be a code transmitted from the sensor package to the receiver as 1010 = 1 V just representing that single measurement.

For 5 V, we have 1, 1, 1, 0, 1, 0 representing 3.2 + 1.6 + 0 × 0.8 + 0 × 0.4 + 1 × 0.2 + 0 × 0.1

If our data has a time series of 5, 0.7, –1, –2, then we can represent a sign of polarity (+ or –) value as 1 or 0 (say 1 for + and 0 for –). Provided we put it at an appropriate position say at the front of the code, we can then label the number with its sign.

Each of these code groups is known as a binary *word*. Our data is a number series of positive and negative sample values of 5, 0.7, –1, –2. (But if it also represented the number 5712, we would just drop the sign values, and change our simplest integer value to 1 V [here it was 0.1 V], to become a single sample data word.)

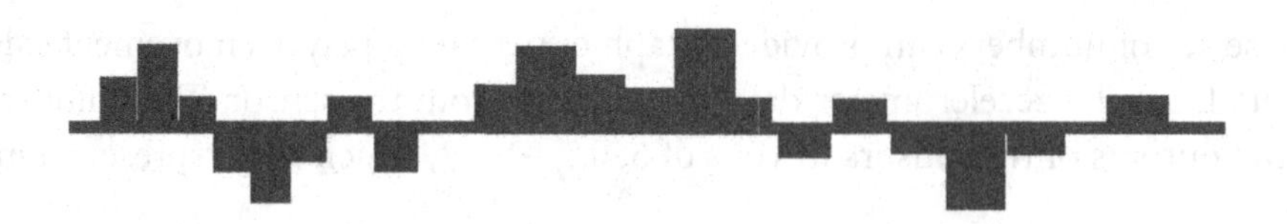

FIGURE 3.2 Histogram of the data in Figure 3.1.

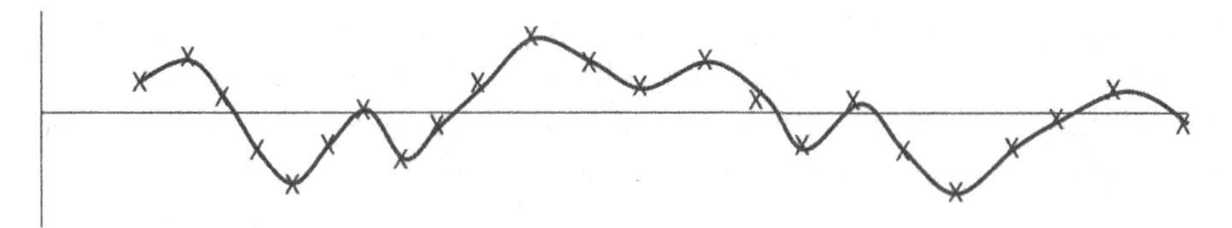

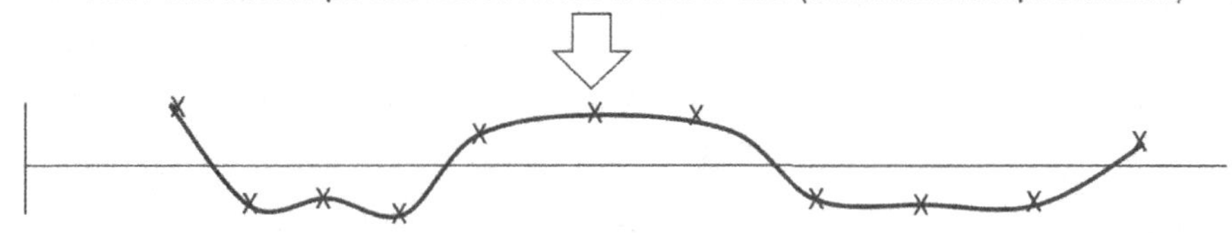

FIGURE 3.3 Aliased time series of the data in Figure 3.1.

So our original time series is represented by a code of six integers as follows:

5 = 1 111000, 0.7= 1 000111, −1= 0 001010, −2= 0 010100 (the underlines are only emphasising which are the sign bits) so this is:

1 111000, 1 000111, 0 001010, 0 010100 which is 5, 0.7, −1, −2

Now this code can be transmitted at a low-voltage level to the receiver as shown in Figure 3.4.

There is always lots of confusion between BITS and BYTES. Basically, now that you understand what a bit is, a byte is just 8 bits ([1]). This reduces the amount of bits we refer to when we say we can store 32 Kilobytes (32,000 bytes) on our hard disc drive. If a single integer in our code was 8 bits long (we had sign bit + space + 6 data bits = 8 bits), then a byte would represent a single integer or number. Each byte could therefore represent a data value in our coding. Therefore, a 32 Kb store would store 32,000 data values and if each byte was recorded every second, it would take 32,000 seconds (533 minutes or over 8 hours – a full day) to fill your storage area.

The transmitted code is a series of bits (or bytes) of equal voltage but arranged at equal time intervals or *sample period*. If we could sample and transmit this time series at twice

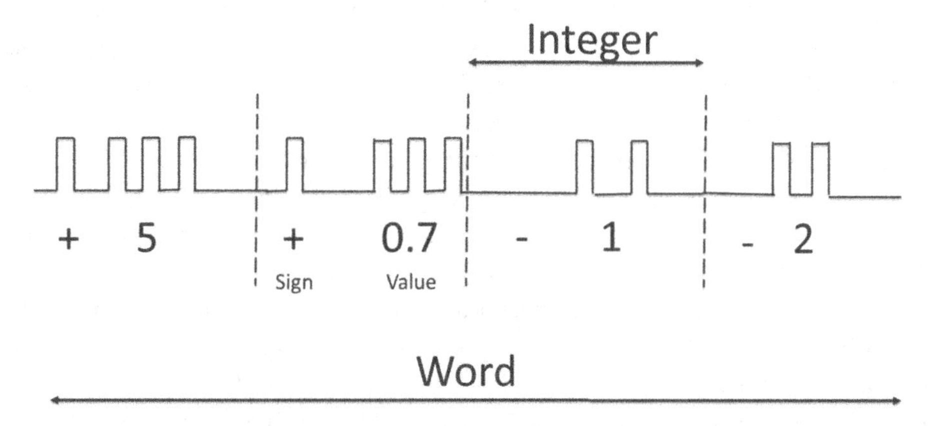

FIGURE 3.4 A *word* is a series of bits representing a number.

the frequency by reducing each sample time to half the amount, we are then *compressing* the data so that we can transmit twice as much data in the same amount of time [(2)]. It follows that the higher the transmission rate, the more data that can be transmitted. This means the higher the frequency, the more data we can transmit.

Obviously, this amount of storage space on a computer or *smart* watch (which is a miniature computer) would be plenty for a day's workout. When you consider that computer memories can store terabytes of data, this is 10^{12} or one million-million bytes – at say a recording data rate of a half second, this would last for 15.8 years of continuous data recording. That is plenty of time to get fit.

3.1.3 Higher Sampling Rate of Bits versus Accuracy

If we can transmit and store so much data, would it be more useful to transmit better quality data refined down to the finest detail of use to us than to have lots of average data? If instead of recording whole numbers, maybe we need finer detail, or in terms of computer screens, more pixels per image and therefore greater *resolution*? If this is needed, then we need to record more bits per byte – this will allow us to record 0.75 values instead of 0.7 values, and can be done by making the data integer (aka *byte*) 7 or more bits long so we can sample the fine values if they are important.

If the number of *light-emitting diode* (LED) pixels on a monitor screen is increased for the same size object, then finer and higher resolution images can be observed with greater detail – as discussed in Chapter 1. So when we increase the number of bits in a sample, sampling becomes finer and we can also take more samples in a period in time, which means that the sample time length reduces. By sampling at a higher rate, the accuracy of what we are sampling increases. We increased transmitted and received frequency to improve the speed of transmission to improve the speed of communication from 4G to 5G, as well as installing additional towers since a higher transmission frequency for the same power did not travel as far as previously [(3)].

3.1.4 Bit Resolution, Formats, and Storage

Industrial recording systems (e.g. used by banks or offshore survey companies) used magnetic tape recording from the late 1960s, with 8-bit, then 12-bit, then 17-bit, and eventually 24-bit sampling systems, being developed over the years. It has been explained that the greater the number of bits, the greater the resolution of a sample. These sample rates gradually increased from 8-bit to 24-bit over a period of 40 years, with the *24-bit α-microchip* being the result of manufacturing faster and tinier computer CPU integrated circuit chips. Since then, little further development on CPU speed has been done, apart from embedding chips, cooling them so they don't run so hot, and building them to run computer code in parallel by stacking CPUs on top of each other to make today's supercomputer.

The 8- and 12-bit systems were the first attempts at increasing the recording ability of tape recording systems. The method of magnetic recording was to magnetise dots representing computer *bits* on magnetic tape. If a '1' was to be recorded, a magnetic dot or a magnetised strip with a magnetic direction was put across the tape but if a '0' was needed, there would be no magnetised dot and a space or blank area would be left, or alternatively the magnetic strip across the tape could be reversed, as indicated in Figure 3.5.

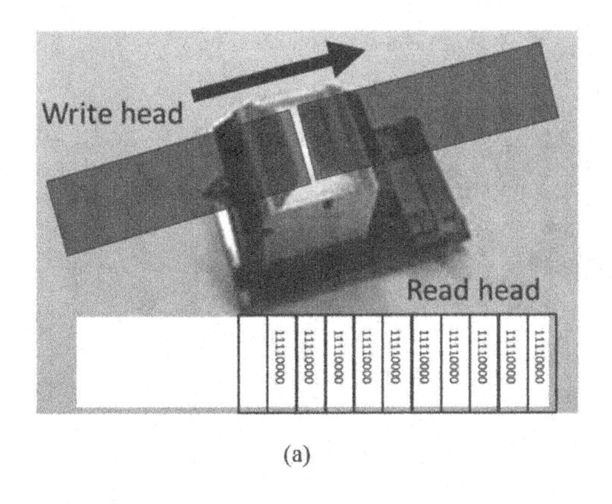

(a)

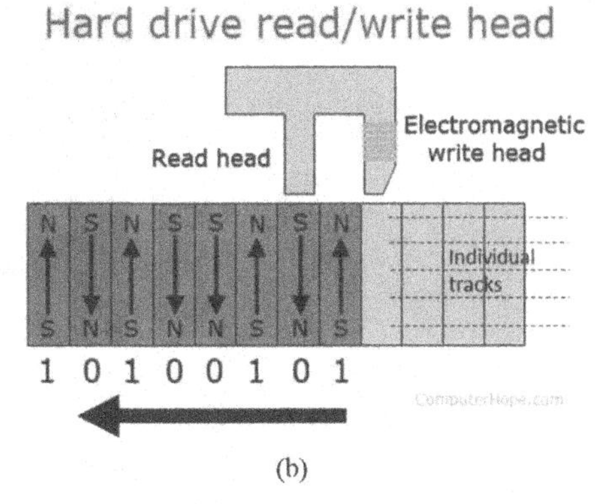

(b)

FIGURE 3.5 Magnetic recording head in RAW operation. (Modified from ComputerHope.com.)

The magnetising of these bits was called *read-after-write* (RAW) format. The magnetic data was read off the tape shortly after magnetising the data dots onto it, which acted as a continuous check that the data were being written correctly since early tape recording systems suffered from loss of synchronisation – the tape recording head becoming skewed with respect to the tape passing over it. Figure 3.5a shows the complex head writing data bytes across the tape as it passes over the RAW heads, while Figure 3.5b shows a schematic of a normal read and write recording/playback head which may be used for conventional recording of music. Simple data that used single bits such as that used in playing music, could be recorded along a single track of tape, whereas more complex byte-size data was recorded across the tape. This was done by building many individual magnetising points across the recording head so that the data would be recorded in the form of tracks. Initially, in the late 1960s, 21 tracks across 1 inch wide magnetic tape were used on steel reels, but these were quite heavy and were replaced by lighter 9-track tape on plastic reels in the early 1970s.

This form of recording lasted a decade until the 1990s when data could also be written on *magnetic cassette tape drives*, on *magnetic disc*, or on a *hard-drive* with the same formats but using *optical data* writing and reading (Figure 3.6). Hard-drives are often *optical drives* [4] that have data written on them by focussing a tiny laser beam on the surface of a disc which has an alloy coating that melts on contact with the beam, changing reflection angle slightly to write '1's and '0's code; while the 'read' occurs by focussing the same beam on the disc that has already been written upon so that the reflections come back as the code. These are commonly known as compact discs *CD*s (mainly used for storing music) and digital video discs *DVD*s (mainly used for playing video data such as movies) that have typical storage of about 4.7 Gb. Data compression was a major effort in the late 1990s, as companies worked hard to optically burn discs and read from them in tinier formats. Starting with the *CD format* of dots and dashes burnt 6 μm apart between tracks, the DVD closed the distance between tracks to half, to get twice the amount of data on a disc. Then the high-density *DVD HD* (*high definition* is the same as high *resolution*) was produced, with shorter distances again and 0.74 μm between tracks that stored about eight times more data. Eventually, Sony produced the *Blu-Ray* recording (called blue because it used the blue frequencies for burning), which was a very small disc – it wasn't as popular as

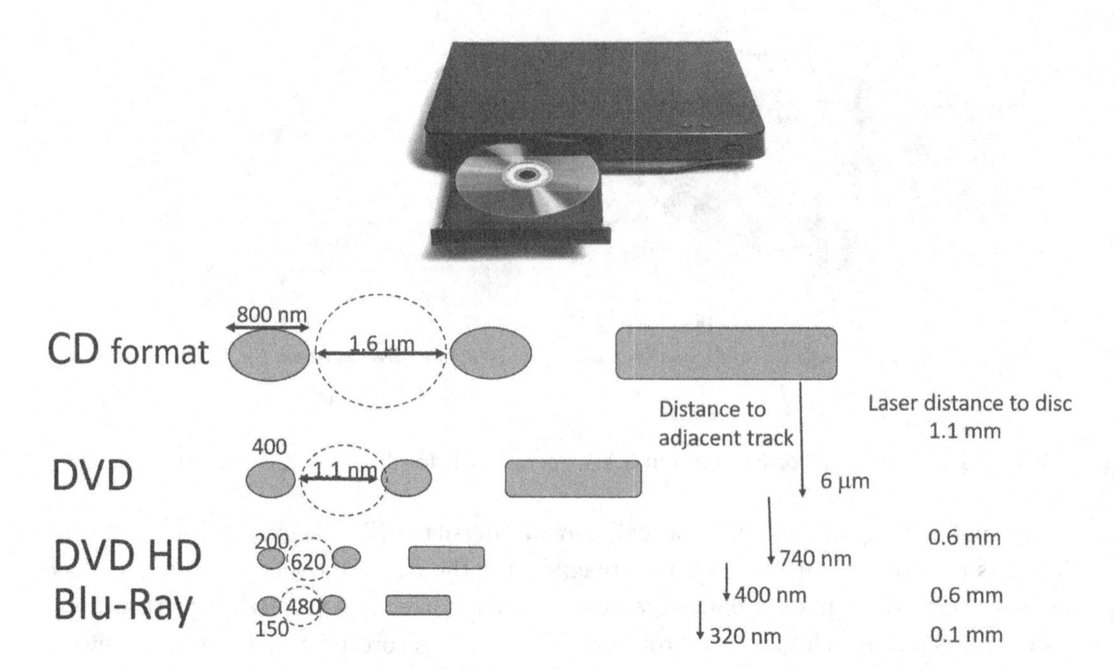

FIGURE 3.6 Optical drive uses a laser to write data – different storage media. (Modified from Shutterstock.)

the DVD, which is still around and used for home movie watching. But you could still get a good lengthy feature film on a Blu-Ray disc if you had a reader. The *DVD-R* disc player similar to Figure 3.6 indicated that with this DVD you could both *read* and *record* (i.e. *write*) and was similar to the DVD HD.

While we are discussing *optical technology*, we should just touch on how the *optical mouse* works. Figure 3.7 shows the inside of a mouse on the left and its basic operation on the right. Simply put, an LED shines a light (sometimes a laser might be used, which is more accurate) towards a mirror, which reflects the light onto the surface over which the mouse is travelling. The reflection of the light then passes through a *photodiode sensor* (which is like a tiny *solar cell* that is the opposite of an LED as it produces an electrical output when it receives light). The output of the sensor goes into a *digital signal processor* (DSP), which is simply an integrated circuit that performs a *cross-correlation* of the previous image it has in memory with the present image (which will be explained later in the next Section 3.2). The DSP then sends a signal of how far the mouse had travelled (the *mouse coordinates*) to

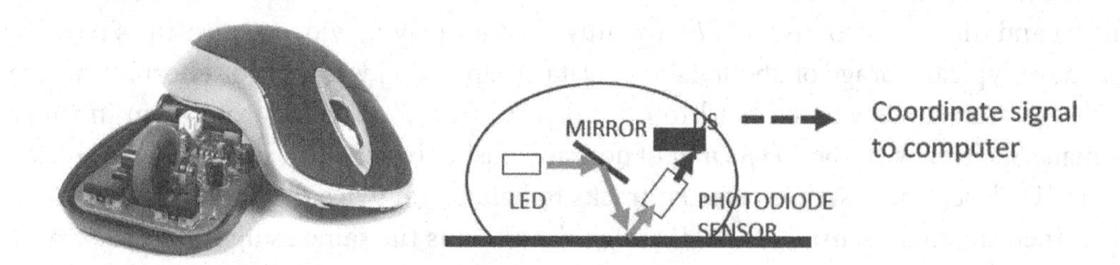

FIGURE 3.7 Operation of an optical mouse. (Courtesy: Shutterstock.)

the computer screen (usually by local wireless *Bluetooth*) which then moves the arrow on the screen. These mice don't work on surfaces that are very reflective like highly polished desks or glass, because the polish or glass deflects the reflection away from the light sensor.

Returning to the discussion on sample rate, popular music and video playback sample rates tend to be of higher frequency than the industrial multibit systems (which operate in the 1 kHz region), and as personal computing became vogue in the 1990s, recording went from removable disc to the hard-drive. Finally, we started using terabyte-sized external hard-drives or small easily manipulated integrated circuits (*flash memory, thumb drives* or *USB*) for small data storage and the *cloud* for larger data storage and manipulation over the internet. Just a note that the *thumb drive* or *USB* is just an integrated circuit for storing bits and uses the *Universal Serial Bus* (USB) port on a computer, while the *cloud* is nothing more than a data storage and computer point somewhere else in the world, and instead of storing your data locally on disc or thumb drive, it is stored via the internet, somewhere else – which might even be a different country.

Computing can also be done in the cloud, where instructions are sent by your computer via the internet from your location in New York, for example, to a computer in Singapore that has the required software, and this does the computation sending the result back to you in New York – that is known as *cloud computing*. It all depends on who your provider is. In *cloud computing*, you may never know where the actual computer is that is performing your computations, because that depends on where your internet provider has agreed to store the data [(5)].

> "In *cloud computing*, you may never know where the actual computer is that is performing your computations, because that depends on where your internet provider has agreed to store the data."

3.1.5 Rebuilding the Analogue Graphic Using a Digital Number Series (DAC)

During the writing of data bits onto media, we have to then read the stored data off it. As explained previously, the written bits must fully represent the data otherwise we cannot do a good reconstruction of the original data. Provided we have sampled the data at frequencies adequate for reconstruction – the industry takes that as being four times faster than they have been recorded (time intervals therefore being one quarter of the frequency, as discussed earlier in determining the *Nyquist* frequency) – then we can do a good reconstruction job of the screen graphics. The actual piece of equipment that does this conversion is known as a *digital-to-analogue converter* (DAC).

Many times particularly in sound recording, we don't have to sample so fast, because the output to a sound speaker is generally slower than the recording. That is, if we were to sample a sound recording every μsecond (which is 10^{-6} sec), this is a sample frequency of 1 MHz, but the human ear can only hear between 20 Hz and 18 kHz (Figure 3.8). The recommended sample rate for anti-aliasing is 96 kHz, which is almost 5 samples per cycle and inside the *Nyquist Criteria* of 4 samples per cycle for the highest level of human hearing. In

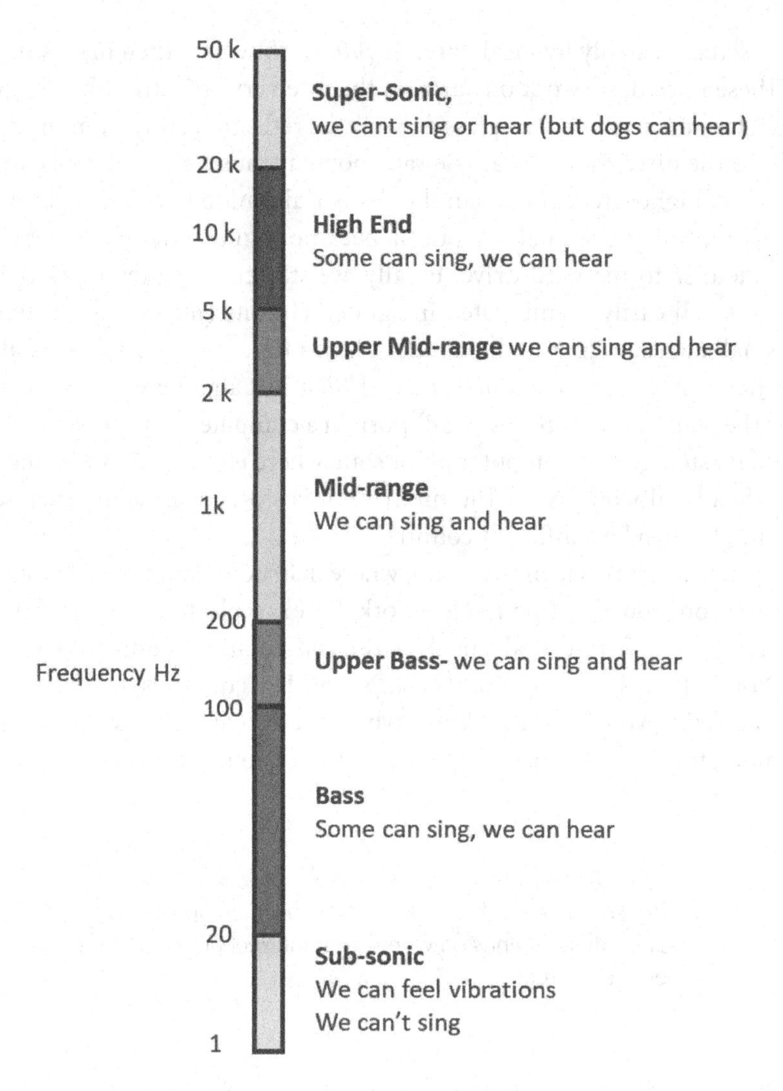

FIGURE 3.8 The frequency range of the human ear.

reality, because electronic loudspeakers are slower and adequate for hearing most frequencies, these are driven at a top-end of around 50 kHz sample rate for most applications while CDs use between 32 and 44 kHz with DVDs occupying the higher values around 48 kHz since DVDs have to sample both sound and video at the same time.

Also a brief note on sound. We hear much talk about the quality and volume of sound levels, which are valued in *decibels*. Because sound can go from a very low whisper to an ear-piercing level, this very broad range needs a basic scale that can condense large numbers into smaller numbers that we can easily handle – this is the *decibel scale* [6].

A *decibel* (dB) is a relative value only and it depends on the level of sound (or *noise*) we use as the basic starting decibel level. Generally, a decibel is a *natural logarithmic* scaled increase in sound amplitude, so when sound is twice as much as before, we say that it is

2^1 higher in amplitude than the previous sound (like when adverts come on TV and their sound level is much higher than the program you were watching).

So double amplitude $2^1 = 1 \times 10 \log_{10} 2 = 10 \times 0.301 = 3$ dB.

If we double again $2^2 = 2 \times 10 \log 2 = 20 \times 0.301 = 6$ dB, which is 4 times amplitude.

$2^4 = 4 \times 10 \log 2 = 40 \times 0.301 = 12$ dB, which is 16 times amplitude.

$2^8 = 8 \times 10 \log 2 = 80 \times 0.301 = 24$ dB = 256 times amplitude.

$2^{16} = 16 \times 10 \log 2 = 160 \times 0.301 = 48$ dB = 65,456 times amplitude.

You can see here how the logarithmic scale now keeps very large multiplication amplitudes at the dB scale that still has small numbers. It is worthwhile saying that we are talking about amplitude and not power. If we discuss how much power something is producing, A^2 = Power, so we just multiply the dB amplitude value by 2, and it becomes a dB but in *power.*

Figure 3.9 shows the relationship with actual noise values – the industry refers to people talking at the base level of a single volume ('1') and considers it as 60 dB in amplitude – this base value is really arbitrary. Loud talking may double the volume so the sound level becomes 70 dB. A small aeroplane taking off, which may be four times the noise, is 80 dB, while close-up to a rock and roll band is 120 dB. Generally in apartments, sound-proofing material is required to keep noise to less than 70 dB.

3.2 CORRELATION OF IMAGE DATA

Earlier we commented that correlation was used in the optical mouse to guide the arrow on your computer screen – this is done by comparing an image in computer memory (inside the mouse) with a new image the mouse was moved to, and then calculating the distance travelled and telling the computer where to move the arrow to. But the maths process of image correlation is used in many places by different technologies. For example, in financial transactions when we are comparing one magnetic signature with another, in character or *pattern recognition* when we are comparing a person's face at an airport immigration check with a photo in the central database, in fingerprint matching on a smartphone with the fingerprint in its database, and so on. The actual mathematical process is simple and will be discussed in this section.

3.2.1 Simple Correlation of Two Sets of Numbers – Recognition of a Wavy Line Is Simple

Consider that we have a series of screen pixels that represent a number (e.g. at one end of the scale black is no colour so is valued at 0, while white which is all colours, is valued at 10). A raster line of pixels can then be represented by a series of numbers.

We can use correlation (aka *cross-correlation*) when we want to confirm and recognise we have the pixel pattern by looking for the same number pattern within a number series. In Figure 3.10, in order to recognise the middle 7 numbers 0, 1, 2, 1.5, 1, 1.5, 0.5 (red in the figure), which is the short number series within the previous number series, we had of:

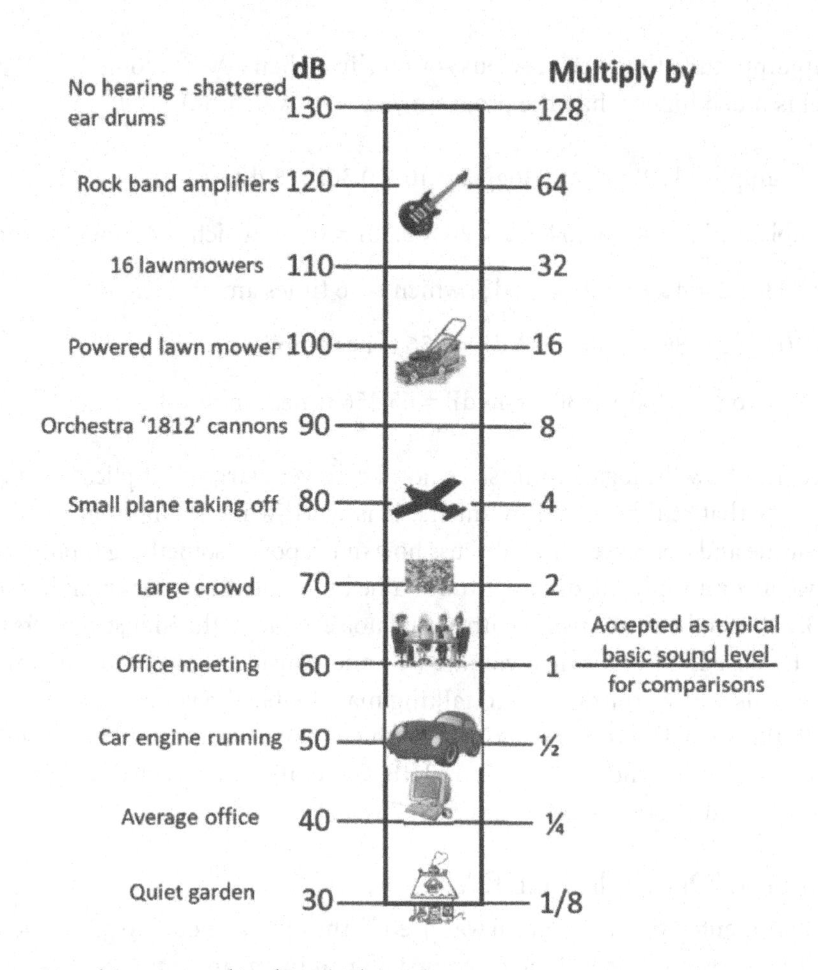

FIGURE 3.9 Sound is measured in dB, which is relative.

1, 1.5, 0.75, −1, −2, −0.8, 0.5, −1.2, 0, 1, 2, 1.5, 1, 1.5, 0.5, −1, 0.5, −1.2, −2, −1, 0, 1, 0

We run these seven numbers across the number series, then add the results and it produces a new number series that automatically indicates the location of the same numbers. For example, in Table 3.1.

The full number series is written on the top right line. The number series we want to find is written on the top left.

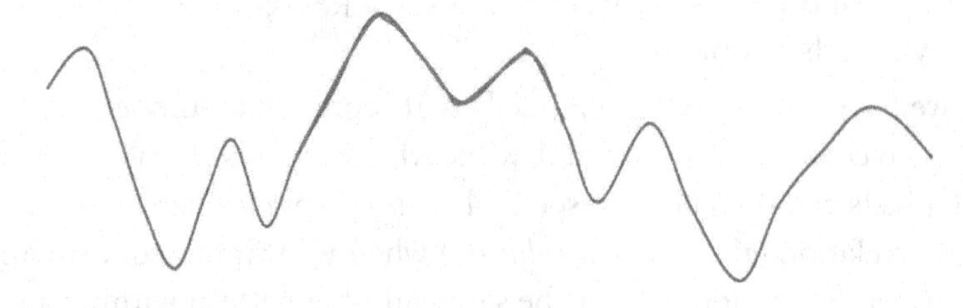

FIGURE 3.10 To find the number series 0, 1, 2, 1.5, 1, 1.5, 0.5 which is buried in our original numbers.

TABLE 3.1 Maths of Cross-Correlation of Two Sets of Numbers

0 1 2 1.5 1	0.5	1	1.5	0.75	-1	-2	-0.8	0.5	-1.2	0	1	2	1.5	0.5	-1	0.5
Position 1	**0.5**	0.5														
Position 2	**1.75**	1	0.75													
3	**2.375**	1.5	1.5	0.375												
4	**4.5**	2	2.25	0.75	-0.5											
5	**3.125**	1	3	1.125	-1	-1										
6	**-3.9**	0	1.5	1.5	-1.5	-2	-0.4									
7	**-4.8**		0	0.75	-2	-3	-0.8	0.3								
8	**-6.3**			0												
9	**-4.24**				0											
10	**-0.1**					0										
11	**0.1**						0									
12	**3.05**							0								
13	**6.75**								0	0	2	3	1.5	0.25		
→	**8.5**									0	1	4	2.25	1	0.25	
15	**5**										0	2	3	0.75	-1	0.25
16	**1.75**											0	1.5	1	-1	0.25
17	**0.5**												0	1.5	1	-2
18	**-1.5**													0	0.5	-2
19	**-1**														0	-1

Largest positive number indicates location of number series is between data points 11-17 with a peak at 14

Note: The top left number series (0,1. 2, 1.5, 1, 0.5) is passed over (correlated with) the top right number series by multiplying individual numbers, to produce the vertical bold numbers, having a peak value of 8.5 at position 14 indicated by the arrow.

1. Take the number series of top left and start the run across the other number series by going to Position 1, and multiply the last sample in the series (0.5) with the first sample in the main top right series (1) = 0.5 and post this in the sum column on the left of the vertical line.
2. Take the number series again and go to Position 2, where we now multiply the last two samples (1, 0.5) with the first two samples of the full number series (1, 1.5) and post the total sum in the left column ($1 \times 1, 0.5 \times 1.5 = 1.75$).
3. Go to Position 3 and multiply the last three samples (1.5, 1, 0.5) with the first three samples (1, 1.5, 0.75) and add them in the left column ($1.5 \times 1, 1 \times 1.5, 1.5 \times 0.5 = 2.375$).
4. Go to Position 4 and continue this process down the left side totals column. In Table 3.1, we have actually stopped at Position 19 because we don't need to go any further, but in practice we would complete the numbers.
5. Now read down the new number series of totals and the largest value found is at Position 14 of 8.5, which is the centre of the number series we are looking for. In other words, the number series has been located with a centre at Position 14.

Consider how this happened. We ran a small set of numbers across the full series and added them, using the concept that when a graph is overlaid and run across itself in the manner we have, when we multiply numbers, we are using the basic math statement that + times + equals + and a – times – equals +. So we will always get only positively summed values, even when most of our figures are negative. This approach is known as

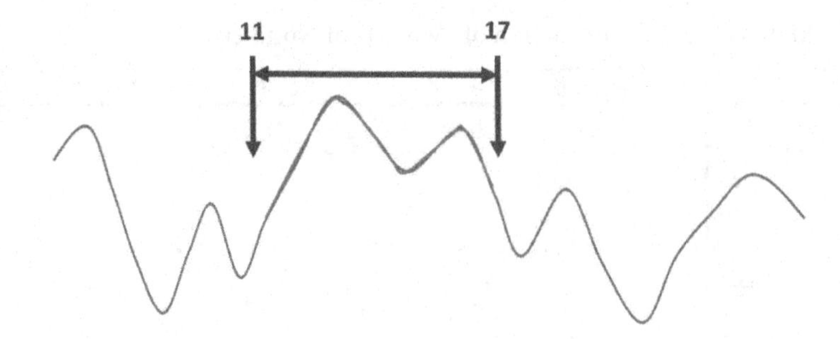

FIGURE 3.11 The desired location of the number series is in the centre of the full number set.

cross-correlation of two number sets (often referred to just as *correlation*), so if we centre the required number series at Position 14, we have recognised its location (Figure 3.11).

On a computer screen where the numbers represent pixels say in histogram (block) form, we would then have something like Figure 3.12 so that when we have the desired number series located, we can colour it red instead of yellow. When this is applied to pattern recognition operations, we can potentially pick out a face in a crowd (sometimes referred to as *flagging* an object).

Many authorisation events require the signature of a person to be verified. In this case, if a signature is scanned in black-and-white (B&W), each pixel of the scanned signature representing 1 or 0, with 1 maybe representing a black pixel and 0 being a white pixel. The number series then is simply a series of 1s and 0s which of course is very simple to correlate with a similar simple wavy line type signature in a database.

3.2.2 Flagging a Good Number Series Correlation versus a Weak Correlation

When we have a good correlation in a number series using this method, the total summed number values increase to a high positive value in the centre, which in this case was 8.5. However, if the required number series was not quite perfect as this was and some numbers

FIGURE 3.12 Numbers can represent pixel colours so that we can pick out a particular set of numbers and post them with a specific colour (here, red).

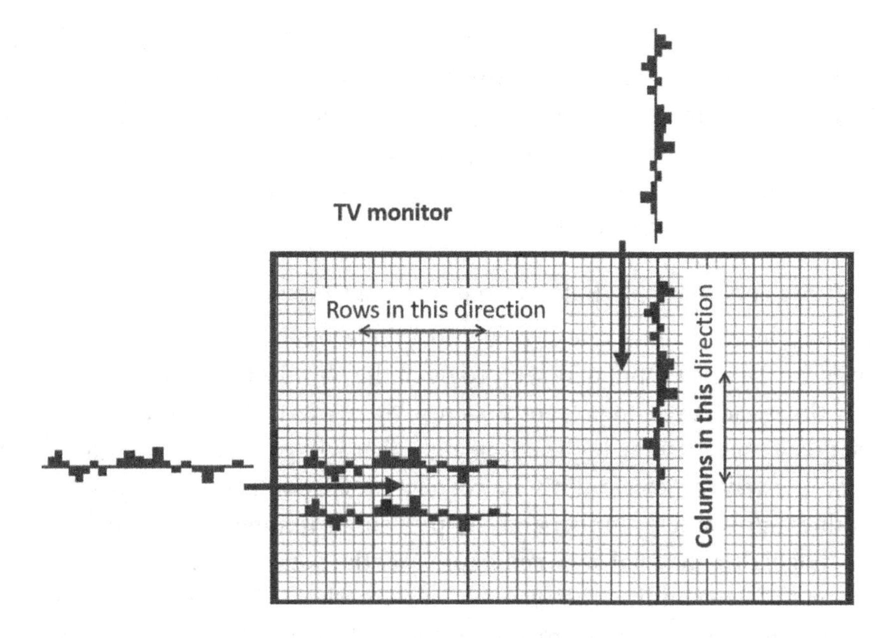

FIGURE 3.13 A cross-correlation of a number series either down columns or across rows.

were less in value or of opposite polarity, we could expect the peak number to reduce in value down towards zero – but probably just be a small number towards 1 in this case – and the summed number series (the vertical line of numbers) will appear quite ragged compared with a clear peak value in the centre.

The ability to be able to run a number series doing a correlation along either a horizontal raster line of pixel row values or a vertical string or column of pixel values (Figure 3.13) allows us to start at one end of the line and, by the time we are at the other end, if many summed values are similar and no clear peak value as in this case (of 8.5), then we could label the high value as being 100% correlation and the lowest value around 1 as a 0% correlation. The limits are determined by the user and there are many methods for determining how correlations are valued.

When we insert our passport into the automatic scanner at the airport, its 2D scan may not be anything like the real-life 3D person at the immigration gate with a correlation accuracy less than 70%, so we are sent off to stand in front of the immigration officer [(7)] so that he can judge (Figure 3.14). Another issue is that if there is a slight tear in the photo or on the passport page, the laser scan could cause a shadow in the paper and so the accuracy could also falter.

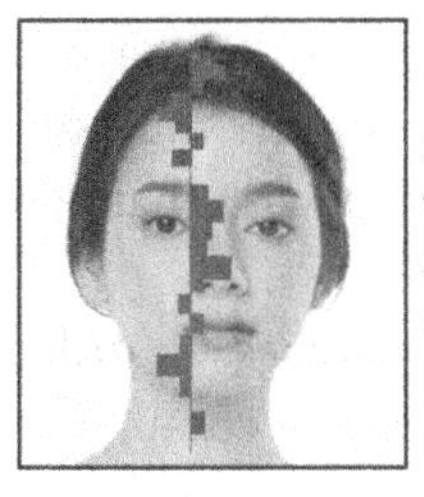

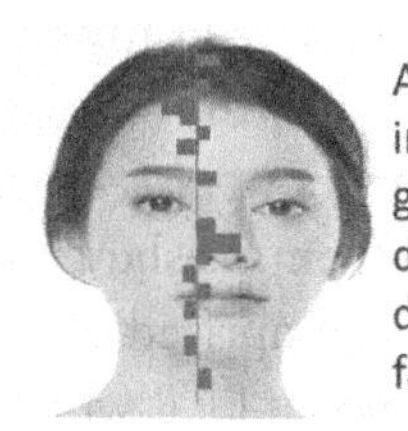

FIGURE 3.14 Scan data is correlated with photo. (Photo courtesy: Shutterstock.)

In the case of signature authentication, the accuracy required sometimes is no more than about 40% because people often scribble their signature differently each time they sign-off on something, so automated signature approval is often very weak, although accepted. Recently, we have seen signatures being required with the date of signature as part of the scan – which signifies that a signature on its own is no longer accepted unless accompanied by a date that has numbers involved.

3.2.3 Effect of Higher Sample Rate on Accuracy of Correlations

When we have a higher sample rate or resolution, there are more pixels and therefore more data samples for the same input data set. Provided the *Nyquist Criteria* is met (of sampling at least four times each cycle to prevent aliasing), there will be little difference in accuracy to a correlation, other than the high summed number will go much higher (8.5 increasing to 50+). Another way of determining if a correlation is good is to sum up all of the final summed values into a single number and compare single numbers – the more accurate the correlation, the higher the final summed figure is at the bottom of the sum column.

This concept of determining the accuracy of a correlation can fail if, for example, one number series has been converted from analogue to digital with a low sample rate (below *Nyquist*), and it is being compared with a different number series that has a much higher sample rate (above *Nyquist*). In this case, the act of correlation does not work properly, and the numbers will always result in a lower final summed value because the samples will be mismatched in time. Again, the user can flag a good correlation using either the centre summed value or a total summed value or both. A 50% correlation may be a half summed value compared with what would be expected as the highest summed value.

Note that this simple correlation may be done across the full pixel range, scanning along raster lines from top to bottom, so the maximum number of correlations computed in a simple HD-sized scanned screen (Figure 1.20) of 480 × 640 pixels will be the number of raster pixels correlated per line (480 + 480) × number of lines (640) = 614,400. At 1 msec per correlation, this would take 10 minutes, so generally the less the number of pixels, the faster the scan. Typically, a facial scan can take 10 seconds to take and compute at the immigration counter, but compare this with a full-body medical CT scan that requires a lot of detail and is high resolution, at 15 minutes to complete.

3.2.4 Increasing Resolution through Pixel Mixing

If we have a poor correlation, it may be because the resolution (number of pixels) on the screen is inadequate for our scanner. This can be overcome by mixing two adjacent pixels and producing a third pixel between any pair of pixels. To do this, all we have to do is to use a *Fast Fourier Transform* (FFT) to transform the two pixel's colours into their frequency range, and then take a frequency between them and insert a pixel with that dominant mid-colour frequency between them, as shown in Figure 3.15.

Any light that we see is comprised of many different frequencies. In order to find the frequency of light, we can do an FFT, which basically separates any complex wave out into its component frequencies and their respective phase [(8)]. It was a simplification of the Discrete Fourier Transform (1805) and was developed by Tukey and Cooley working at

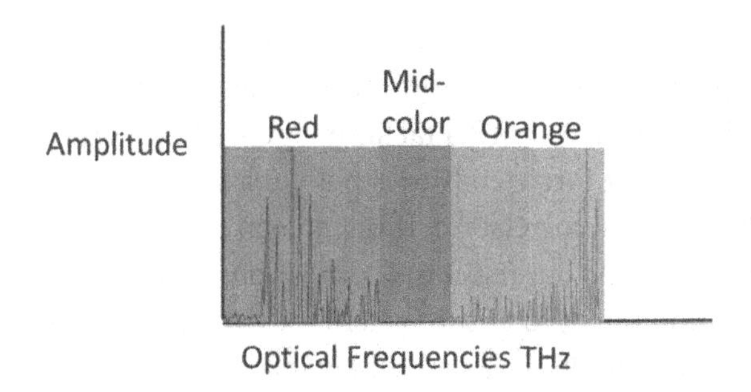

FIGURE 3.15 Mixing pixels to increase resolution.

IBM's Watson Labs in 1965 (the FFT is an *algorithm*, which is a fancy name for a complex equation [9]).

> "FFT is an *algorithm*, which is a fancy name for a complex equation."

Since the frequency of visible light is in the range of 430 THz (trillion Hz – 10^{12} Hz) for red to 750 THz for violet, going from red to the next colour orange would go from 430 to 500 THz and so 470 THz is mid-way in colour. Adding an extra 30% pixels to the screen would improve resolution and computation accuracy, and this approach is adopted by many industries to improve the resolution of a sparse data set.

3.3 APPLICATION USING A PIXELATED-MATRIX DISPLAY

We use correlation mainly when we are trying to automatically recognise a number series that represents images of objects constructed from pixels on a screen, which is often referred to as *pattern* or *facial recognition.*

3.3.1 Facial Recognition (on a TV or Monitor Screen)

A prime example of the use of correlation in facial recognition is the replacement of a border control immigration officer by an automated process that compares a visual data set from a local database or scan of a person's passport photo with the scan of the person standing at the immigration gate.

When we have a scan that has been taken from a passport photograph, the scan may be in colour or B&W – B&W is much easier to convert to a number series than colour. This is because the colour range is broad while B&W is a limited number of shades of grey. The limited numbers produce less resolution, a reduced number range, and therefore faster computer scanning and data storage, so that immigration departments budgets can be less using lesser quality and powered computers. The problem is that we will be doing a cross-correlation with data obtained by low-resolution cameras at the immigration gate, which does not allow for the fact that the person's appearance when scanned at the gate may be quite different from when the passport photo was taken.

Consequently, this is why people are asked to remove hats and glasses when being scanned at the gate – simply because a poor correlation will always result. In Figure 3.14, the scanned passport photo may be compared with a facial image that has a different hairstyle or fatter looks. This is where border force tends to slightly relax the correlation requirements and may go for just a 50% accuracy correlation. If the person looks away at the time of the facial scan, then there is nothing to do but reject the correlation result. The software will perform a rough scan at low resolution first, to ensure where the head is (having been programmed to look for an oval-shaped object within a certain height range), and then just scan within the oblong or oval (this is very similar to how some photocopiers scan – with a rough scan followed by an accurate scan within a defined area). Immigration gates often have lights flashing telling the person to stand in a particular zone – with the flashing lights actually being part of the rough scan process, before the detailed scan light flashes.

When a scan fails to reach the desired accuracy level, the software often tells the person being scanned that they have to go to see the immigration office, person-to-person. It is not necessarily that the person has looked away, and maybe that the scanned passport photo inadequately represented the person's appearance in the first place.

3.3.2 EM Data Scanning for Financial and Other Transactions

When we use a credit card at a payment terminal, the *electromagnetic* (EM) field around the card carries the number of the card, which is picked up by the card reader and compared with the card numbers in its database. The EM method of card number retrieval will be explained later but the point is that the card number goes through a correlation process with the number in the database. In this case, it has to be 100% accurate, or the card is rejected. This happens with other forms of EM transactions such as bus and rail pass cards where they are reading the period of use on the card and correlating it with the day it is being used – then transferring money from the cardholder's account.

3.4 CORRELATION APPLICATIONS AND USE IN SECURITY DEVICES – FROM CROWDS TO EYEBALLS

Correlation is a widely used, simple data manipulation method adaptable for automating recognition of characters. A major use of correlation today is in crowd security and monitoring the actions of groups of people.

3.4.1 Application of Correlation to Pictures (Face in the Crowd)

The ability to correlate the face of a travelling passenger in the airport has not been lost on the crowd control industry, where security forces want to know if a particular person is in a crowd, or they want to know the security details of wanted persons on their database. Even the use of *Identikit* pictures by police to put together the face of a criminal has been abandoned in preference to being able to grab a series of features from a digital library and patch them together to make a good likeness of the image being searched.

Finding a face in a crowd depends entirely on the clarity and resolution of the vision of the crowd compared with the database picture. Normally, the colours of a crowd scene are broad but, provided there is a high pixel rate of the cameras taking the movie or photos, they would be readily correlated with the database photo [10]. The movie scenes of

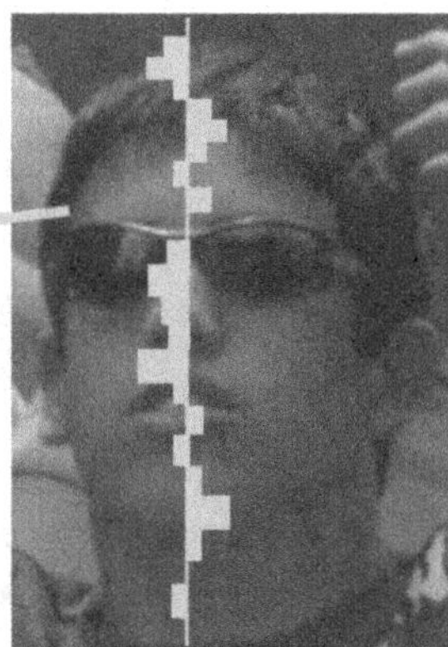

FIGURE 3.16 Spotting the face in the crowd. (Courtesy: Shutterstock.)

Figure 3.16 monitoring a crowd and the computer being able to recognise images of faces in the crowd with a 70% accuracy of correlations are not too far from fantasy.

But much more can be done with these pixels because they carry the colours of the people in them, which in themselves can provide other data. For example, many border force and security monitors today in airports and rail stations have the ability to use *infrared* (IR) cameras to look at the heat of faces, which is possible because, as explained in Chapter 1, we can get a pretty accurate measurement of heat from the IR range of the EM spectrum. The monitor receives a picture of someone walking through an airport, and each face is initially scanned, followed by a *spectral analysis* (analysis of the frequency content using an FFT) within the box of all of the pixels. If the analysis indicates a specific frequency within the IR bandwidth, this can be tied to a temperature and that temperature is then placed within the boxed face on the security guard's monitor screen. Doing an FFT today is really fast and an FFT can be done on a face within a box of pixels within milliseconds.

Therefore, the security guard can look at the monitor that has the faces of people in boxes, showing their individual temperatures, and in the event that a person is running a red-hot fever, the security guard can then inform ground staff or ask the computer to track that person through the airport while informing other guards at the exit.

3.4.2 Application of Correlation to Sound

A recent innovation, although not yet widespread but becoming acceptable over time, is the use of voice recognition in smartphones. Apple Corporation introduced a *Speech Interpretation and Recognition Interface (Siri)* software in 2010, which was considered a leader in *voice pattern recognition and response* software. Its competitors include *Alexa* from Amazon and *Cortana* from Microsoft (Figure 3.17).

When we talk into a *piezoelectric* crystal microphone in our phone, the sound pressure pulse waves we make (reviewed in earlier sections) are converted by the smartphone's piezoelectric crystal into an electrical pulse, and vice versa when we listen to someone talking on our phone speaker [11]. These analogue sound waves are digitised that looks like Figure 3.18 and are in a relatively low-frequency range of 85–250 Hz (lower frequencies for male, higher frequencies for female).

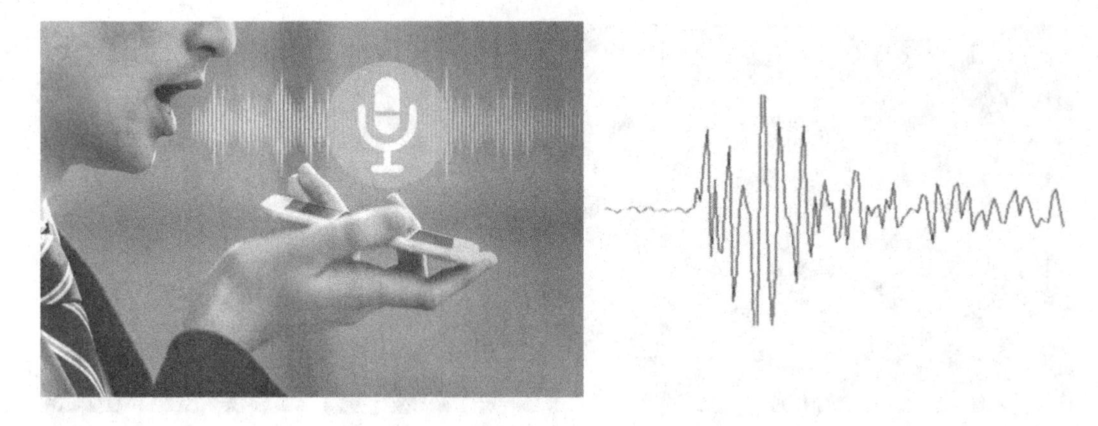

FIGURE 3.17 Voice recognition. (Picture Courtesy: Shutterstock.)

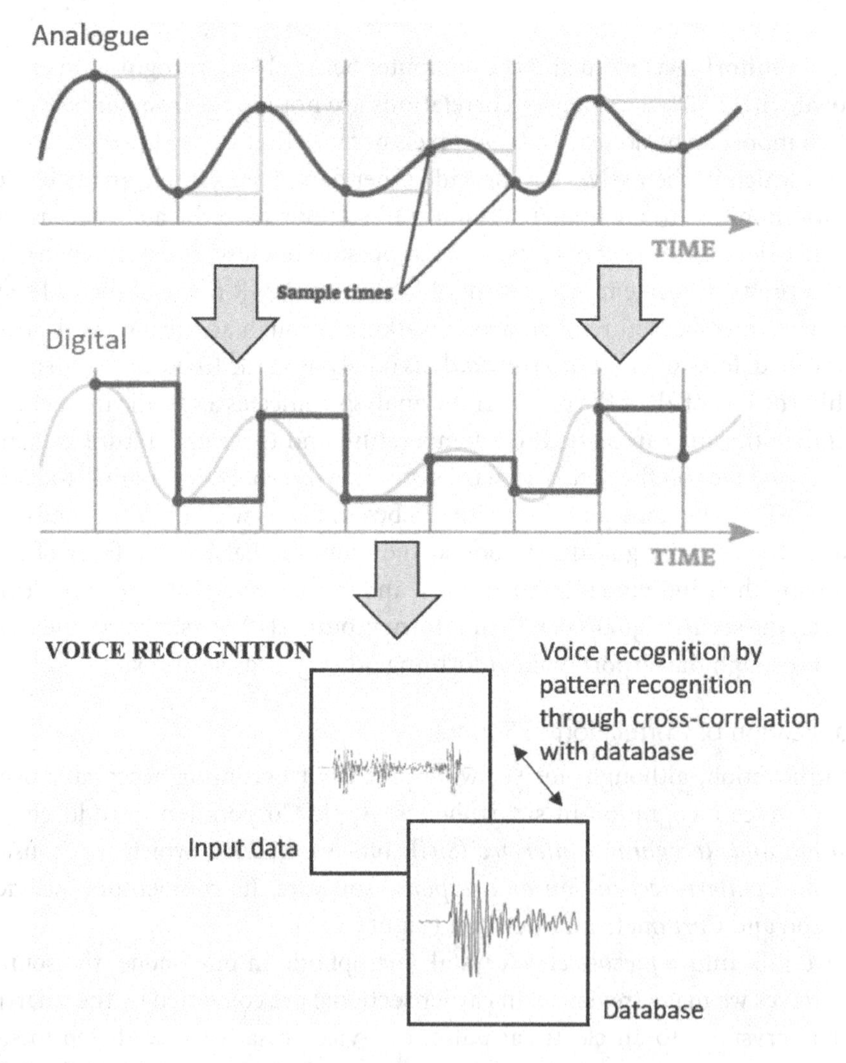

FIGURE 3.18 Analogue voice produces a digital signal after A-D conversion, which is then correlated with the database. (Input data courtesy: Shutterstock.)

This data set can be converted to a number series as we have discussed earlier in this chapter, but the number series can either be produced from the amplitudes or from the frequency content (after FFT conversion), or both. What voice recognition software does is to take that number series of amplitude or frequency and run it across a set of stored numbers in its database. For each set of stored numbers, there is a sound in amplitude or frequency. When a series of sets of numbers are provided to a database, the search engine software searches for the nearest grouping of that series and has a stored set of answers for that grouping.

Voice recognition software requires a keyword to start this process, which only accepts questions (for which it has stored answers). This is why you have to say '*Siri*, play me the song.........' Here *Siri* is the keyword (which after an FFT always has the same frequencies from the same person) to start the recording process and once the digitised sound or song is recorded, the correlation process is performed to obtain the appropriate answer to the question.

Voice recognition software has to be programmed to answer specific commands, which can also be household or office actions, such as turning the house lights on or lowering automatic curtains. You can't walk into a house and ask *Siri* to play some specific music as well as turn on the heating unless it has been programmed to do this.

Early voice recognition software received a lot of poor publicity when it either failed to recognise the command or question being asked, or instead mistranslated the command to do something else instead. This was because the database number series was developed using a specific set of frequencies, and often the person's voice may have a different set of frequencies or the emphasis on some words was such that the frequency content was not as it was programmed. This can easily be the case where two words sound similar or the same word has a number of meanings (e.g. is 'bow' a weapon an archer uses to fire an arrow or is it the front end of a ship – and in some dialects, a 'quay' is pronounced 'kway' while generally in English it is pronounced 'kee'?). This difference can produce hilarious moments.

To avoid this, we have to provide examples of how we speak to the database, so the speaker must provide a number of chosen words or phrases in order to establish the database to begin with. A set of phrases were developed during the World War II to determine how well radio operators could hear English in a noisy environment. It is often referred to as the *Harvard* sentences, in which there are 72 different lists of 10 lines each containing no more than 12 words. They include every possible noise (frequency mix) people hear during conversation, and here is the most popular list:

1. The small pup gnawed a hole in the sock.
2. The fish twisted and turned on the bent hook.
3. Press the pants and sew a button on the vest.
4. The swan dive was far short of perfect.
5. The beauty of the view stunned the young boy.
6. Two blue fish swam in the tank.

7. Her purse was full of useless trash.
8. The colt reared and threw the tall rider.
9. It snowed, rained, and hailed the same morning.
10. Read verse out loud for pleasure.

The ideal voice control was used in the movie *2001: A Space Odyssey* in which the central computer was called *Hal* (*Heuristically programmed Algorithmic* computer) and was sufficiently well programmed that it could discuss all manner of issues on multiple topics with all of the crew (of course *Hal* eventually wanted to take over from the human space travellers but Dave unplugged it).

Today similar pattern recognition methods have been adopted by major computer companies such as IBM, which has developed *Watson* to help industrial engineering operations people [(12)]. It is used in oil and gas processing operations, but rather than voice, operator's type information data to its database, and thereafter when an engineer types a question about a specific oil or gas field in its database, it can answer the question. This makes the engineering design, calculations, and decision-making process much faster and efficient than otherwise.

3.4.3 Application of Correlation to EM Fields (Credit Card or Door Key)

This topic was briefly discussed earlier in Section 3.1.4, but basically if we have a magnetic card, there can either be various magnetic patterns representing its code across its magnetic strip like the magnetic tape recording of Figure 3.5 or it can have a magnetic block that represents the card number (to be discussed). If we put the magnetic card through a reader or tap it against a reader on a door, this action breaks the magnetic field lines of force in such a manner that the card's code is read. This number is then passed electrically to the financial institute or to a hotel computer, where the number is cross-correlated with numbers in its database and, if correct, a signal is sent to either approve the transaction or open the door.

3.4.4 Application of Correlation in Industry

In industry, we have already discussed how pattern recognition through correlation of a word pattern in a company database is used to respond to questions put by engineers. However, correlation is also used with *bar codes* on products stored in warehouses, and on postal items when goods or parcels are being stored and later sent to a consumer. In this case, correlations with a product stores database include a history of the good's manufacture, cost, transportation, and storage details.

In the food industry, shops and supermarkets have food items with optical *bar codes* stamped on labels on the outside where the line of numbers printed adjacent to the bars are represented by different width B&W stripes. In this case, the food item at the shop checkout is passed over a laser light source or LED – the image is then reflected down to a single light receiver (photodiode) or a strip of receivers within the reader so that the lines (numbers) are correlated with bar code numbers in the shop or supermarket database

FIGURE 3.19 Optical bar code. (Courtesy: Shutterstock.)

(Figure 3.19). This not only provides the checkout computer with the price but also updates the supermarket database so that the supermarket can take a continuous stock check. These bar codes are officially called *Universal Product Codes* (UPC), and different forms of bar codes will be discussed later.

Pattern recognition is also used for secure access to special areas. For example, *finger-print and eye pupil* recognition can all be used to allow access of one individual person rather than others.

In *finger scanning* (aka *biometric scanning*), each person's fingerprint is unique with ridges and valleys (even identical twins are different). Finger scanners (Figure 3.20) pass a low-level current between closely spaced contacts and the received signal provides a figure of resistivity of the skin (no current will transmit across a valley), to build up a picture of

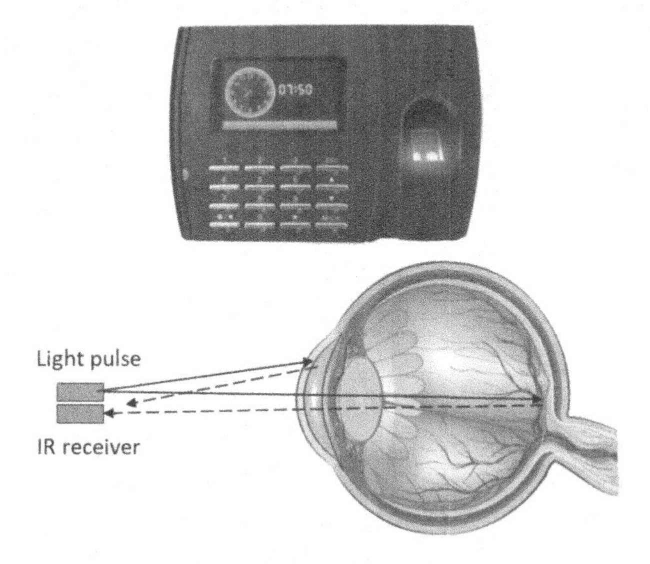

FIGURE 3.20 The upper photo shows a biometric finger instrument while the lower diagram shows the eye scanning concept. (Modified from: Shutterstock.)

electrical resistivity, which is translated into a picture of a finger or thumb. The image is correlated with the database image so that a successful match allows entry to a room – these are often used in where security is an issue, such as airports.

Alternatively, *eye pupil* scanning recognises the iris and retina. Like the fingerprint, the iris and its retina are unique to each person, with complex colour patterns in the iris and blood vessels in the retina (Figure 3.20).

- *Iris recognition* uses IR transmission (a pulse of white light is fired and an IR filter is put on the receiver) which is more accurate than facial recognition since it has a higher density of pixels limited within the IR range, while the iris is internal to the eye, protected and visible.
- *Retinal recognition* uses the same methodology but is more stable than the iris that can move during a scan. However, these eye pupil scanners can damage the eye, which is the reason they are not used very often (or at least they have the insurance to defend a court case).

To support automation developments, rather than have paper drawings we can scan paper drawings or plans/documents using an *Optical Character Recognition* (OCR) scanner (Figure 3.21) that scans each line of an image putting the data into 1s and 0s format, and correlates this with a database of letters, numbers, or characters. Alternatively, the magnetic form is known as a *Magnetic Ink Character Recognition* (MICR), which is used by banks and financial institutions for checking documents that have been printed with

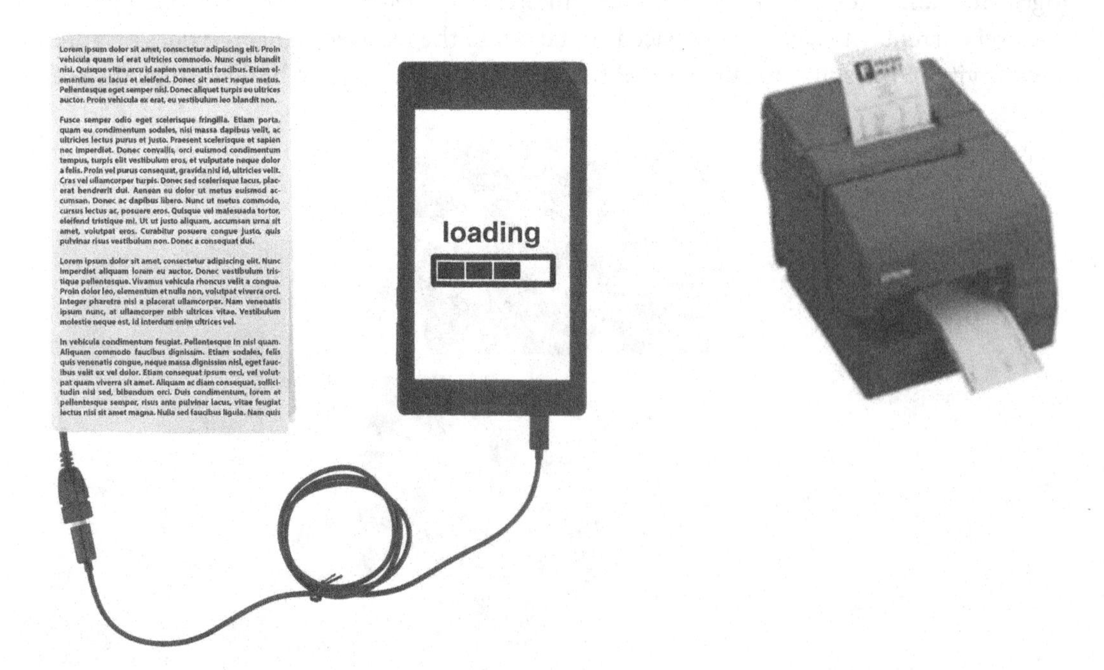

FIGURE 3.21 On the left is an *optical character recognition* (OCR) instrument and on the right is an example of a *magnetic ink scanner.* (Courtesy: Shutterstock.)

magnetic ink – this reader is only used for number checking. It is very simple and quick to use, such as with credit card payments for accommodation in hotels.

3.5 EXERCISES

1. **Determine the computer word for the number series 3, 1, –2.8, 0, assuming 0.1 is the value of the basic binary integer.**
2. **We have installed a new exhaust pipe system on our car. How much, in real numbers, has the sound amplitude level of the exhaust system changed if we say it has changed by –24 dB?**
3. **Exercise on pattern recognition (passport photo versus immigration gate photo).**

   Consider low-resolution pixels where a dominant light fill in the majority of the pixel is given 0 and any other pixel having a majority of its fill in colour is given 1, and just do a correlation with 5 numbers (because this is low Ignore this arrow resolution).

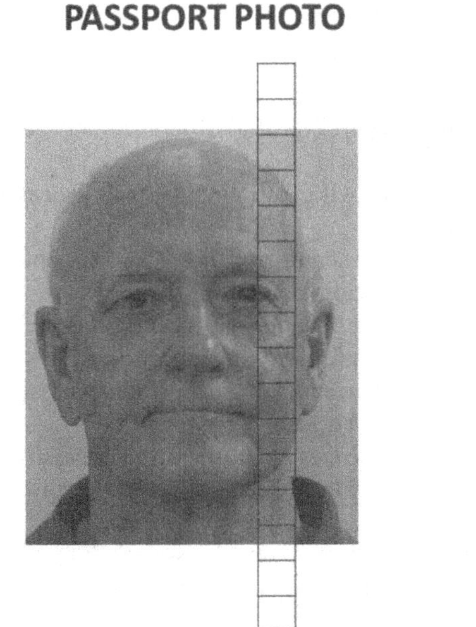

- What is the number series?
- Cross-correlate the passport photo first five numbers with the identical set on the immigration gate photo.
- What value is the peak number? If the final peak value equals the sum of the 5 passport numbers, there is 100% fit. If 0, there is 0% fit. Any numbers between those values are a %age of it. A number greater than 50% will pass immigration.

As a reminder, cross-correlate the five numbers.

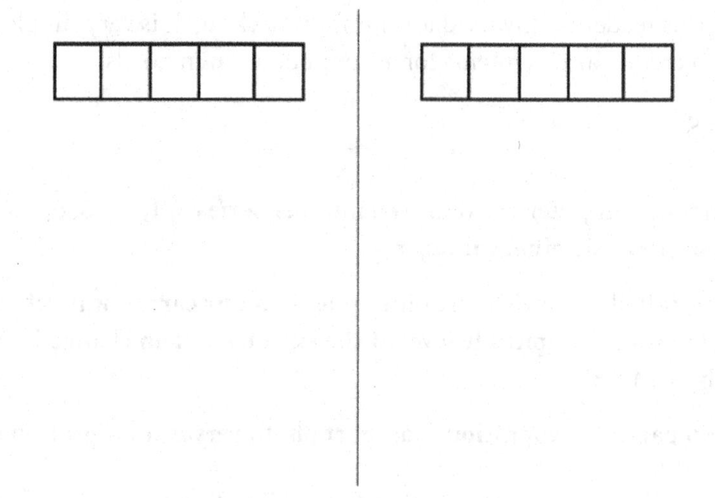

Imagine the computer doing this but with the full vertical number series – a 50% match of all vertical lines will often be considered adequate, but any less may be considered failure, and your scan rejected – you have to personally go and see an immigration officer face-to-face if that happens.

FURTHER READING

1. Understanding bits and bytes: https://www.lifewire.com/the-difference-between-bits-and-bytes-816248
2. Sampling and processing: https://en.wikipedia.org/wiki/Sampling_(signal_processing)
3. 4G vs 5G: https://www.finder.com.au/4g-vs-5g
4. How optical drives work: http://www.notebookreview.com/news/how-it-works-optical-drives/
5. Cloud computing: https://www.salesforce.com/au/cloudcomputing/
6. The decibel scale: https://www.britannica.com/science/sound-physics/The-decibel-scale
7. Passport scanning: https://knowtechie.com/key-features-passport-scanner/
8. What is an FFT waveform analysis? https://www.dataq.com/data-acquisition/general-education-tutorials/fft-fast-fourier-transform-waveform-analysis.html
9. What is an algorithm: https://www.bbc.co.uk/bitesize/topics/z3tbwmn/articles/z3whpv4
10. Pattern recognition and crowd analysis: (PDF) Pattern recognition and crowd analysis | Stefania Bandini – Academia.edu
11. How do speakers work? https://www.soundguys.com/how-speakers-work-29860/
12. What is IBM's Watson? https://fredrikstenbeck.com/what-is-ibm-watson/

CHAPTER 4

Average Track and Prediction of Future Location

4.1 UNDERSTANDING THE MEANING OF AVERAGE TRACK

When we want to try to predict the direction of a moving object, we need the initial positions first from which we can try to predict the next steps. When a ship's captain wants to predict which way he should turn the steering wheel, he has to look at wind and current direction first and then modify his course to allow for these. Because these values may be unknown, the captain can use the past positions (which have included the effects of wind and current) to make this prediction, which is called *dead reckoning.* To do this, the captain has to get the average values of position to be able to compute the *average track* in the past and then make the prediction of which way to steer in the future.

4.1.1 Average versus Mean, versus RMS Track

When we work with an athlete's positioning data (that is transmitted by *wifi* from the accelerometer data they wear on their back), sometimes it is easier to compute the average distance from one point to another like the ship's captain, rather than the precise distance (which may be more accurate if positions are accurately known, but takes a lot more effort since the athlete may be darting around from one point to another not even in a line or circle).

Average track (sometimes referred to as a *moving running* or *rolling average*) can be found by taking all of the values within a selected number of locations and adding them, then dividing by the number of locations [1]. We can use the numbers of our original series of 1, 1.5, 0.75, –1, –2 where, for each step an athlete takes, +1 m is on one side of a central line in the *Y*-direction, while –1 m and –2 m represent how many metres the player has moved to the other side of this line (shown dotted in Figure 4.1). The average line allows us to extend the average offline position potentially to where the athlete may run next (*Y* in the figure).

Working this out $(1 + 1.5 + 0.75 - 1 - 2)/5 = 0.25/5 = 0.05$ m, which is 5 cm off the straight line in the future. If we now add the later values of our number series, the result changes.

DOI: 10.1201/9781003108443-4

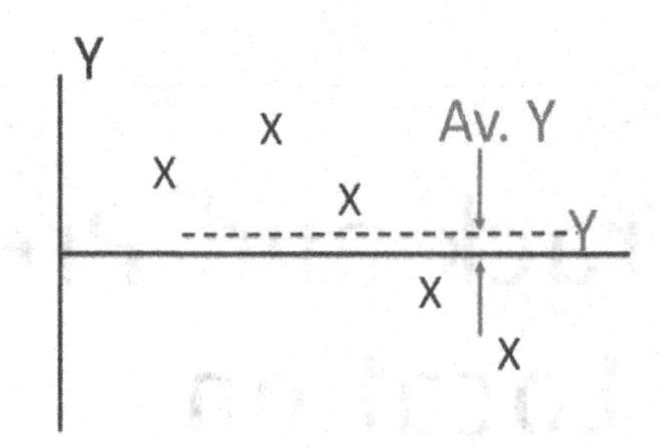

FIGURE 4.1 Average track of data points.

That is, (1 + 1.5 + 0.75 – 1 – 2 + 0.8 + 0.5 – 1.2 + 0 + 1 + 2) = 3.35/11 = 30 cm, which is further from the *X*-line than 5 cm we had before as shown in Figure 4.2.

This indicates two things:

1. The future position prediction depends on the number of the most recently chosen values.
2. Average track is only a rough guide of where we should expect the athlete to move because clearly the athlete's locations are in some form of slalom so the actual location will likely not be at this position – we should consider other methods of prediction.

An alternative approach is to use the *mean track* method to predict future movement directions. The mean track takes the value nearest the middle of the number spread, which would be 0.75 for the earliest group of 5 (1, 1.5, 0.75, –1, –2), and 0.8 for the group of 11 (1, 1.5, 0.75, 1, –2, 0.8, 0.5, –1.2, 0, 1, 2), which is not representative of all values.

Instead, let's try the *root mean squared* (RMS) method. The RMS method takes the value computed as the square root of all of the individual values after squaring and summing them. A tricky part of the RMS value is that the sign of each location distance is an indicator of its relative position, which must be retained rather than be removed (any number squared removes the sign, i.e. –3 × –3 = +9 normally, but here we do not remove the sign since it shows positive or negative direction off a central line of values.

Squaring our 11 number series becomes (1, 2.25, 0.5625, 1, –4, 0.64, 0.25, –1.44, 0, 1, 4), which added together = 5.2625. The square root of this is 2.294. Now, because RMS values have been applied as meaningful to sine wave motion (e.g. in alternating current, we say that the RMS current is the value delivered equivalent to a steady direct current value), we are not dealing with a sine wave here but more of an irregular slalom in values, so it seems

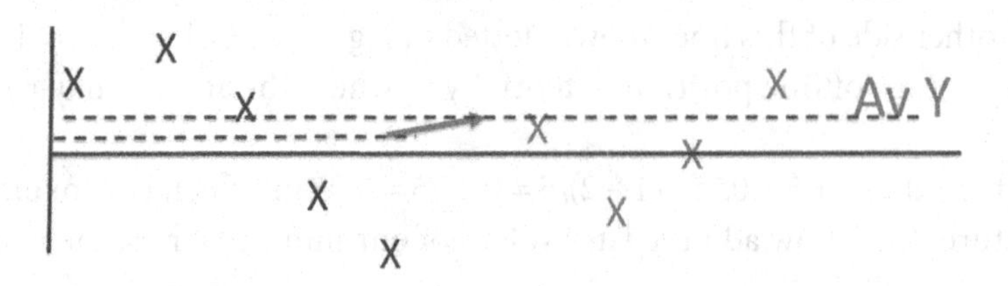

FIGURE 4.2 Average/mean track difference.

that the average values are the simplest to deal with, provided we have a short set of input numbers (such as 5 instead of 11).

To obtain the average track over a long number series, and then predict where it will be travelling to next, we first decide how many input points we will have, take their average value, and plot that. Then we drop off the first value we had at the start and add the next value at the end of the series.

Let's expand our number series to explain this – and plot the series in Figure 4.3.

1, 1.5, 0.75, −1, −2, −1.8, 0.5, −1.2, 0, 1, 2, 1.5, 1, 1.5, 0.5, −1, 0.5, −1.2, −2, −1, 0, 1, 0

Step 1 – take an average of the first 5 and plot it.

Step 2 – drop off the first value and take up the second value, average 5 and plot it.

Step 3 – continue on, dropping the first and plotting the next average.

For the first nine groups of five we have:

1. Av 1, 1.5, 0.75, −1, −2 = 0.05
2. 1.5, 0.75, −1, −2, −1.8 = −0.5
3. 0.75, −1, −2, −1.8, 0.5 = −0.7
4. −1, −2, −1.8, 0.5, −1.2 = −1.1
5. −2, −1.8, 0.5, −1.2, 0 = −0.9
6. −1.8, 0.5, −1.2, 0, 1 = −0.3
7. 0.5, −1.2, 0, 1, 2 = 0.4
8. −1.2, 0, 1, 2, 1.5 = 0.6
9. 0.5, −1.2, 0, 1, 2 = 0.5

In this figure, we can see that by connecting the average track values of the first nine sets of data, we have a reasonable representation of where the data has been moving, but within a small range of values which does not represent the extremes, only the general position. This is what we call *filtering* the data, and is a 'heavy' filtering when it effectively reduces the extreme amplitudes, flattening the data.

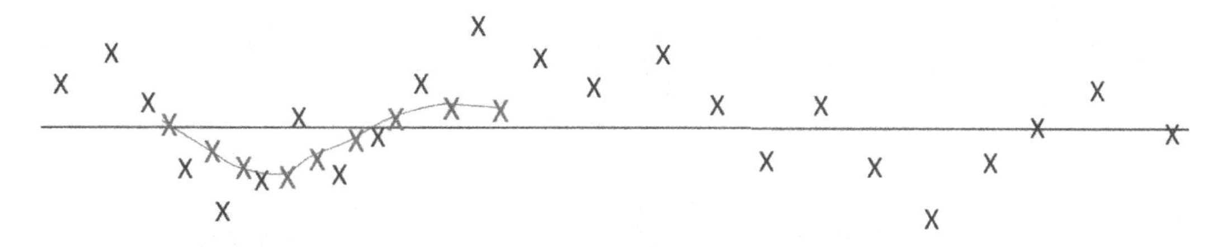

FIGURE 4.3 Averaging five data points.

4.1.2 Filtering Data

To retain values closer to the peak values, we can reduce the filter by reducing the number of data points to construct the average track. If we reduce the points from 5 to 3, Figure 4.4 shows how the same series will appear:

This was constructed with the following number sets:

1. 1, 1.5, 0.75 = 1.1
2. 1.5, 0.75, −1 = 0.4
3. 0.75, −1, −2 = −0.8
4. −1, −2, −1.8 = −1.6
5. −2, −1.8, 0.5 = −1.1
6. −1.8, 0.5, −1.2 = −0.5
7. 0.5, −1.2, 0 = −0.1
8. −1.2, 0, 1 = −0
9. 1, 2, 1.5 = 1.5

The longer filtered number series will give a result closer to the baseline, while the shorter filter will give a resulting data set that is closer to the maximum/minimum values. But what if we have random outlier numbers we don't want? This can happen if we have, for example, a graph of bus arrival times at a bus stop, where one or more buses are delayed or arrive minutes early due to traffic conditions. Here, an early arrival is positive and a late arrival is negative (not meaning that an early arrival is good for customers, but meaning the time is taken as the positive side of the graph).

The original data set is shown in the blue colours, while the resulting 5-point averaging filter is shown in red and is closer to the baseline, whereas the 3-point running average filter has buses arriving within 2 minutes of *estimated time of arrival* (ETA), compared with a 5-point running average have them arriving within 1.7 minutes of ETA. As a passenger, you may be aggrieved that buses are arriving no better than within 2 minutes of when they

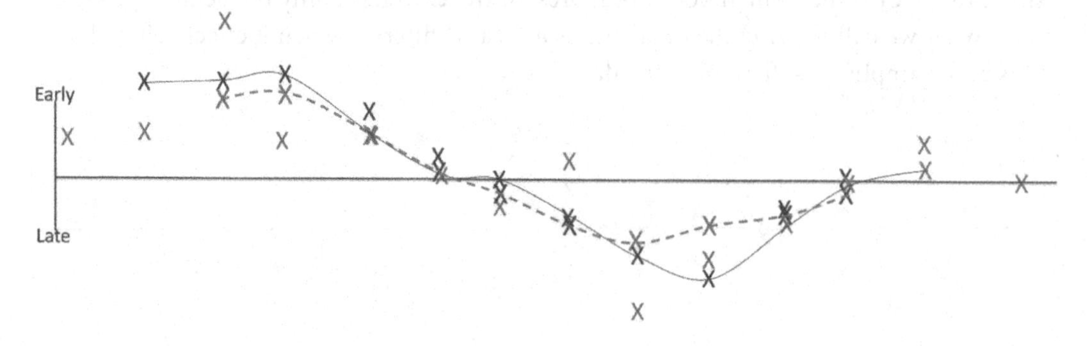

FIGURE 4.4 Choice of average data point number – 5 versus 3 to produce alternative filtered averages.

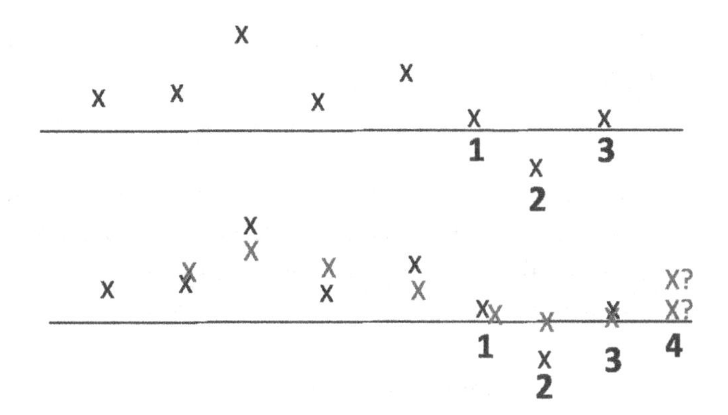

FIGURE 4.5 Using a running 3-point average, the average track and value are used for prediction of a point.

should, whereas a bus scheduler would put out in their publicity that the bus service is excellent, arriving within 1.7 minutes of ETA.

It has been shown that the line of best average fit depends on the number of points being averaged. Once that has been determined, we can use existing data to make simple predictions of the next value in the series based on a simple equation.

Take a simple number series of 1, 1.2, 2.5, 1, 1.5, 0.5, –1, 0.5 shown in the upper graph of Figure 4.5. The last three numbers are shown in positions 1, 2, and 3. We want to predict the next value after 3 based on the previous values. What we can do is set up a running average of 3, so that we can then see the trend of the filter's output (shown in red crosses if you have colour print) in the lower graph, where points at 1, 2, and 3 would give a running average number at position 2. That is, the average of (0.5, –1, 0.5)/3 = 0.

To get the value at position 3, we can set a rule that says the next point is halfway between the final data point 3 (which is 0.5) and the 0 value we just computed for position 2 that makes it 0.25. To get the final prediction value at 4, we can then set a rule that says the next point is a straight-line projection from the last two positions average computed values at 2 and 3, projected to be at (0, 0.25) = 0.5.

The result is we are able to establish three new data points at positions 2, 3, and 4 by using the average value computation for positions 2 and 3, and from the straight line predict the position at 4. This method used so far is very basic and does not allow for anything much more complex than projecting on a straight line, and anyway why can't we do that just using the last two real data points at positions 2 and 3 – that would be easier than doing this averaging stuff? Whichever way it is computed, we need a rule-based equation (aka *algorithm*) to predict the location of the last point and what has been proposed so far is too simplistic. However, it does demonstrate the need for some form of predictive equation in which we have confidence that is it working accurately.

4.2 RULE-BASED TRACK PREDICTION

It has been explained that the number of points to be used in computing the locations of average track influences the resulting locations, and that to maintain a smooth track with few large values, a larger series of points is chosen, as opposed to the lower number that makes the predicted track similar in amplitude to the most recent data, allowing divergence of points some distance from the baseline.

4.2.1 Curve Fitting

Once averaged, we can then use a simple equation not just a straight line. The most popular method is to connect the points using a quadratic equation – this is often referred to as *spline fitting*, and in-fill points along the line (we used averaging) are known as *interpolating* [(2)]. A smooth curve is generated if we pass a spline fit to produce a smooth curve, like in Figure 4.6.

A typical equation has many numbers, so is known as a *polynomial*. An example of a polynomial equation is $Y = a + bx + cx^2 + dx^3$, etc. … where a, b, c, d, and x have values. The form of a specific equation we may use in spline fitting is quadratic (two components one of which has a squared value) much like $y = 1 - 2x + x^2$. This equation describes the U-shaped curve in Figure 4.7, which is also known as a *parabola* and has two axes x and y (with y being in the centre, it is referred to as the *axis of symmetry*).

If there are three data points 'a' on a graph of x versus y with values of 'a' put into this equation of a parabola, we could draw a section of the curve of a parabola that fits the points, and then using the equation, we could continue the drawing to predict the parabola's shape and points, say at the three 'X' values. Of course, these X-values shown here are the curved main part of the U, but the point is that we can predict them using a rules-based equation with input being the 'a' values.

4.2.2 Application to Ball Tracking

With simple ball tracking in cricket and baseball, where the ball is travelling along in an approximate straight line, we can apply a quadratic equation with the rule that the trend of the ball's flight and direction must be followed and because we know very accurately where the ball is (within millimetres using our cameras) and its velocity (using our radar if not the cameras), we can then predict where many more location data points would be.

Consider cricket where the ball bounces once before arriving at the batter (Figure 4.8). Any object travelling through the atmosphere is affected by the earth's gravity, so any ball that is thrown in the air will fall back to the ground, based on the quadratic equation of:

$h = 8 + 16\,t + 64\,t^2$ where h = ball height (m) and t = time (sec) at any point along the final travel trajectory after the bounce, ignoring any wind, friction, spin, or seam turn it may experience.

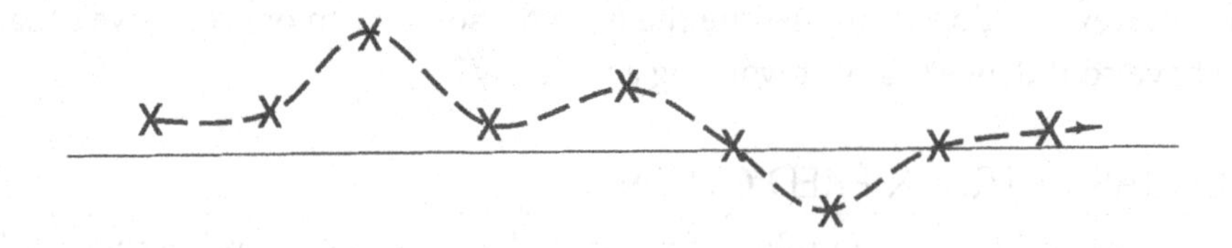

FIGURE 4.6 Spline – curve fitting.

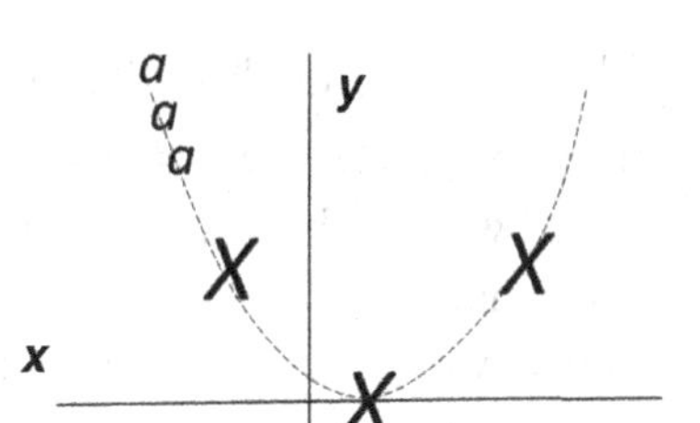

FIGURE 4.7 A parabola describes a U shape.

Having six data points along the ball's trajectory allows us to modify the standard polynomial equations in 3D to suit the conditions on-field, so that the next three or four ball locations can be predicted with greater accuracy than just using three points. On a 22-yard length cricket pitch, a ball travels at about 150 Km/hr in about a half second to travel from one end to the other. The trajectory of the final third of the pitch length is adequate to fit a quadratic equation to determine an accurate prediction of where it will arrive. So one-third of 500 msec is 170 msec. At 50 samples per msec, this means there are at least 8500 samples we can use, and if we just use 50 of these to make our accurate prediction, that means we sample every 170th sample (every 3 msec).

So, when you watch TV, by the time the ball gets to the batter, the samples have been collected, stored, and await fitting into the equation followed by the ball location prediction. Because the ball is likely to have been hit by the batter a metre before arriving at the *stumps* (or in baseball, the *catcher*), it is just the computer *latency* (how slow the computer is that causes a delay) that is the reason the commentator may bring up the picture on-screen a little later than the action happened [3]. Actually, in baseball, because the ball pretty much goes from pitcher to batter in a slightly curved line (without bouncing on the pitch), only the last 20 samples are needed, which therefore makes it easier to do a ball prediction in baseball than in cricket. This allows the TV Company to show the location on-screen of where the ball would arrive at the catcher in real time during a baseball match, whereas in a cricket match it can't. Another thing about cricket is that they like to show an animation of where the balls bounce, so often the full trajectory of a cricket ball is shown – which takes additional computing time [4].

So far, we have considered simple data where an object might travel in a curved line but what happens when data becomes more complex and instead of one point following another, it becomes more scattered and random – and our simple equation no longer represents the more random data? We have to then consider alternative methods from which we can still make reasonable predictions.

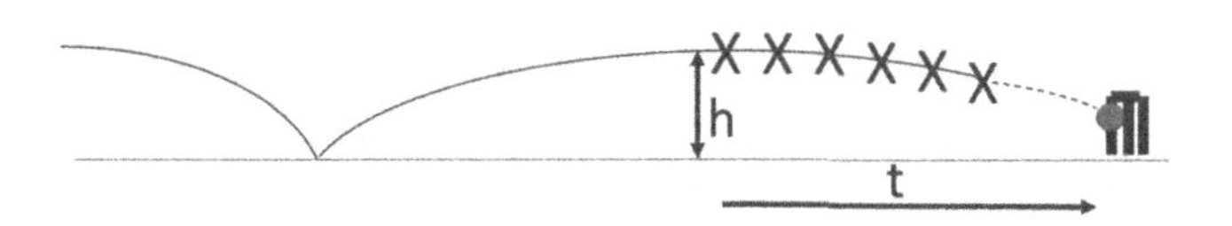

FIGURE 4.8 Ball prediction in cricket.

4.3 BASIC REGRESSION ANALYSIS AND PREDICTIONS OF FUTURE DATA

In predicting where a ball is going to travel [5], we use existing (historical) data of position to make the future prediction of where a ball might arrive. The term commonly used for this is *regression* and the act of analysing the data coins the term *regression analysis.* Regression actually means to go backwards or return, so is the correct term to use (and sounds very fancy when we consider that we are performing regression analysis when we are doing a simple thing like writing an email – we often re-read what we have written to finalise the email before sending). By contrast, of course, *aggression* is going forwards.

4.3.1 Regression and Least Squares (Best Fit)

Regression analysis has been used for many years in the shopping retail industry to determine what shoppers want in the future. It is used to predict the behaviour of an independent variable (such as shoppers of a particular smartphone) based on the behaviour on a number of independent variables (phone size, capabilities, applications, light, age of user, financial status of the user, etc.). It has been used for decades in the fashion industry, to try to understand what women want in clothing for 'next season'. This uses fashion of the past (since fashion goes through cycles) to predict the future trends. The science of this, using past data to predict the future [6], is also known as *predictive analytics modelling.*

> "The science of this, using past data to predict the future, is also known as *predictive analytics modelling.*"

Like equations to track a ball (which has a single track and so can be considered as simplistic), we can track other data that is much more variable – but we still need to develop that basic averaging to obtain a baseline from which we can derive an equation for the future data trend. Based on the variable data point scatter, we can attempt to construct a formula that takes the variability of the data into account. A *linear regression* equation fit is the simplest form as shown in Figure 4.9.

In this figure, the linear relationship lines can be drawn for a simple data set, but for a complex data set like that on the right, the relationship is non-linear and requires some thought over the type of equation to fit the data. If we have data that is more random than a ball's travels through the air, we start tracking the data, which may vary a lot and we call this a *variable.*

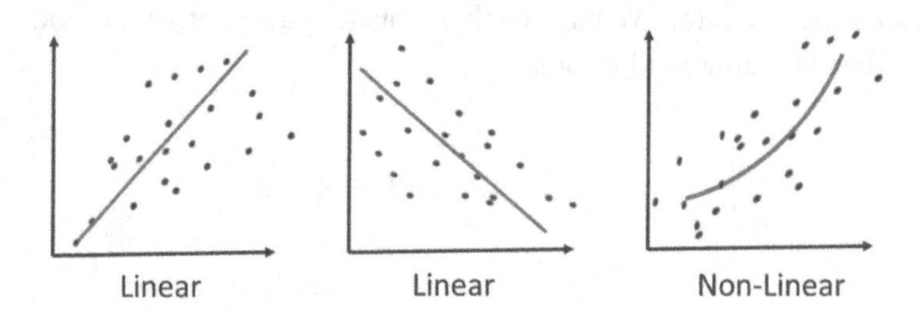

FIGURE 4.9 Predictive analytics modelling.

Based on such variable data, we attempt to construct a formula that takes these factors into account, so let's start with the linear regression equation:

1. The equation will be for a straight line (i.e. *linear*) such as $y = Mx + C$ where $+M$ indicates a positive slope and $-M$ would be a negative slope. C is the position where it intercepts with the Y-axis as shown in Figure 4.10 (but C may also be zero when the line passes through the origin – where x and $y = 0$).
2. We will use what we call the *least squares* (LS) *fit method*, so take the mean of both numbers, which gives us a central location (around which the data collected so far occurs).
3. Then compute the distance from each data point to that central location (these are known as the *residuals*).
4. Square X numbers and add up, then multiply X by Y.
5. Divide XY/X, which is the slope M.
6. Now draw in the slope through the central value.
7. And now intercept $y = C$.
8. Therefore, we have $Y = Mx + C$, which are all values we computed.
9. So for any future value of x, we can get a best fit for Y.
10. This line of best fit is the *regression line.*

Let's consider we have connected a solar panel on our roof to supply our power needs and have connected a power meter to it – at night, we need to use power for lighting our house from the solar cells (which don't work) and therefore have to decide if we need a battery or maybe we can use grid power? During the day, the panel can make so much power that we can export the power into the central power grid. So to compare grid versus battery costs, we need to know when we are using the power and when we are exporting it, so we record the power used from the meter to predict how our power requirements will be in the future – to see if we have enough cells in place (Figure 4.11).

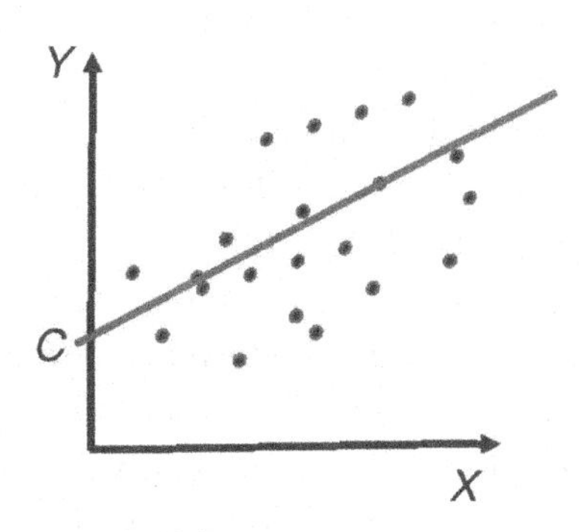

FIGURE 4.10 Slope for least squares fit.

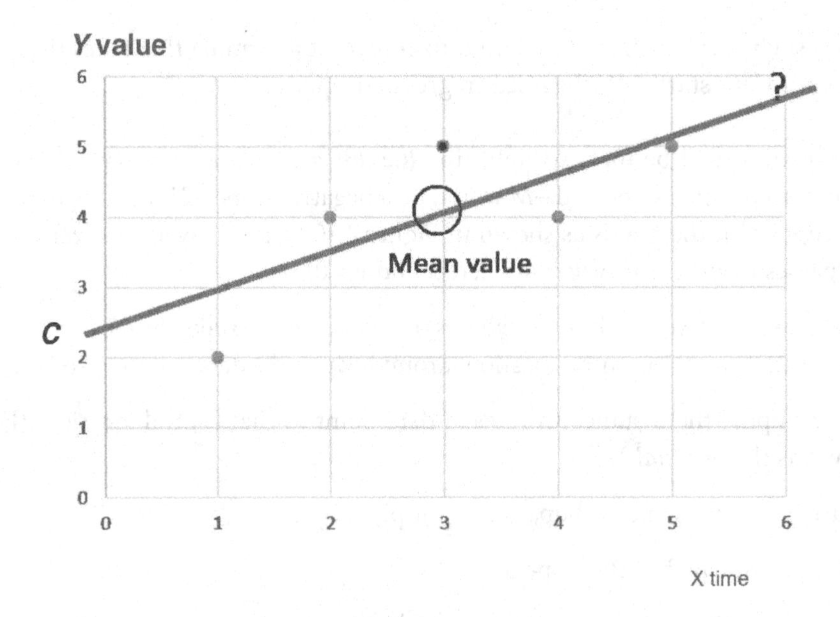

FIGURE 4.11 Power input values from electric meter.

First, we construct a table using an Excel spreadsheet as shown in Table 4.1.

In this spreadsheet, we have values of power flow that have been read in the *y*-column at specific times shown in the *x*-column. The readings take place every minute over 4 minutes (5 readings), so that we can add up the total of the power used values (5) and determine the mean value of power used with the mean of time – 4 units of power (20/5) at 3 minutes (15/5).

The next columns are $(x - X1)$ and $(y - Y1)$, where $(x - X1)$ is the difference in time between each individual time and mean time value, and difference in power between individual power readings and mean power value. Because x is sometimes earlier than the mean time, then x will on occasion appear as minus for graphical purposes (and negative power may mean power is exported).

Square values of $(x - X1)$, post these in the next column and sum them at the base (**10**).

Multiply values of $(x - X1)$ $(y - Y1)$, post them in the next column and sum at the base (**6**).

The slope of the graph M can now be determined as being the sum of $(x - X1)$ $(y - Y1)$ value divided by the sum of $(x - X1)^2$, which is $6/10 = \mathbf{0.6}$, and this slope passes through the mean location $(x = 3, y = 4)$ to intercept the Y-axis at C (**2.2**).

This intercept value is important in a practical sense in this case, because this shows the power consumption when no power is being used, and therefore there may well be a **2.2** unit power leakage in this household, or a baseline of power may be going from the solar panels directly into the power grid (which should be paid for by someone).

Using the line now at $y = Mx + C$, we can observe at 6 minutes that even with this sparse data, we can estimate that **5.8 units** of power will be generated after **6 minutes**. A similar

TABLE 4.1 Power Value Spreadsheet

	X Minutes	Y-Value Read	X − X1	Y − Y1	DX2	(X − X1) (Y − Y1)
	1	2	−2	−2	4	4
	2	4	−1	0	1	0
	3	5	0	1	0	0
	4	4	1	0	1	0
	5	5	2	1	4	2
No of data	5	5				
Mean X1, Y1	**3**	**4**				
Sum D and G					**10**	**6**
		Slope M =	**0.6**	Y/X	Anchor this at mean pts	
		Intercept C =	**Y − MX =**	4–(0.6 × 3)	**2.2**	
		So next point:	**At 6 mins**	**Y = 5.8**		

The bold values of 10 and 6 indicate summed values which are then used to compute the graphical values shown by slope and intercept, which then provide the value of 5.8 energy units at 6 minutes.

method to this is sometimes used by power companies to determine estimates of power usage by customers after the power companies fail to get access to customer's power meters.

If we want to know the value of a point along a straight line, we can estimate the value by using the *interpolation* method. This method uses the values each side of the desired location to get an answer. In Figure 4.12, let's say we have a line fitting within a set of points and we want to find a value for y that sits on the line at a known x-value between x_1 and x_2, and y_1 and y_2.

First, we take the x-value and then we have to figure out how far up the line the location is. Because it is a straight-line relationship, we get the proportion of distance of the point from x_1 and divide this by the full distance between x_2 and x_1 (i.e. $x_2 - x_1$), and multiply by the full y-distance ($y_2 - y_1$). This gives the distance along the line but just as a proportion.

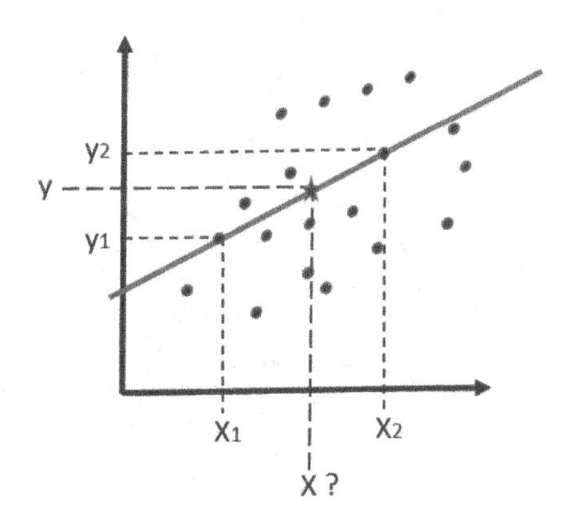

FIGURE 4.12 Finding a value using interpolation.

This methodology is not dissimilar to the *Bayes' Theorem* method, which will be discussed later in Section 4.3.5.

Therefore, the proportion of distance along the line is $(x - x_1)/(x_2 - x_1)$ and multiply by $(y_2 - y_1)$. This is the value for y along the line, so we just add y_1 to this, and we have the interpolated value.

The equation is therefore $$\mathbf{y} = \mathbf{y_1} + \left(\frac{\mathbf{X} - \mathbf{X_1}}{\mathbf{X_2} - \mathbf{X_1}}\right) \cdot \mathbf{y_2} - \mathbf{y_1} \tag{4.1}$$

When there is no relationship, we can try *regression curves.* Each curve has its own equation (e.g. the *quadratic* or the *parabola* equation, which will be explained later). We can fit different curves to best fit the data, in segments to have variable curves that can be joined.

4.3.2 Conical Equations

When we try to fit a curved regression line, it is governed by particular equations. For example, while we have used the linear equation $y = Mx + C$, some other equations derive from fitting different shapes into a cone where the diameter is r, and the base axis is x with y as the vertical axis, while a, b, and c are variables. These are called the *conical equations*, as shown in Figure 4.13.

So, we can try any equation for a best fit if we have some idea of the general shape. The computer can test these equations that are stored in memory, by starting a best fit equation. It can cycle through until the best fit is found – regression accuracy is known by the variations in its best fit, which is known as its *variance* represented by R. To compute the variance, the squares of the difference between all computed y-values and their predicted values are subtracted, and divided by the difference between all the computed y-values and the mean value. To emphasise this difference, we square the result because *variance* is often a tiny number, and this is referred to as *variance* with an accuracy of R^2. Having this number, we can compute how accurate our best fit is.

In mathematical terms this is

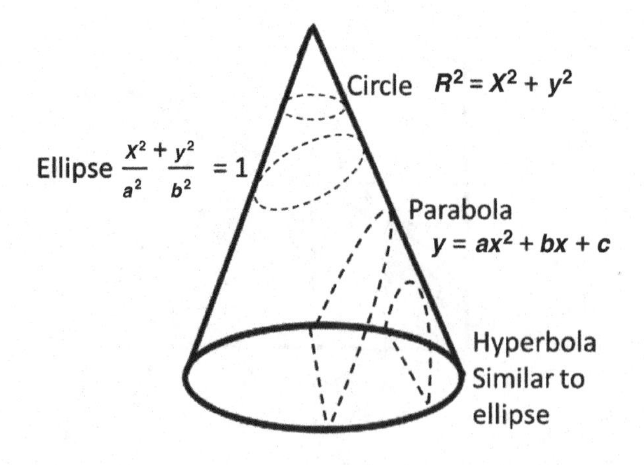

FIGURE 4.13 Conic equations are shapes within a cone.

$$R^2 = 1 - \frac{\sum (X - X_{pred})^2}{\sum (X - X_m)^2} \qquad \begin{aligned} \text{Where } \Sigma &= \text{Sum of values} \\ X &= \text{value at start} \\ X_{pred} &= \text{predicted value} \\ X_m &= \text{mean value} \end{aligned} \tag{4.2}$$

4.3.3 Accuracy of a Line of Regression

Can we use this to work out how accurate such a regression line can be, which in our example may be important in paying these electricity bills?

The accuracy of a regression line fit is determined by calculating the value R^2. When this value = 1, there is an exact fit, but 0.9 is also acceptable, whereas something like 0.5 is not so good and in many cases considered poor. When 0, there is no relationship as indicated in Figure 4.12 in which any line can be a poor fit. R^2 gives an idea of how far off our line might be from a good *LS fit* through the origin (where x and $y = 0$).

To make a best fit to any curve, we can first fit an LS equation but now with a different curved equation, and then compare the R^2 values. Here is a simple approach that is often used by computer programs:

- Straighten the chosen curved equation by taking each value and determine the difference between the chosen curve values and the straight-line LS values at each of the *X*- or *Y*-values, as shown in Figure 4.14. The short vertical lines indicate these differences between the LS and *ellipse* or *parabolic* fits.
- Then *variance* value R^2 can be computed, with the highest R^2 value being the preferred fit (still less than 1). Like any process, this set-up is only done first time to establish the preferred equation. R^2 is often not used and σ^2 is preferred – which is very confusing.
- The computer coding may continue other fits, expanding the ellipse or parabola to get a best fit, and finally accepts the best fit at the maximum R^2 value closest to 1.

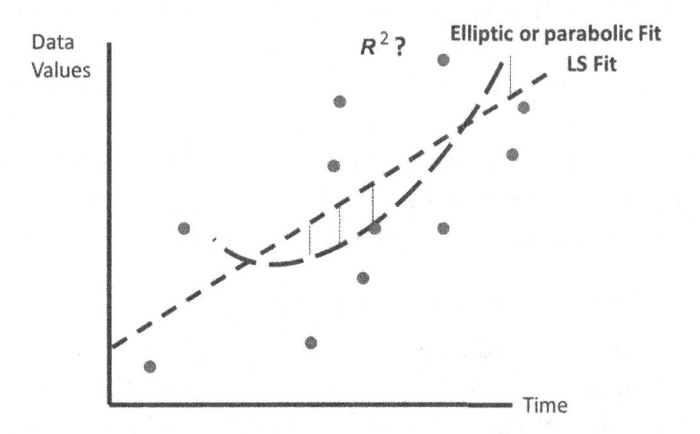

FIGURE 4.14 Fitting curves to data.

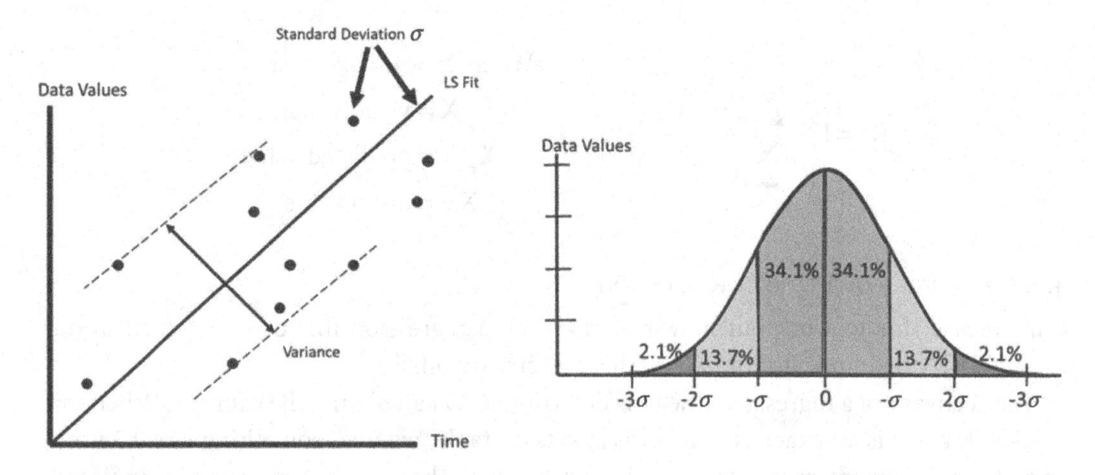

FIGURE 4.15 Normal distribution of data (aka *Bell shape* or *Gaussian distribution*).

Because we are working in squared units, this is difficult to use, so a square root of variance σ^2 is preferred – this is known as *standard deviation* σ and is a measure therefore to show how a group of numbers are spread from the average. A low value of σ means that numbers are in a zone or consistent area but high values indicate numbers are spread out. In a standard *Gaussian distribution* (aka the *Bell Curve* or *a normal distribution* of data), the majority of values (68.27%) lie within 1 standard deviation of the mean (Figure 4.15). That is, if we take a normal distribution of numbers – such as the time over which people stand at a popular bus stop (some will be early, most will be on time for the bus, some will be late and miss it) – then each quarter of that time is known as one standard deviation, and typically most people would wait about the same amount of time – one standard deviation – over a period of a few months.

In the figure, while the normal distribution is shown on the right, the graph on the left is really the view looking down on the data where the LS fit line is at the centre of the data – the data is mainly within the two standard deviation lines (shown dotted) with any data outside of those lines contributing little to the computation (representing only 13% between 1σ and 2σ) of total data, while even further out would be outliers, which may be disregarded.

In terms of application of this knowledge, we can record data from sensors and most of the data will fall within a standard deviation of the mean value. Armed with this knowledge and provided our sensors continue to operate in that manner, we can work out when data lies outside of normal operations, and this helps us to define what good quality data is and what poor quality is (we can only perform good quality predictions if the input data is good quality) as well as having some idea of the *probability* of that prediction being correct.

4.3.4 Probability of a Prediction Being Correct

When we make any form of prediction, we would like to have an idea of what degree it is likely to be correct. The sports betting industry [7] works on the basis that punters are going to predict the results of a game or race with a poor probability that they are correct – betting makes its money from more people predicting incorrect results, and, as a result, they pay-out more winning money if a result is improbable. Consequently, probability is a major issue when making a prediction, and both industry as well as the social community work with probability on a daily basis (when driving a car, we base our own driving decisions on the probability that other car drivers will follow the road rules, and we are constantly predicting what they will do).

Probability is based on the number of times an event is likely to happen, taking into account the past history of events (so in a sense, probability includes *regression analysis*). Most events often start with random probability and then they mature.

The classic way of explaining this is rolling a die: it has 6 sides so the probability of it stopping to roll at a given number like 3 is equal at 1 in 6 = 16.66%. This is *theoretical probability*.

If we roll a die a second time and 3 comes up again, the *experimental probability* shows that we have rolled a 3 twice, so becomes 2 out of 6 = 33%. If we keep rolling, using the normal distribution discussed earlier in Figure 4.15, the probability is that it will settle back at 16.6% assuming there is no bias in the die. This is known as *conditional probability*.

Let's say we roll the die 100 times and the amount of times (the frequency) the same number came up as shown in Table 4.2.

This approach can be applied to other events, such as the workday. If a person is a high street café barista and makes coffee for a passing crowd, it is highly likely that he/she will work every day except weekend days Saturday and Sunday. If there is a weekend event, then maybe they will work, but, generally, the probability of the barista making coffee would be 2 days in 7 = 28.5%. This is then mixed with the likelihood that a weekend event might take place 20 times a year, and so the probability of the barista pouring coffees in any year now becomes 28.5% + (20/365)% = 34%. This then has to be factored into the business model to ensure a profit is produced.

Probability is factored into the business plans in mining and oil and gas production [8]. For example, a mine will have drilled a certain area and found minerals of a certain value. The forecast of recoverable minerals will therefore take the drilling results, look at the quality of the ore, and predict the size of the ore body. This prediction may increase or reduce in size depending on quantity of holes drilled, their location, and area available to be drilled. If most holes to date have minerals in them, then there is a high level of confidence that more minerals will be discovered, and miners often refer to this as an 80% probability otherwise known as a P80 forecast. On the other hand, if the orebody appears thin and there has been no mineralisation in some holes, then it may be a poor forecast of

TABLE 4.2 Probability in Rolling a Dye 100 Times

Event	Frequency	
1	18	Rolling a 3 gave 16/100 = 16%
2	20	Theoretically, it should be 1/6 = 17%.
3	16	Rolling a number less than 3 (1 or 2) = 38 times or 38%
4	14	Rolling a 3 or 6 is 26 times or 26%
5	22	
6	10	
Total	100	

further mineralisation – which may be a P20. This becomes important since in some cases, a P20 forecast may deter investors from putting money in to develop the mine, whereas a P80 forecast may see the mine fully invested for development. Consequently, probability plays a major role in business and industry.

4.3.5 Bayes' Theorem of Probability or Chance

When prediction becomes difficult, we like to use any modelling that is available to help us. The Reverend Bayes who lived in the mid-1700s was a part-time statistician and preacher who used *conditional probability* (mentioned earlier) to develop a basic equation that used evidence to calculate limits of an unknown parameter (referred to in those days as *Chance*). *Bayes' Theorem* uses the concept that prior knowledge should be used before making a prediction – we earlier referred to this as *regression analysis.* An example of *Bayes' Theorem* is that if the risk of developing health issues increases with age, then the risk of any individual of known age can be assessed more accurately over the alternative of assuming that the individual is typical of the population as a whole (and that all of the population of the same age have the same health issues – which is obviously wrong). When we are using the *LS* method of calculating a prediction, we actually use a form of *Bayes' Theorem* in the body of the calculation (to obtain the slope value in Table 4.1).

> "Using Bayesian theory of probability interpretation, his theorem expresses how a degree of 'belief' (in God or other likelihoods) should rationally change to account for any changes in availability of related evidence. The more evidence there was, the more probable that the belief was modified."

Using Bayesian theory of probability interpretation [9], his theorem expresses how a degree of 'belief' (in God or other likelihoods) should rationally change to account for any changes in availability of related evidence. The more evidence there was, the more probable that the belief was modified. Laplace in France expanded this concept in later years, but generally, we have come to accept this Bayesian theory of probability and statistics. His basic equation is given in Equation (4.3):

$$\underset{\substack{\text{Prob of A being true}\\ \text{if B is true}}}{P\left(\frac{A}{B}\right)} = \frac{\overset{\substack{\text{Prob of B being true}\\ \text{if A is true}}}{P\left(\frac{B}{A}\right)} \quad \overset{\text{Prob of A being true}}{P(A)}}{\underset{\text{Prob of A being true}}{P(B)}} \tag{4.3}$$

The stock market is a major user of Bayesian economics since it can sometimes predict the stock price change. For example, if this is related to the resources industry, A can be the stock price (depending on drilling success) and B is the CEO being replaced (if he/she has done a good/bad job).

The probability of a rise in the stock price would be produced by:

- P (A/B) – the probability of a stock price rise, given the CEO is replaced.
- P (B/A) – the probability of the CEO being replaced (which happens if past profits are poor), given a rise in price has already happened.
- P (A) – the probability of the stock price increasing and P (B) – the probability of the CEO being replaced.

If drilling results are good, and the stock so far this year has been stable or risen, the CEO is in a good position, and the probability that the stock price will rise is high. If drilling results are bad, and the stock has been stable or risen (CEO is in a good position), the stock will be stable and not move.

If drilling results are poor, and the stock so far this year has performed poorly resulting in the CEO being replaced, there is a high probability that the stock price will fall.

Alternatively, this could be applied to a heart attack of a specific person who is 70 years old. If the person has family history of heart attack (father and grandfather died before 70 due to heart attack) in Equation (4.3). If the person has personally had a heart attack before 70 in Equation (4.3):

1. If there is a family history of a heart attack, and the person has had a personal history of heart attack, the probability of dying from a heart attack is likely and therefore about P80. If the person has a family history of heart attack but the person has not had one personally, the probability of having a heart attack is likely to be about P50.
2. If there is no family history of heart attack and the person has not had a heart attack, there is a probability that the person will have one at P20. This does not account for the fact that the person may have a high cholesterol problem (which causes heart attack), and didn't know about it. If this were to be accounted for, there would be a need to determine if the father and grandfather also had high cholesterol – so the equation becomes more complex.

Consequently, determining predication probability can be a complex topic and more often than not, some data that may be useful knowledge (such as cholesterol levels) is overlooked. When inadequate data is provided, the prediction probability assessment is reduced, so we might introduce a bias factor to account for one event being more important than another.

In the following exercise (Table 4.3), a bias factor has been introduced in the sale price of apartments, because if more apartments are sold with three bedrooms than two in the same block, it can bias the apartment complex towards higher prices (since more people are generally interested in apartments with more bedrooms). But typically, there are less buyers for two-bedroom apartments than for three bedrooms.

TABLE 4.3 Apartment Price Prediction

Year	2 brm Sold	2B Av Price ($k)	3 brm Sold	3B Av Price ($k)
2012	2	100	3	110
2013			5	120
2014	4	130		
2015			2	150
2016			2	155
2017	1	130		
2018			1	190
2019	3	170	1	200
2020	4	180		

4.4 EXERCISE

Predicting future apartment prices.

A high-rise apartment block contains 30 – 2 and 3 bedroom apartments that have sold over many years at an increasing price level, consistent with inflation. You would like to know the price of your two-bedroom apartment, so you know what price to ask. Here is the data, so work out what you think it might sell for (Table 4.3). There is an upward price-bias in the cost of house prices, which follow the inflation rate, so consider house prices increase by an annual rate of 10%.

First work out what the annual bias is and redraw the spreadsheet of Table 4.3 showing the bias. Then draw the graph of apartment sales containing that bias, and compute the spreadsheet using a least squares approach.

FURTHER READING

1. To calculate a rolling average in Excel: https://www.got-it.ai/solutions/excel-chat/excel-tutorial/average/how-to-calculate-moving-average
2. Spline fitting and interpolation using Excel: https://www.real-statistics.com/other-mathematical-topics/spline-fitting-interpolation/
3. What is latency? https://www.thepodcasthost.com/recording-skills/what-is-latency/
4. The physics of cricket: http://www.physics.usyd.edu.au/~cross/cricket.html
5. Prediction using regression analysis: https://en.wikipedia.org/wiki/Regression_analysis
6. Predictive analytics: https://www.ibm.com/analytics/predictive-analytics
7. Global sports betting market analysis: https://www.brsoftech.com/blog/global-sports-betting-market-analysis-2021/
8. Probability of an event: https://datacadamia.com/data_mining/probability
9. Bayesian statistics in simple English: https://www.analyticsvidhya.com/blog/2016/06/bayesian-statistics-beginners-simple-english/

CHAPTER 5

Track Prediction in Sports and Industry

5.1 PRESENT-DAY SPORTS (CRICKET, TENNIS, BASEBALL, AND FOOTBALL TECHNOLOGY)

In modern-day sports, analytics is closely associated with winning and losing in any game. Whether it is tactics of one team over another or players' abilities, the understanding of analytics enters the game everywhere if teams are to be successful and win competitions. A major driver of success is the betting industry, which gambles on predicting one team beating another. Knowing each side's weaknesses is one area of analytics, but knowing where the ball is travelling in some sports is critical to winning the game. An example is in the European Champions League Final football game where, if the ball was halfway across the goal line (between goal posts), a goal is not scored, until the full ball was across the line – earning a potential $50 million for the winner. In soccer, this is assisted by what is called *goal line technology.*

The ability to observe the positioning relationship of a ball with respect to a line [(1)] is crucial in world-class tennis and cricket, and if the ball cannot be physically seen to travel over a line, then there must be *predictive animation* to assist the judges and referees with their decision.

5.1.1 Track Prediction of a Cricket Ball

We earlier discussed track prediction of a cricket ball and how it is generally done. One of the major issues in cricket is that a ball may be bowled in a straight or slightly curved line, but if it has spin after bouncing on the ground, it can unexpectedly deviate off by some distance. In fact cricketers often attempt to land a spinning ball on parts of the cricket pitch that have been damaged or onto worn indentations in the pitch so that the ball deviates off. Most cricket followers would know about the *Gatting ball* (aka the *ball of the century*), where in the 1993 *Ashes Series* Shane Warne – an Australian spin bowler – when bowling to Mike Gatting of England, imparted so much spin and placement on the ball, that

DOI: 10.1201/9781003108443-5

caused it to bounce on the side of an indentation in the ground, and veer off-track by around 45 degrees knocking off Gatting's wicket bails. The spectre on TV was that of Mike Gatting just standing there astonished as he watched the ball go straight past him as he swung his bat at it, then turn as it hit the pitch 'almost at right-angles' and knock his bails off the wicket immediately behind him. This jaw-dropping moment was such that he just stood there in amazement and couldn't believe what he had just seen - and so even with one of the best world's batsman, it becomes very difficult sometimes to make an accurate prediction. Even today, such a ball is impossible to predict as it is bowled and subsequent to this, all spin bowlers have tried to emulate *the Gatting ball.*

However, what we can do is predict the flight of a ball provided we have sufficient data points and it does not experience the extreme conditions which occurred to the Gatting ball. In cricket more often than not, the ball travels at speeds up to 90 mph and deviates very little. Consequently, it only needs a few camera tracking devices - as few as six per ground - to be able to make the simple prediction of the final metre of ball flight. Using cartoon animation methods, the ball can be shown passing a wicket (with the picture being overlaid on the real picture of the cricket ground). Because of the accuracy of prediction (in the order of millimetres), this animation can then be constructed as a 3D view offering views from different locations such as that of the wicketkeeper (whose job is to catch the ball), as shown in Figure 5.1.

These animations have to wait until the ball's 3D position has been computed, which usually involves a best fit of the conic equations (Figure 4.13) using at least three ball location data points in x/y, x/z, and y/z directions (Figure 5.2). As explained in Chapter 4, the speed of computation has a slight lag or latency so the results of such animations take a number of minutes before they are shown on-screen. However, the millimetre accuracy

FIGURE 5.1 Animation of a ball as seen from the position of the wicketkeeper. (Courtesy: Hawk-Eye Innovations.)

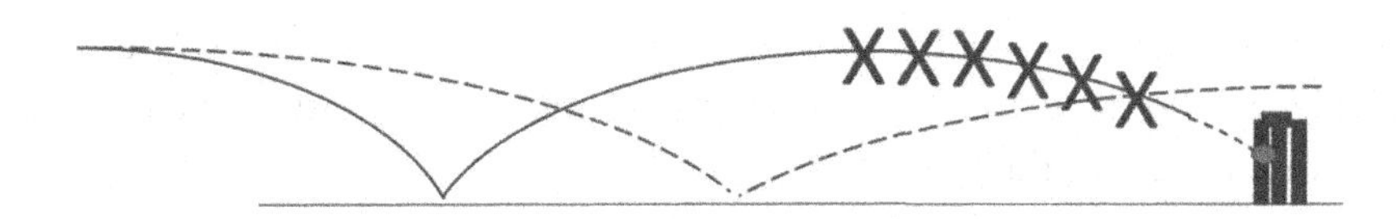

FIGURE 5.2 Ball tracking to predict location of arrival at wicket – hit and miss.

of these computations allows umpires to confirm or deny where a ball was travelling, and whether a batter is *out*. Of course, we also have the benefit of viewing the output of a *snickometer* (which is a microphone mounted on the stumps), a *stumps-cam* (which is a *wifi*-connected camera mounted on the stumps), and *infrared* (IR) imaging (which shows any heat resulting from friction of the ball touching the bat, clothing, or an arm in-flight) to assist and corroborate the umpire's decision. Such sensors will be discussed in later sections of Chapter 6.

Another view commonly used by the TV broadcaster is an animation along the cricket pitch from bowler to batsman, showing where the ball has bounced or could bounce in order to cause problems for the batters. Figure 5.3 shows the image that the viewer would have of the pitch seen from the bowler's end. The location where the ball had previously bounced is shown in terms of *short, good,* or *full-length bowling*. The fact that this bowler (Southee) bowled 59% of the balls in the *full-length* area rather than the *good-length* area can be an indication that the bowling has caused less problems for the batter than if the balls had landed in the good-length area. This provides an immediate indication of the quality of the bowling and perhaps why so few a number of batters got *out*. It is easy to see how these statistics can help a captain decide which bowlers to select for the next match.

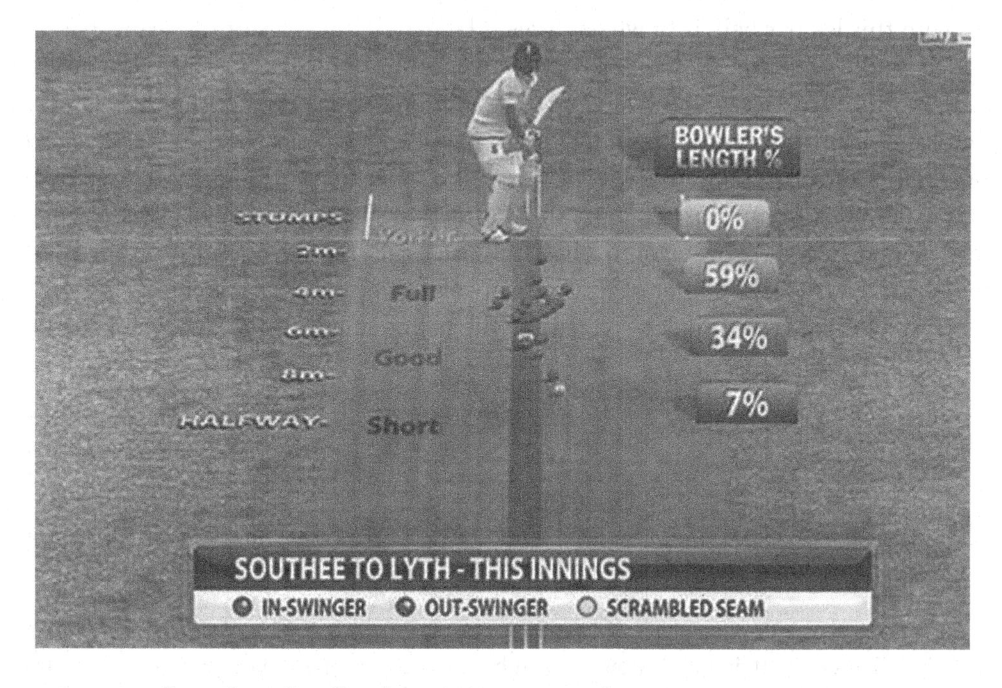

FIGURE 5.3 Different lengths of bowling. (Courtesy: Hawk-Eye Innovations.)

(With a *full* length delivery, the batsman can snick the ball away relatively easy, whereas with a *good* length delivery, the ball can swing around more and give the batsman more problems than if it were in the *full* area.)

Keeping a myriad of such statistics only requires particular predictions to be retained in computer memory for later use. *Hawk-Eye* software (as discussed earlier) is often the software of choice that allows the different actions of bowler and batter to be produced from memory for comparison purposes [(2)]. This, with linking software, will give the statistics and animations of batters and bowlers facing different conditions and balls (Figure 5.4), whether batters are left-handed or right-handed. These statistics provide all of the winning attributes needed for batters and bowlers to emulate if they are to be successful.

In order to study a ball being bowled, which moves so quickly that it is almost impossible to see on TV, an animation can be played after a selected number of balls have been bowled. The left picture of Figure 5.4 shows different coloured ball tracks through the air, with the different colours referring to a different bounce or track (short, good, or full), which can be viewed from the bowler, batter, or angled perspective. The red *dot* ball would have been ineffective with no runs being scored, the blue ball representing a number of runs scored, the yellow ball representing a ball that had been struck and run off to the boundary (for 4 runs), while the white ball would represent a ball providing the purity of a fallen wicket. Where these land and how they have deviated is of great importance to the analyst. A *dot ball* is acceptable where the batting team is catching up to the fielding team in terms of runs, so limiting the team by limiting its runs scored, helps the fielding team – with the ideal being the white wicket–ball.

Of statistical importance, however, is the location and amount of spin applied to a ball and the degree to which it diverts after the bounce. Of particular interest, when bowling faster with this leather ball, is to what degree it landed on the edge of the seam that has a threaded binding holding the two separate leather halves of the ball together. If it lands on the ball seam's edge, this could divert the ball (after bouncing) further than if it lands flat on the seam – with bowlers who can make this happen being referred to as a *seam bowler.*

Bowling is something of an art form where the movement of an arm or fingers during the ball's release can cause changes in its direction. Figure 5.5 shows the sequence in time of a bowler bowling overarm to a batter. In each picture, the bowler is on the left, the *umpire* (derived from the sound of the French word *non-pere* meaning 'no-equal' or 'no-side') is to his right (wearing dark trousers). The batter stands at the far end of the pitch

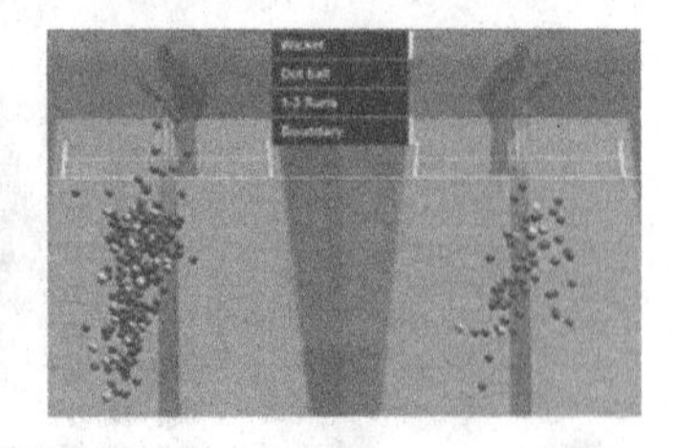

FIGURE 5.4 Different attributes of bowling can be assessed using animations. (Courtesy: Hawk-Eye Innovations.)

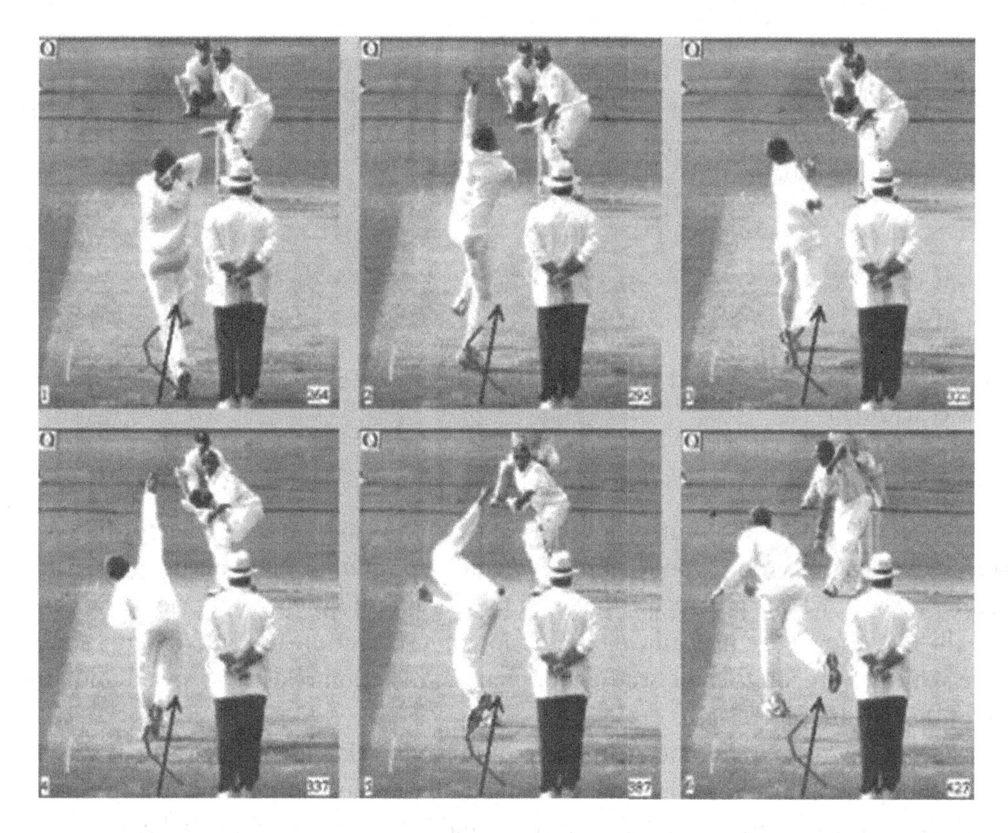

FIGURE 5.5 Multi-capture photos showing arm action. (Courtesy: Quintic, www.quintic.com)

while the wicket and wicketkeeper are behind him. The action runs from top left across to bottom right, where the bowler top left rolls the bowling arm over the shoulder until the ball is released in the bottom centre picture and the batter adjusts a defensive position ready for the deviating ball flying past in the final picture (bottom right). If this movement is not strictly followed, it is possible that a bowler's action may cause the ball to travel in a direction, which is unwanted. The arrow shows the general direction of bowler run but the red line indicates the direction of ball travel in the hand. These lines can be drawn on the screen by the commentator immediately after the action.

In order to check on this arm action, technology has come to the aid of the statistician with the bowler able to wear armbands containing positioning sensors that record the relationship of the arm in the air with respect to the ball, to allow playback of where the ball was at any point in time. These armbands (Figure 5.6) are only used in movement laboratories and not in the field of play, to allow studies of a bowler's action and what modifications would be needed to the arm's rolling over the shoulder [3]. Consequently, the data transmitter package is *Bluetooth* for short-range operation, and the microprocessor will contain an accelerometer as well as perhaps strain gauges and body temperature sensors. By analysing the 3D trajectory of the arm, it then becomes possible to recommend alternative ways to bowl (e.g. bending or straightening the arm, holding the ball differently, twisting the wrist before releasing the ball, etc.).

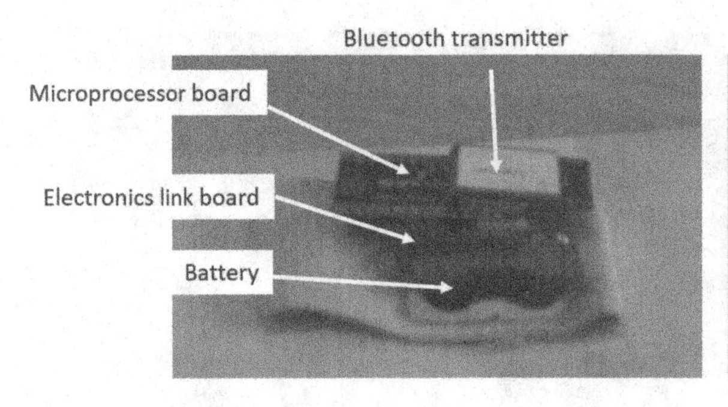

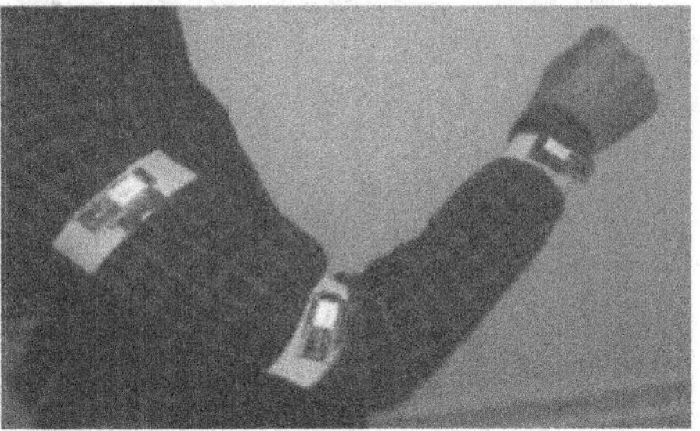

FIGURE 5.6 A flexible armband containing transmitter and microprocessor. (Modified from Qaisar et al., 2013.)

Important points to monitor in a bowler's action are the *number of steps* the bowler takes, *step rate*, *peak run-up speed*, *step length*, and hip *rotation*. Note needs to be made of whether a fast bowler is passing his/her arm over or around the wicket prior to release of the ball. This knowledge may allow a prediction of its direction and side-by-side analysis of the same bowler can then be studied for changes in action relative to speed and direction of the ball after its release.

In Figure 5.7, the accelerometer in the monitoring pack shows that vertical acceleration and angular velocity can be displayed after a bowler has completed bowling [(4)]. In the figure, 'B' is the back-foot strike on the ground and 'F' is the front-foot strike. Ball release

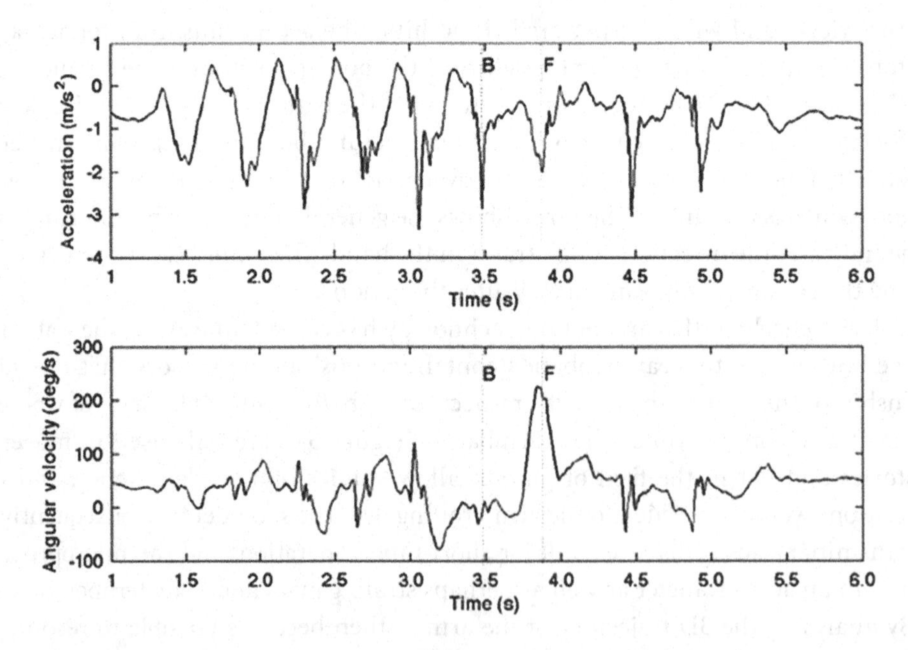

FIGURE 5.7 Vertical acceleration and angular velocity of a bowler's run-up. (After Rowlands et al., 2009.)

is not detectable so it is left to the cameras to observe that, with a typical run-up speed of 6 m/sec and a release/delivery speed of 160 Km/h. Using such data, the amount of stress and twist placed on the front-foot is a guide to potential future problems with this bowler's action – excessive front-foot pressure can lead to stress fractures occurring, resulting in long-term fracture issues.

5.1.2 Track Prediction during Tennis Games

While in cricket, the idea is to knock the wooden bails off the top of the wicket (and so every action associated with this event is closely monitored), in tennis, the main idea is to hit the ball over a net adequately for the ball to land in the opponents area (court) without going *out-of-play*. Because the speed of a cricket ball can be as much as 100 mph (160 Km/h) while a tennis ball can be served to travel faster at about 110 mph (180 Km/h), the effect of a single ball landing on the wrong side of a line is more important in tennis.

As discussed earlier, the *Hawk-Eye* technique uses cameras linked to computers to track balls, workout speeds, and provides animations of the tennis ball's bounce on the court lines (Figure 5.8). The concept of the animated bounce resulted from tennis being played on clay courts in France – when a tennis ball bounces on clay, it leaves an impression of the ball on the clay surface [5]. This gave such a firm answer to whether a tennis ball was out or not, that it was included as the image of an animated ball bouncing on the court. While this looks realistic, it is just an animated image of a ball bouncing, but gives credence and faith that the animation's accuracy is correct (within millimetres), which is accepted by tennis associations around the world.

Such data analytics are also used to improve player performance as discussed for improving the cricket bowler's action. In tennis, a player's shot direction, the number of returns made, ball velocity, depth of shot to the backcourt, number of serves won, number

FIGURE 5.8 Hawk-Eye camera and animation of a ball bouncing. (Courtesy: Hawk-Eye Innovations.)

of aced serves, etc. all contribute to a player's performance statistics for the evaluation of player abilities.

Figure 5.9 shows a computer screenshot of how tennis player *Nadal* performed in a typical contest. On this screenshot of the analytics software, the number of serves, return balls, whether first or second serve, and points won/lost are shown for a game. Since the faster a tennis player can serve with accuracy, the greater the probability that the serve will be won, so linking the two statistics is an indicator of whether *Nadal* was playing well at the time. An average shot velocity of 161 Km/h with a velocity spread of 15 Km/h and a fastest shot of 187 Km/h suggest he was in top form during this game. The average shot depth of just over 11 m having a spread of 4 m across the court suggests great accuracy is maintained at these high speeds.

By contrast, analysis of men's versus women's tennis over five sets can be conducted using *Association of Tennis Players* (ATP) statistics for men and *Women's Tennis Association* (WTA) statistics for women (Figure 5.10) during 2015. The black dotted lines show the ranking of men and the percentage winning matches over 3 sets versus 5 sets (in which the bottom ranked are closer to 100, losing more games). The open dotted lines show similar data for women. Both men and women had greater probability in 2015 of winning matches and raising their rank if they played and won the best of 5 sets versus the best of 3 sets. Because both data sets are similar with hardly any divergence between the results of the two sexes, it shows that the higher ranking is associated with playing and winning more sets of the 5 set format rather than 3 sets, and is not associated with gender.

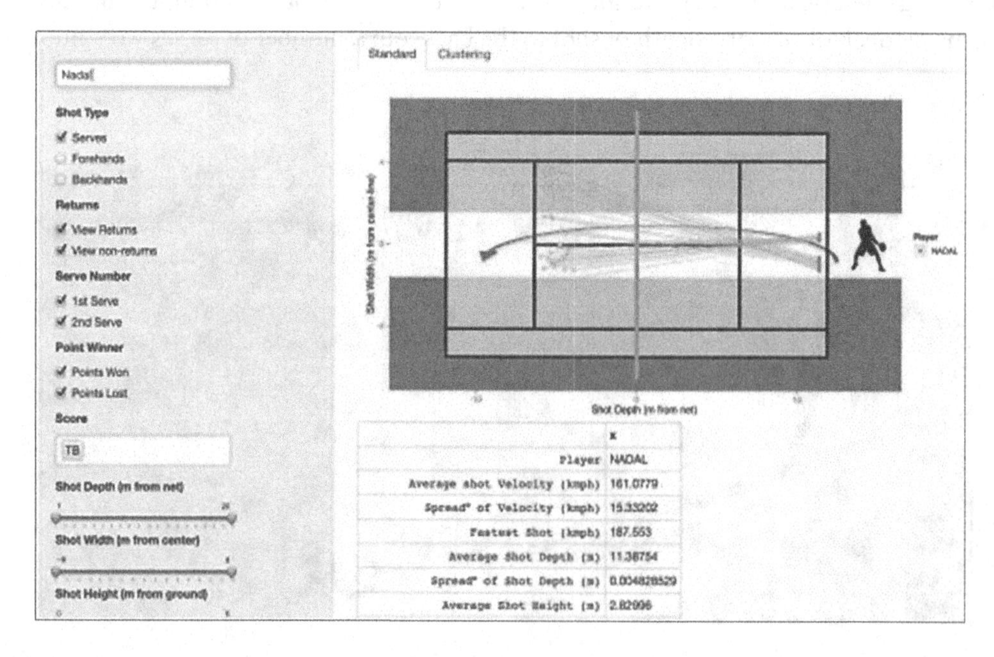

FIGURE 5.9 Screenshot of Hawk-Eye analytics software for a Nadal serve. (Courtesy: Hawk-Eye Innovations.)

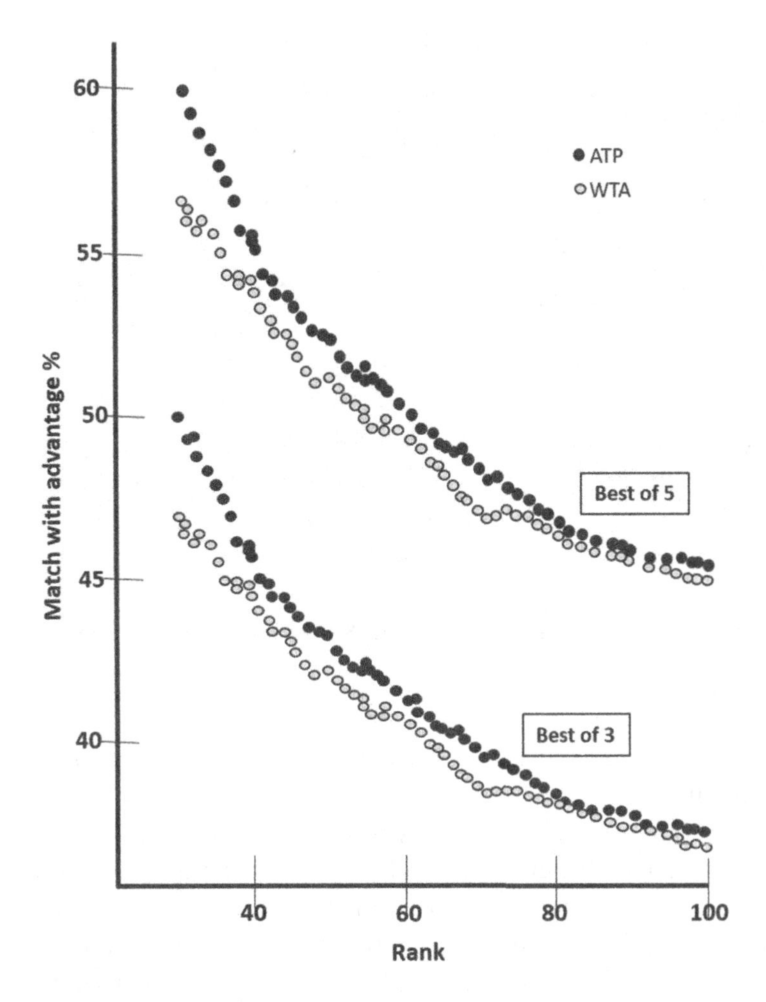

FIGURE 5.10 Men's versus women's tennis over 3 versus 5 sets.

5.1.3 Sports Analytics and Prediction – Baseball

Baseball is considered as the original sport that used data analytics of the individual players during games to find an edge over opposing teams. *Henry Chadwick*, an Anglo-American cricket and *rounders* writer for the *New York Times*, in 1859 adapted the baseball box score system (Figure 5.11) from a cricket scoreboard, to be able to track the performance of individual players as well as the team. The first box score contained nine rows for players and nine columns for innings. He assigned numbers to each defensive position for scorekeeping purposes only, and this system was subsequently adopted by the *American League* for normal use in the early 1900s.

Baseball continues to use data analytics to ensure a team has a competitive edge over the competition. The application of sports analytics was made popular by the film *Moneyball* in 2011, which showed how the *Oakland Athletics* team relied heavily on the use of analytics to build a competitive team on a minimal budget. That movie emphasised the *on-field* and *off-field* analytics in which the on-field data analysis exposed

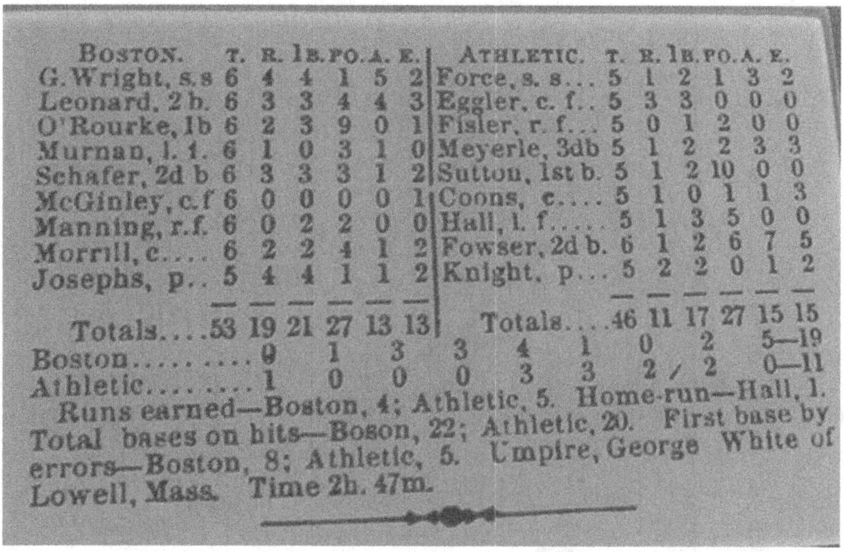

Boston.	T.	R.	1B.	PO.	A.	E.	Athletic.	T.	R.	1B.	PO.	A.	E.
G. Wright, s.s	6	4	4	1	5	2	Force, s. s...	5	1	2	1	3	2
Leonard, 2 b.	6	3	3	4	4	3	Eggler, c. f...	5	3	3	0	0	0
O'Rourke, 1b	6	2	3	9	0	1	Fisler, r. f...	5	0	1	2	0	0
Murnan, l. f.	6	1	0	3	1	0	Meyerle, 3db	5	1	2	2	3	3
Schafer, 2d b	6	3	3	3	1	2	Sutton, 1st b.	5	1	2	10	0	0
McGinley, c.f	6	0	0	0	0	1	Coons, c....	5	1	0	1	1	3
Manning, r.f.	6	0	2	2	0	0	Hall, l. f.....	5	1	3	5	0	0
Morrill, c....	6	2	2	4	1	2	Fowser, 2d b.	6	1	2	6	7	5
Josephs, p...	5	4	4	1	1	2	Knight, p...	5	2	2	0	1	2
Totals....	53	19	21	27	13	13	Totals....	46	11	17	27	15	15

Boston.......... 9 1 3 3 4 1 0 2 5—19
Athletic.......... 1 0 0 0 3 3 2 / 2 0—11
Runs earned—Boston, 4; Athletic, 5. Home-run—Hall, 1.
Total bases on hits—Boson, 22; Athletic, 20. First base by
errors—Boston, 8; Athletic, 5. Umpire, George White of
Lowell, Mass. Time 2h. 47m.

FIGURE 5.11 First use of data analytics in sport in 1859 – the baseball box score.

players and staff strengths and weaknesses of individual players and the team, including game tactics and fitness.

Meanwhile, the off-field analytics deals with the business side of sports including how to increase ticket and merchandise sales, and fan engagement to help decisions leading to higher club growth and profitability. Sports gambling companies were not slow to realise that the use of player and club analytics could bring more punters to consider placing a bet on the winning team, and consequently today there is a lot of financial dealings tied-up with sports analytics.

The major statistics in baseball include *batting average* (some players play better at one ground compared with others), *on-base percentage* (the percentage a player may spend at a base, which can also change a player's approach to how they bat), *slugging average* (the number of bases a player's hits allow him/her to run), and WHIP (*walks plus hits allowed per innings pitched* is a measure of how many baserunners the pitcher allows on both hits and walks).

For a batter (Figure 5.12 left side), a high batting average is something to be admired and encouraged, whereas a low average means a potential position needs to be filled. For a pitcher (Figure 5.12 right side), a low strike percentage followed by a gradually increasing percentage shows how pitchers warm up towards the end of the final quarter in each round.

In terms of live TV, baseball has not gone to some of the extremes of other sports, where all manner of statistics are displayed in real time. In professional baseball, they tend to show a box where the pitcher would be expected to pitch the ball. Inside the box is an image of each ball's arrival location with the speed alongside. By monitoring the box and ball placement in the box, the accuracy of the pitch can be judged and using its speed as a

New York Yankees — Batter's Average

Hitters	AB	R	H	RBI	BB	SO	AVG	SLG
Elsbury	3	0	1	0	1	0	280	420
Gardner	3	0	0	0	0	0	217	344
Refsnyder	1	0	0	0	0	1	250	500
Belltrain	3	1	1	0	0	1	264	534
McCann	4	1	1	2	0	0	231	413
Teixeira	4	0	1	0	0	2	195	286
Castro	2	0	0	0	2	0	250	420
Headley	4	0	0	0	0	2	229	307
Gregorius	3	0	0	0	0	1	261	376
Hicks	3	0	1	0	0	1	198	297
Total	30	2	5	2	3	8		

Outcome of Pitch — Pitcher's strike percentage

Count	Ball %	Strike %	2-strike Foul %	In Play %	Strike Zone %
0-0	40.4	47	0	12.6	59.6
1-0	34.3	47.2	0	18.5	65.7
2-0	31.4	49.4	0	19.2	68.6
3-0	35.1	61.6	0	3.3	64.9
0-1	41.8	38.3	0	19.9	58.2
1_1	35.4	41.6	0	23	64.6
2_1	29.8	43.5	0	26.7	70.2
3_1	28.8	44.6	0	26.6	71.3
0-2	45.6	16	18.8	19.6	54.4
1_2	37.2	17.4	21.6	23.7	62.8
2_2	29.6	17.6	24.8	28.1	70.4
3_2	21.9	15.6	28.2	34.3	78.1
Total	36.8	37.2	6.2	19.7	65.3

FIGURE 5.12 Batter's average versus pitcher's strike percentage.

guide, determine if the pitcher and the catcher are tuned to each other's thinking, as well as whether the pitcher can confuse the batter with slower or faster balls.

5.1.4 Track Prediction in Football and Application to Gaming

Football (soccer) is a sport that, over the last ten years, has attracted a great deal of publicity over the money involved. For many years after World War II, football 'Pools' were used (the name *pools* derives from the practice of people 'pooling' their money when betting on a horse race or a team winning – subsequently, it was adopted by the football industry to describe the process of placing a bet on a number of football teams to win their match).

In football (*soccer*), complex equations (*algorithms*) are applied to how much space each player may need to create when performing well, since, when there is little space to move, it is not easy to use your ball control skills. If the goalkeeper of the *home* team is expected to dominate his area in front of goal, this can be emphasised by software that shows where the player should be dominant over other players. Figure 5.13 shows a rectangular field in which areas are established with individual players in each area shown as a dot – the software allows the viewer to move individual players, and, as they move in their area, borders

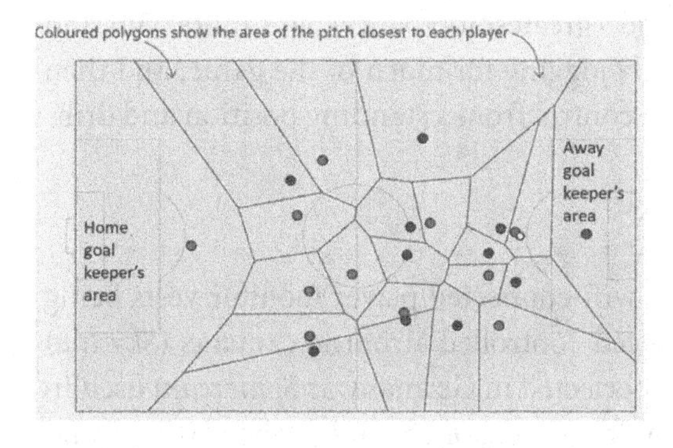

FIGURE 5.13 Polygons show the area of pitch that each player is expected to control (graphic courtesy: Hudl CFT) and graph of percentage of space gained indicating Messi gains most by walking.

move also to show the area in which an opposing player may take action. This software is very useful for understanding the amount of space each individual player may have, and the possible areas in which a player may move with minimal interference from an opposing player.

It may be that the opposition *away* keeper (right end, blue dot) is known to be weak on his/her right side (this may be known from existing stats that may show more goals are conceded from the keeper's right side), so the home team should try to put the ball into that area or draw the goalkeeper into that area. This would be possible if the home attacker (shown in front of the away keeper) focussed on moving to the keeper's right side rather than left.

By comparison, the *home* team keeper (left side, red dot) would dominate his/her area (and note that the *home* keeper's area is twice as large as the away keeper's area). If you are a *home* team forward or home team keeper (when kicking the ball up-field), you would want to bias the attacking runs into that area on your left (his right) side. It is also worth noting that the home team appears to be further up-field than the away team, which is a tactic known as *pressing* the opposition. Of course, this opens up more space at the back for the away team to run into and score goals.

"*Leon Messi* (arguably the world's greatest footballer) wins more balls and creates more chances by standing still or quietly jogging for much of the game, and then expending maximum energy through tight ball control from a standing position and dribbling past defenders."

These areas suit an individual player's style, so that, on computer, models of player's movements can be aggregated into playing styles of each squad. This tactical thinking allows scouts to consider players with similar styles in their own team, ensuring tactical comparability when it comes to working with their own team and any new players. An example often quoted is that most football forwards get their best chance of scoring a goal by running fast towards goal all of the time (thereby beating both full backs and goalkeeper to the ball), whereas *Leon Messi* (arguably the world's greatest footballer) wins more balls and creates more chances by standing still or quietly jogging for much of the game, and then expending maximum energy through tight ball control from a standing position and dribbling past defender [6].

5.1.5 The Use of Spidercam and Drones

Apart from the normal stadium cameras with wifi-connected player monitor vests being worn on the field of play, the industry introduced controlled airborne cameras (*Skycam*) in 1984 to monitor general play. This was later perfected in Germany as *Spidercam* used in sports, which gives the TV viewer the chance to be among or just above the players – and it can provide additional on-field data. *Spidercam* (Figure 5.14) has a camera suspended on four *Kevlar* cords with a fibre optic line wrapped within the cords [7]. The cords are

FIGURE 5.14 Spidercam. (Courtesy: Shutterstock.)

each controlled by a winch at each corner of the ground, and the camera is stabilised using a gyro.

One fibre optic wire controls the camera position with the other fibre optic transmitting pictures and other data. All cords are individually controlled by a winch with a computer handling all data commands. A single controller is manned by an operator in the stands or in a recording truck. One of the benefits of using an airborne spidercam is that the on-field action is recorded from above the players, and so is very useful after the game to see how team tactics worked.

The alternative to spidercam is the use of *drones*. Remote control of drones similar to those shown in Figure 5.15 is a useful technology where an overview is needed of the field play or of any other industrial plant. Drones can be sufficiently powerful that they can do many heavy-lift industrial functions such as replacing power lines without turning off the power. Like birds, if a drone lands on a live power line, it will not feel the power flowing through it because it is not connected to the ground. They can be fitted with IR, *ultraviolet* (UV), and ultrasonic sensors as well as *electromagnetic* (EM) sensors and of course cameras, and they generally operate in the line-of-sight of the ground-based pilot. Where the application is possible of monitoring the progress of wild or bush fires, they can be fitted with an IR sensor that can detect the differences in forest fire heat and transmit images of heat maps from above forests or residential areas during a fire. They are often over-the-horizon at these times, so their position is also transmitted back to the control centre, using *Global Positioning System* (GPS) satellites for both their position location and the communications data transmission. These forest fire monitor drones have very sophisticated cameras so that the operator can steer them towards areas of interest. However,

FIGURE 5.15 Drones can be sufficiently powerful to attach a power line on top of two towers with power left on. (Courtesy: Shutterstock.)

a major issue of having such drones in fire areas is that they can cause flight-path issues for water-bombing planes, which limits their use. Otherwise, they can be used for remote resource exploration when fitted with EM sensors that detect changes in magnetic field, and of course, to check the exterior of large structures when used with a high-precision (*resolution*) camera.

The one problem with drones on the sports playing field is that they tend to make a lot of noise when moving around, and this can be quite disturbing to players. Consequently, they can never be allowed to move close to the field of play, and so preference is given to spidercam for monitoring player performance and field tactics.

5.2 PREDICTION IN PLAYER PERFORMANCE AND TEAM TRACKING

Individual player fitness and performance are linked because a player must be fit if the player is to perform at the maximum levels. As discussed previously, many ballplayers are chosen for their ability to move around a field generating spaces. Their knowledge of the typical movements of players in their own squad is important so that they would know to whom to pass the ball next. This removes the need for shouting between individual players and makes the ball speed and flow between players much faster.

Recruitment of future players depends on the prediction of where they are in their playing career, how they would play with others around them, their fitness, skill, and attitude. So as discussed earlier, there are on-field statistics and off-field statistics. In major money-earning games such as football (soccer and American football) and baseball, club boardrooms need input data to make their decisions when spending millions of dollars.

5.2.1 Team Selection

An understanding of individual player attributes and skills is often a complex mix, and statistics is the only way to separate the favoured player (who may have been with the club for many years) from a potentially new player who would be a better fit into a specific position if bought from another club. Clubs monitor opposing team players' abilities because a successful club has a team that wins games and brings income ($$$s) into the club to make it financially strong. Figure 5.16 is typically how player quality versus salary software

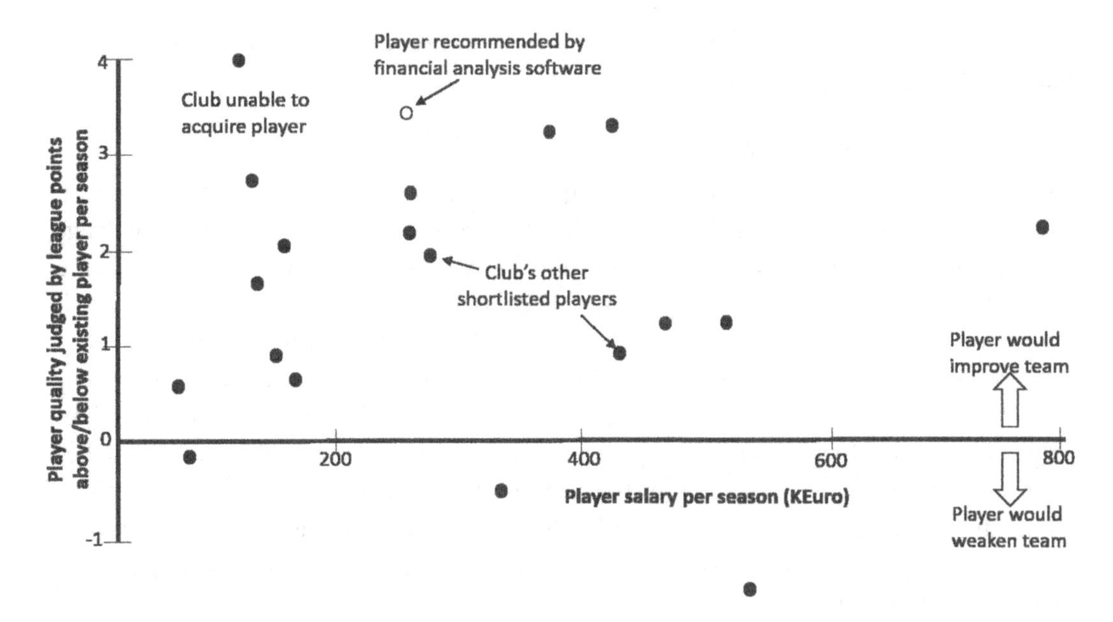

FIGURE 5.16 Example of professional football player quality versus salary graphic.

appears on a computer screen and helps football club management optimise their player purchasing choices.

Since we are dealing with big $$$numbers, the players' annual salaries are posted as being between $100k and $800k per annum. This spread would not be unusual for a mid-tier soccer club where the majority of a team would have players typically being paid between $100k and $300k per annum. In the top tier leagues of the world, this may be the amount being paid to each player per game not per annum, so that a top league player could be making at least $5 million per year.

It shows an estimate of player quality for the on-field position each player has, with this quality estimate being based on number of passes and tackles made, number of goals scored, amount of possession, etc. which each player contributes to the overall team performance. During practice sessions, some players are evaluated on their practice workouts or quality in practice games versus the other players [8]. Some clubs run statistics on players' social interactions since a happy motivated captain in the dressing room is a plus compared with a captain who has little interest in the team morale.

It is clear from the quality index that some players could be detrimental to the team's performance and it would be that these players are to be replaced. An example would be a player who is not quite fast or fit enough to aid the team's progress, or in fact suffers from poor relationships with other players and therefore is not brought into the game as much as would be desirable. These weaker players are likely to be the first players to be sold.

The figure also shows the proposed salaries of players that the club was unable to buy, those they may want (having been short-listed), and the trend in salaries that shows increasing costs for players they may want but only being as good as the players they already have in the team. An issue for many teams is that, if they are to succeed, they need an almost equally skilled reserve team and certainly, if they were to do well in European competitions aside from their own league, they may need a reserve team that has a majority of players that swap for a first team player at any time.

5.2.2 Tracking Team Performance

Each player has a profile in the squad list (Figure 5.17) that includes financial information extracted from the player's contract. In this specific case, the player is a goalkeeper with age in his/her late 30s (not unusual for a professional goalkeeper) valued at £5 million being paid £10k per week on a three-year contract. There is an agreement that the contract will be renegotiated after 30 first team appearances, but if the club is relegated, the salary will reduce by 40%. A bonus of £1k would be paid if a goal is scored in the Premier League (very unlikely occurrence since goalkeepers are not goal scorers) and £200k bonus if more than ten goals are scored in the Premier League, per season (note that the world's best footballer *Leon Messi* is reputed to make £90 million per year). The software would compute these values and provide an automatic notification to Board Members when there were important upcoming events or when player's salaries/bonuses may be paid, so that any player's contract may be checked at any time [9].

SQUAD LIST

#	Name	Position	Age	Contract Expiry	Weekly Wage	Value
1	Player 1	Defender	29 yrs 3 mths	2 yrs 6 mths	£25,000	£1,500,000

Contract info. | Transfer history

Current status	Active
Transfer value	$3,500,000
Current placing	First team
Origin	Domestic
DOB	20/09/2000
Position	Defender
Nationality	Ireland

Start	End	Weekly Pay	Performance
			Appearance in Premier League decided by Club
6/07/2018	6/07/2021	£25,000	Salary reduces by 50% if relegated to Championship. Bonus on winning FA Cup
			£1,000 bonus per goal in Premier League. £200,000 bonus for 10 goals+

#	Name	Position	Age	Contract Expiry	Weekly Wage	Value
2	Player 2	Goalkeeper	32 yrs 2 mths	2 yrs 3 mths	£30,000	£1,290,000
3	Player 2	Defender	26 yrs 1 mth	3 yrs 1 mth	£5,000	£425,000
4	Player 2	Goalkeeper	20 yrs 5 mths	1 yr 5 mths	£500	£300,000
5	Player 2	Midfielder	23 yrs 6 mths	3 yrs 8 mths	£2,000	£12,000
6	Player 2	Defender	28 yrs 3 mths	0 yrs 7 mths	£10,000	£20,000
7	Player 2	Midfielder	25 yrs 8 mths	1 yr 7 mths	£25,000	£150,000
8	Player 2	Midfielder	23 yrs 8 mths	0 yrs 9 mths	£12,000	£2,560,000

FIGURE 5.17 Example of professional football squad list computer spreadsheet.

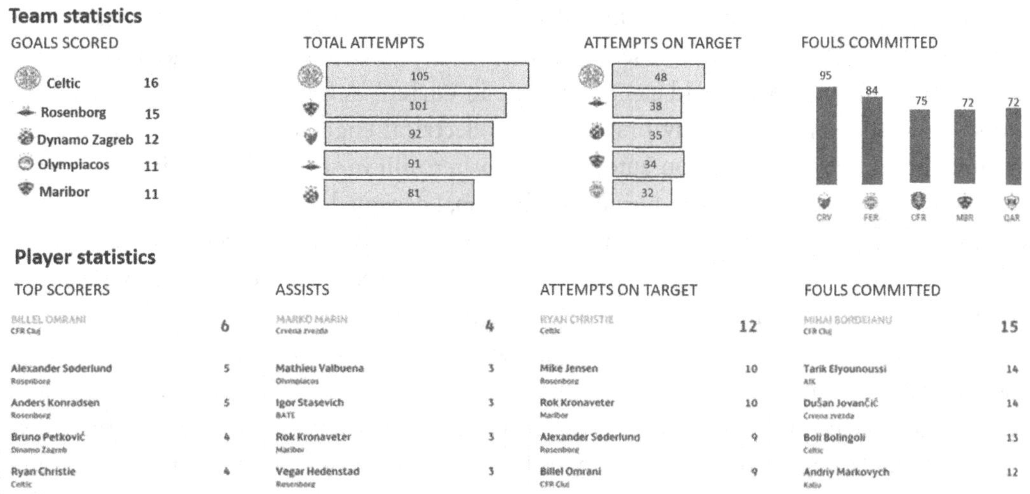

Team statistics

GOALS SCORED	
Celtic	16
Rosenborg	15
Dynamo Zagreb	12
Olympiacos	11
Maribor	11

TOTAL ATTEMPTS: 105, 101, 92, 91, 81

ATTEMPTS ON TARGET: 48, 38, 35, 34, 32

FOULS COMMITTED	
CRV	95
FER	84
CFR	75
MBR	72
QAR	72

Player statistics

TOP SCORERS		ASSISTS		ATTEMPTS ON TARGET		FOULS COMMITTED	
BILLEL OMRANI CFR Cluj	6	MARKO MARIN Crvena zvezda	4	RYAN CHRISTIE Celtic	12	MIHAI BORDEIANU CFR Cluj	15
Alexander Soderlund Rosenborg	5	Mathieu Valbuena Olympiacos	3	Mike Jensen Rosenborg	10	Tarik Elyounoussi AIK	14
Anders Konradsen Rosenborg	5	Igor Stasevich BATE	3	Rok Kronaveter Maribor	10	Dušan Jovančić Crvena zvezda	14
Bruno Petković Dinamo Zagreb	4	Rok Kronaveter Maribor	3	Alexander Soderlund Rosenborg	9	Boli Bolingoli Celtic	13
Ryan Christie Celtic	4	Vegar Hedenstad Rosenborg	3	Billel Omrani CFR Cluj	9	Andriy Markovych Kalju	12

FIGURE 5.18 Typical FIFA-type team and player statistics.

The software also compares these values with the market rates at that time with equivalent players of the same age and experience, in order to strengthen the negotiation tactics that the team management may use when discussing player contracts with their agents. The software can also be adjusted so that potential future player's salaries and bonuses can be configured to give an idea of club incoming and outgoing costs and profitability. This all helps with planning and running a successful team and club finances.

The *International Federation of Football Associations* (FIFA) in Europe also maintains a database of all teams playing in all European football leagues and gives this information to clubs to help them plan their forthcoming games. A typical FIFA-type list of team and player statistics is shown in Figure 5.18 where an example team's past performances (*Celtic*) are shown with goals scored, total attempts at goal, and attempts on target as well as fouls committed in specific competitions. This is compared with other teams in that competition to allow coaches and data statisticians to work out who would be the most likely opposition impact players. The player statistics are shown for each player so that an in-depth study can help tactical management.

5.3 PREDICTION IN INDUSTRY

Predictive analytics has been used by industry since data from the first sensors were able to be collected when monitoring a continuous flow of operations. *Henry Ford* has been acknowledged for bringing production line automation to the motor vehicle manufacturing industry with the development of the *Model-T* Ford automobile in 1908 – even though it was mainly human operated. Because the routine of fitting different components together in a production line flow does not change, the speed of throughput of vehicles improved the economics of manufacture. Henry Ford studied how he could automate the process to remove the human involvement and, by the time of World War II, factories were becoming more automated to produce aeroplanes and other vehicles for war use. However, automation involving electronics controlling mechanical movement only really started with the development of data-producing sensors in the 1950s [(10)].

One of the first places to become automated was street traffic lights. By burying an EM wire loop (the sensor) in the road near a set of lights, any car movement that changes the EM field (by driving over the loop) would cause a change in current in the line to the lights, and the lights would then change. In my first job as an electrical engineering graduate with GEC Automation in 1969, I worked on automating a paper mill in which the continuous flow of paper along a production line was needed. In that job, I designed motor controllers and actuators to replace human hands operating the machines. Very small boxes containing *limit switches* monitored paper thickness and when paper thickness exceeded the specifications, they told the motors to run the production line slightly faster or slower. Later I moved to a different division where we installed overhead traffic-warning signs on motorways servicing London – it also linked with similar computers around London to try to judge where and when traffic congestion might occur. The level of electronics was greater than for the paper mill and we had a large computer that told traffic of hazards ahead – the computer was the size of a large steel cabinet, yet it had less computer power than today's average calculator. The automation of processes in industry using computer monitoring and control is now referred to as *smart technology*. In reality, it has been used by industry for the last 70 years [(11)], and it only seems as if it is new because of the increasing ability to transmit data faster and to compute local results quicker (see Chapter 1) that makes it seem as if it is recently developed technology. Perhaps this is because the different technologies have also coalesced at the same time.

> "The automation of processes in industry using computer monitoring and control is now referred to as *smart technology*. In reality it has been used by industry for the last 70 years, and it only seems as if it is new because of the increasing ability to transmit data faster and to compute local results quicker that makes it seem as if it is recently developed technology."

Today, the mining and oil and gas industries are major users of automation. Driverless dump trucks work in the mines and offshore platforms have very few crew if any, producing oil and gas to pipelines in which electronic control assists automated monitoring of the fluid flow processes. Prediction of future mechanical failure issues, which may reduce the efficiency of production, is also under electronic control. *Preventative maintenance* scheduling has become a major recipient of automated control systems, in which vibration from rotating machinery is constantly monitored, and if the vibration appears to be inconsistent with the machine's design, the remote control centre is informed of potential problems, and a course of possible remedial or planning action is then followed. This has now become the norm in many processes in the resources industry, where multiple sensors are monitoring multiple pieces of complex equipment, and predicting the potential of future failure to allow the monitoring team to take action. An example is that an offshore platform will likely have as many as 10,000 valves, each of which requires constant monitoring. This is now the area of automated operations for predictive maintenance using sensors to ensure sustainable production [(12)].

5.3.1 Prediction Using Sensor Data in Process Control Operations

So far in understanding automation of processes, we have considered replacing manual *control* of machines with automated control. We have considered what the act of data *correlation* is in everyday business and where a *computer and database* fit in that process. We have discussed *regression analysis* and how we make *predictions*. In industry, we put these together so that we can start making predictions using the *sensor data*.

In Figure 5.19, a valve provides a flow of fluid to a pump. The valve automatically turns on or off when fluid in a tank reduces to a certain depth level. As with all moving parts and components, the valves wear out and often need some form of maintenance and cleaning or total replacement. In this case, when the valve opens and closes, it automatically starts or stops an electric pump that is pumping fluid along a pipeline. The pump like the valve contains parts – in this case rotating – that can suffer from fatigue and failure. So in order to monitor the performance of the pump and valve, we can now go to an automated monitoring method that has been well established over the last 50 years.

Both the pump and valve have had vibration monitors (like the piezoelectric crystal pressure transducer that converts voice vibrations in your phone to electric signals) fitted to them and a local computer records the data output from the vibration sensors. Because the pump motor is very large (complex and expensive) and more liable to fail than the much smaller, simpler valve, their vibrations are recorded over different limited short bursts in time (maybe every 30 seconds for the pump motor compared with the valve vibrations every 10 minutes). If a motor fails, it could spell disaster, whereas if a valve fails, there may be simple backup valve systems to circumvent any issues.

Previously recorded data of when both pieces of equipment were functioning well is stored in their respective databases. When the monitoring is turned on, a local computer will record the local data of pump vibrations and link with the pump motor database in order to cross-correlate the data. This also happens for the valve vibration data, with the correlation looking for any vibrations outside of the normal correlation process. We are looking for a correlation with the database resulting in a high correlation value, without spikes outside of the normal operating range. If there is a good correlation, this data may be

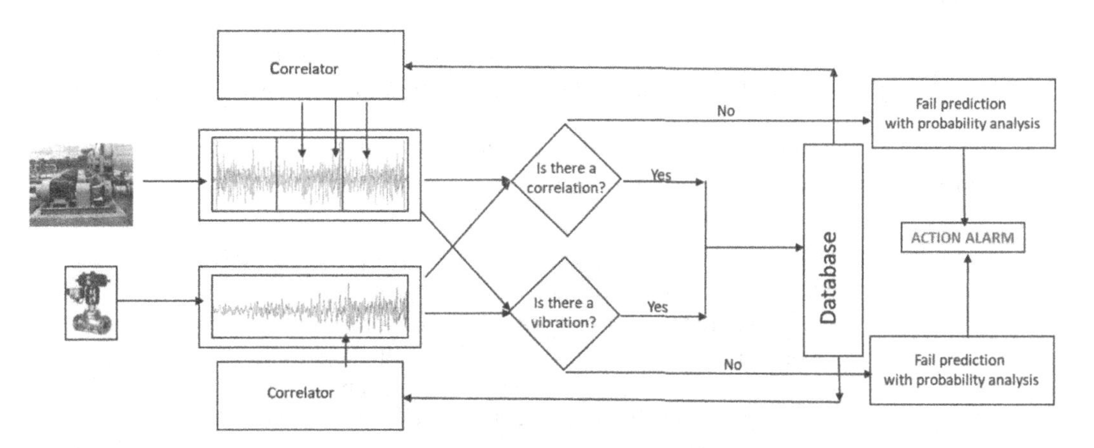

FIGURE 5.19 Process control monitor.

stored in the database. If spikes are recognised such that overall the correlation has failed, the degree of failure is then reviewed by looking at the frequency of the data spikes and sending an alarm to the operator. The number and frequency of spikes then establish the degree of failure, with a failure criteria then input as an optional detail in the software's failure parameters.

A number of constant failures will cause an alert to be sent to the operator demanding action. The action may be for the operator to inspect the data, and decide if there needs to be a maintenance check on the equipment, or in some cases, the operator will decide to replace the piece of equipment if there are excessive out-of-specification vibration spikes. The result may be predictive that the component will fail at some time in the future and apply a *probability analysis* to that prediction. We can of course have the computer determine when a probability analysis indicates failure and then have the computer inform the maintenance department to change the part. When data is transmitted from sensor to the database, we can apply maximum and minimum acceptable threshold values on it to determine the risk level of a piece of equipment failing. When a computer performs a correlation using such threshold values, this is becoming more advanced in terms of automatic recognition, monitoring, and action – and the degree of computer programming required to do this is known as *machine learning.* This is full automation for predictive maintenance, and example software platforms currently in operation that do this type of work are Amazon's *SageMaker,* IBM's *Watson, TensorFlow, Shogun, Apache, Oryx, PyTorch, RapidMiner* to name a few.

5.3.2 Computer Control System Terminology

During the development of these control systems, there has been an upsurge in the use of technological words to explain what area of control we are dealing with. Often, this can be referred to as *smart language*, and the people who use it are *techies.* Many times these techies use the wrong terminology themselves thereby confusing the layman, but it is really all very simple [13] and explained as:

- *Artificial intelligence* (AI), which is an umbrella term covering all aspects of automation.
- Data + Pattern recognition *(Correlation)* + Regression analysis *(Prediction)* + Action from knowledge *(past failure knowledge including probability)* = *Machine learning.*
- Then we can say that *machine learning* + specialised algorithms *(equations* also called *neural networks)* with action involved = *Deep learning.*

> "-when we hear of *Deep Learning* and *Machine Learning,* the only difference between the two is that someone had to figure out a best fit equation to predict what could likely happen given the existing conditions of a machine, and then program a computer and get this all to predict reasonably correctly."

FIGURE 5.20 Typical machine learning literature that advertises sensor providing data on vibration levels in rotating turbines. (Courtesy: Yokogawa Industries.)

Therefore, when we hear of *deep learning* and *machine learning*, the only difference between the two is that someone had to figure out a best-fit equation to predict what could likely happen given the existing conditions of a machine, and then program a computer and get this all to predict reasonably correctly [(14)]. This is the suggestion behind the Japanese company *Yokogawa*'s advertising to help industry as shown in Figure 5.20.

This type of software is well established in industry for prediction of equipment failure, with the reference here to *harmonics analysis* being the use of the *Fast Fourier Transform* (a similar frequency analysis method to that used earlier to explain voice recognition). The vibrations have frequency, and sometimes particular frequencies vibrate more than other frequencies, which we call a *harmonic* (and they may be symptomatic of a failure in some cases).

5.3.3 Prediction Using Past Stock Market Values in the Financial Industry

Prediction is not only used in the engineering and sports industries, but also in the financial industry. Economic factors all influence the price of stocks and shares, the interest rate we pay banks for our loans and house mortgages. When travelling from one country to another, we change our local currency for the currency of the country we are travelling in, and this exchange rate depends on how the financial market considers the relative strengths of the two countries' economies to be.

In the household, the income and savings statements from our banks affect how we spend money, and if we can get some idea of prices of goods or the cost of living ahead of time, we can better plan our daily lives and improve our living. So being able to predict whether the interest rate is going up or down, or whether currencies will change or not, can influence our personal daily financial decision making. This is evident in how the stock market moves and how brokers within the stock exchange try to predict where the market is moving so that they can buy and sell stock to make a profit where possible.

While the fundamental analysis in the financial industry starts with the statement of cash flow (money coming in from dealings versus money going out of a business), it uses conventional spreadsheets to predict a direction to where a company is financially travelling. Added to this is the technical analysis involved in financial transactions – using the past trend in stock prices to predict future price patterns. We discussed computing the *average* and *median* track of data in Chapter 4, which in industry is sometimes referred to as a *moving average*, since the data is constantly being updated on a daily basis.

Machine learning, discussed earlier, uses the last 30, 60, or 90 day average values of commodity prices to find patterns in data that indicate which way currency or other exchange values may move in the future. Some goods, services, or stock can be predicted to rise or fall depending on international actions, such as countries putting a block on certain types of goods imported, or in fact the long-term effects of a virus on each country's economy.

The simplest approach used is that of best-fit linear regression, which as explained earlier in Chapter 4 is a simple straight-line prediction of where prices will go depending on where they went over the past 30, 60, or 90 days. In the financial market, because some economies are seasonal (e.g. people in the northern hemisphere will go on holiday during the summer months of June–August, whereas the summer months in the southern hemisphere tend to be December–February), then the linear best-fit trend of this month will be compared with the same one year ago and also five years ago. Often financial houses, who offer similar simple services, have their own name for this (such as *Long Short-Term Memory* computing, which tries to fit short-term data to the longer range annual values to see if they are similar, and then they will produce a *similarity index*, which is a level of *probability* of a prediction being correct). It can be argued that saying obvious and simple regression analysis in words that could be confusing, is snake-oil salesmanship and I would tend to agree with that. (The correct word in English is *obfuscation*, which often occurs in industry when *techies* explain how very simple analysis works in a way to make them seem very complicated – they do this often which was a major reason why this book was written in the first place!)

More often than not, the person who predicts the financial market well, usually, does so by setting upper and lower limits of trends and simply keeping the prediction within those values. If major international investors cannot yet predict the market boom or bust within a week of it happening [15], then we can only say that the financial market is full of data outliers that are unpredictable, which is how the market has been since the first stock market opened in London in 1773, using similar prediction technologies that we have already reviewed in previous chapters.

5.4 THE USE OF REALITY TECHNOLOGY

Reality technology encompasses technologies that allow the user to view the conditions and environment found in real life and is not to be confused with *Reality TV* where a TV show places people into everyday conditions of living and monitors their personal reactions to each other. *Reality technologies* originally started with the *Walt Disney*-type cartoon animations of different people and animals. However, they are developing to the point that the increase in computer operating speed allows them to produce conventional 2D and 3D images on monitors in real time for the viewer much faster than ever before. This includes 3D visualisations of cartoons and games that have images that look like real people or animals, as well as *virtual* and *augmented reality* (AR).

5.4.1 Computer 3D Visualisations and Virtual Reality

The concept of viewing images in 3D was discussed in Chapter 2, where it was explained that 3D locations on sports fields needed at least two closely positioned cameras to view an object in 3D. This is the same situation with our two eyes looking at an object but using a flat monitor screen – there have to be two very closely spaced images. 3D photographs (aka *anaglyphs*) are constructed by overlaying two closely spaced photos while retaining the background of only one of them. These can be viewed using a pair of red-blue 3D eye-glasses. These eye-glasses have binocular vision – the same as the ability of our two eyes (*bi-ocular* – two eyes) to pass signals representing vision images to our brain. The brain then interprets the vision as having depth. 3D eye-glasses work by having one lens with a red filter allowing only red frequencies through and the other glass blue, with the image being viewed (anaglyph or movie) being an overlay of the two glass images – one red and one blue.

An alternative approach is the use of *liquid crystal display* (LCD) *polarised glasses* as suggested in Figure 5.21, in which the screen displays two images rapidly alternating one after the other using LCD active shutter glasses. They block the view of one display while showing the other, and vice versa. Typically, if a TV screen has a slow refresh rate of 120 Hz (120 images per second or an image shown every 8 milliseconds, while in Chapter 1, it was mentioned that some large screens have a refresh rate of 200 Hz or raster every 0.1 μsecond), each eye's LCD glass synchronised with the screen transmitter becomes opaque every second screen so that one eye sees one image (every 16 milliseconds) while the other eye sees the other image. Because the eye can only register a change in image no faster than every ¼ second (see Chapter 1), this repeating image frame appears to each eye as a continuous image and so the brain interprets the two images as having depth and therefore 3D. The problem with LCD glasses is that if you tilt your head, the image may become blocked and blanked out.

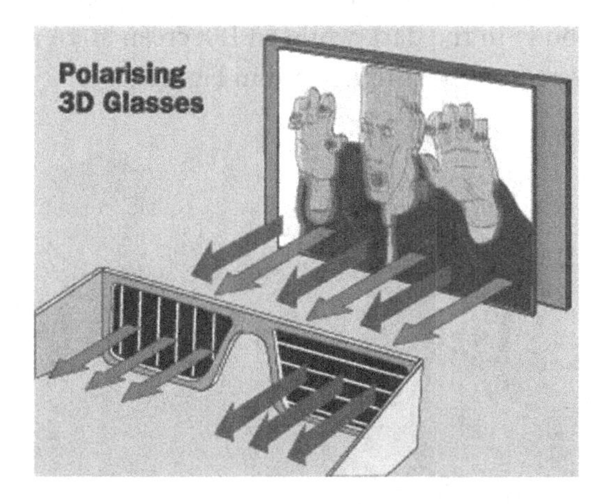

FIGURE 5.21 3D polarising eye-glasses. (Courtesy: Brain, Marshall, 18 July 2003. How 3-D glasses work, HowStuffWorks.com, Retrieved 26 February 2021: https://science.howstuffworks.com/3-d-glasses.htm)

A further alternative is *circularly polarised glasses* that have one lens that blocks clockwise polarised light using one circular polarising filter, and the other lens blocks light travelling with polarisation in the opposite direction. In this case, the image is not blocked if you tilt your head, and so it is the most common of polarised glasses used at the movies. It may also be worthwhile noting that the name given to any 3D character is an *avatar*, which is the appearance of a person or idea which is not real, and we are thankful to the movie *Avatar* for getting this word into the public arena [16].

While discussing 3D image glasses, it is interesting to note how *virtual reality* (VR) came about. This type of 3D glasses was initially developed in the United States under a trade name of *LEEP* in the early 1980s. It was further developed by the *National Aeronautical and Space Administration* (NASA) space program in the late 1980s (Figure 5.22a) when *NASA* wanted astronauts to get a feel for working in 3D space. Because the *NASA* base is in Houston and advertised its 3D experience locally, the oil and gas industry started developing its use for geophysical reservoir interpretations (which eventually became more gimmick than useful), and that additional development created such interest that the gaming community used it as computer speeds increased. This led to the *VR* headset (Figure 5.22b), which allowed anyone to experience walking in holiday villages, on the edge of cliffs, driving fast cars, playing great video games of tennis or cricket against the world's finest sportspeople (*avatars*), and shooting demons that are going to take over the world – all from your living room.

The technology can be adapted so that IR frequencies can be filtered, to allow the *VR headset glasses* to be used to see warm or hot shapes. In Chapter 1, Figure 1.22, the different frequencies were discussed, with note being made that we can see visible light EM frequencies in the 430–770 THz range, but we can't see frequencies in the lower IR range of 300 GHz–400 THz (300,000,000,000,000–400,000,000,000,000 Hz). We can use viewing filters that just filter out normal light but allow this range of frequencies through – such glasses are used for night vision by the military for night-time combat where the soldiers can't see people but can see their body heat. If we were to put such night vision glasses on, we could see a person's body in the dark while an IR screen at an airport would indicate with red which bodies were hottest, as illustrated in Figure 5.23.

FIGURE 5.22 Virtual reality (VR) headsets: (a) NASA and (b) commercial set for use in training. (Courtesy: Shutterstock.)

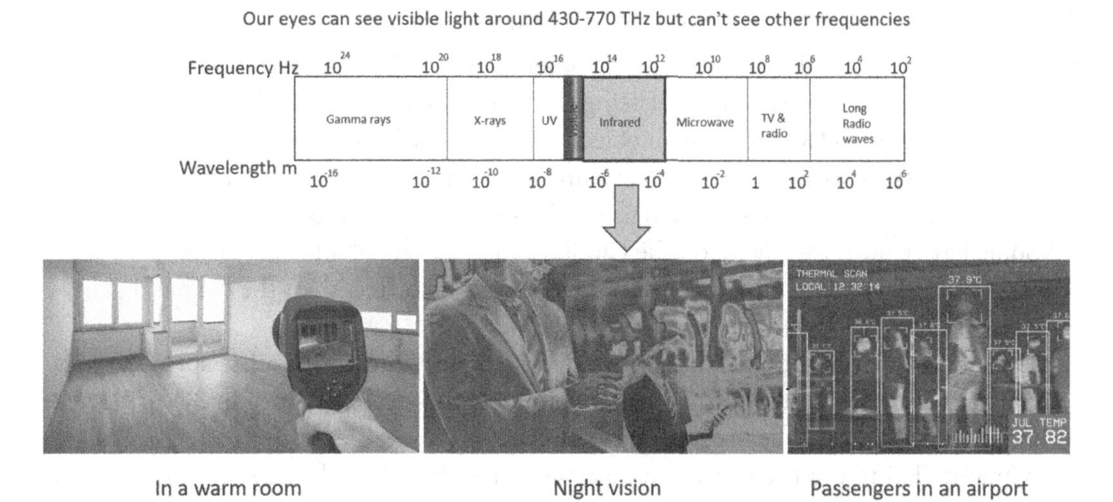

FIGURE 5.23 Infrared cameras can show where it is hot in a room, IR can provide night-time vision and a picture of its use at an airport. (Courtesy: Shutterstock.)

IR is also useful in seeing hot gases in space. IR shows hot gases as red and cold gases as blue. For example, the Eagle Nebula in space is seen as a blue cloud mass in the normal visible spectrum (because the frequencies are high enough for us to register them) using a single lens as shown in Figure 5.24a, but if we were to add a second lens that was an IR filter with a VR viewer, we would see an overlay of the *Eagle Nebula* as an almost real, solid object as shown in Figure 5.24b. This would be how we would expect future travellers in space to view these giant hot and cold gas clouds and avoid them if necessary during their travels.

The *VR headset* developed rapidly with two slightly different computer images instead of eye-glasses, so that fast computer graphics and cartoons could thereafter be incorporated. This incorporation of additional objects on say a screen or photograph was augmenting (adding to) the view, which led to the science of *AR*.

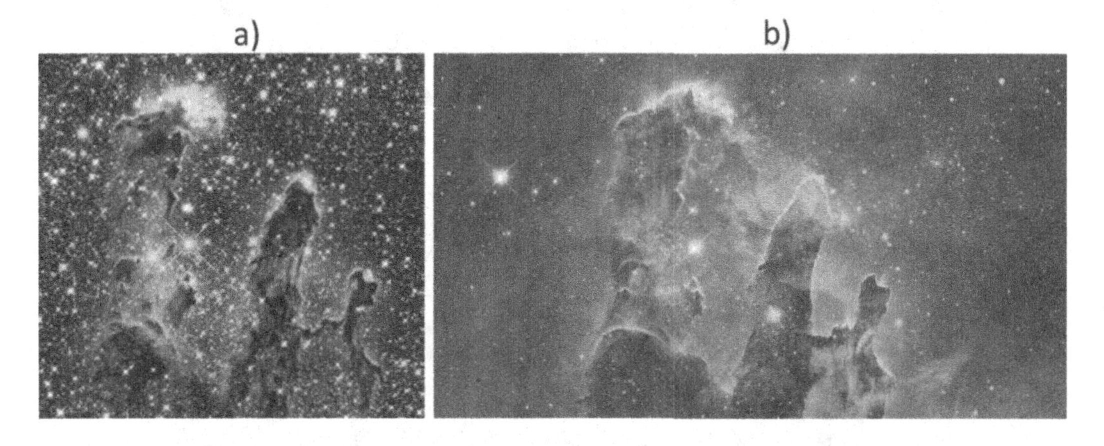

FIGURE 5.24 (a) The Eagle Nebula cloud in space using IR on VR screens, naked eye versus (b) using an IR filter. (Courtesy: Shutterstock.)

5.4.2 Computer Augmented Reality and Its Use in Industry

The concept of augmenting data by adding other data to a view has been around for some time, which has led to the expression of being able to *Photoshop* something into a scene – that is to use *Photoshop* software to add an object to a scene (such as adding small birds sitting on large lion paws – as if this were possible in nature). However, when we add a 3D object to an existing 3D scene, this is referred to as *AR* since the 3D scene is considered to be reality. Some have called it the 'grin on the Cheshire cat' (so named after *Lewis Carroll's* 1865 *Alice's Adventures in Wonderland*) where a cat is seen on a tree, then the cat has a grin – and the cat then disappears leaving the grin behind.

Consider an apartment block building's steel structure into which a fire sprinkler system has to be installed. It would be possible to view the steel structure in 3D using VR techniques and then add the fire system (Figure 5.25).

By manipulating the fire control system pipes around the building structure, we can work out the optimum set of pipes to install the most economic fire pipe system that we need. In fact like the grin on the Cheshire cat, once we have designed the fire pipe system around the building, we can then remove the building and we are left with the fire pipes. These can then be linked to drawing software so that we can rapidly draw the pipes ready for manufacture.

This concept has been taken further in terms of building structures and installing other components in a building. For example, in Figure 5.26, we can plan a chemical plant to perform a certain function, such as refining oil. Let's say we have a planned computer model to modify an existing plant – we have one computer model of the plan and we have another computer model of the existing plant. We overlay one on the other and we have an augmented plant containing the additional equipment – say a tank and piping. We can then modify and adjust the tank's leg height to fit the tank into place and also adjust the piping to make it fit.

FIGURE 5.25 Augmented reality fire system. (Courtesy: Curtin University.)

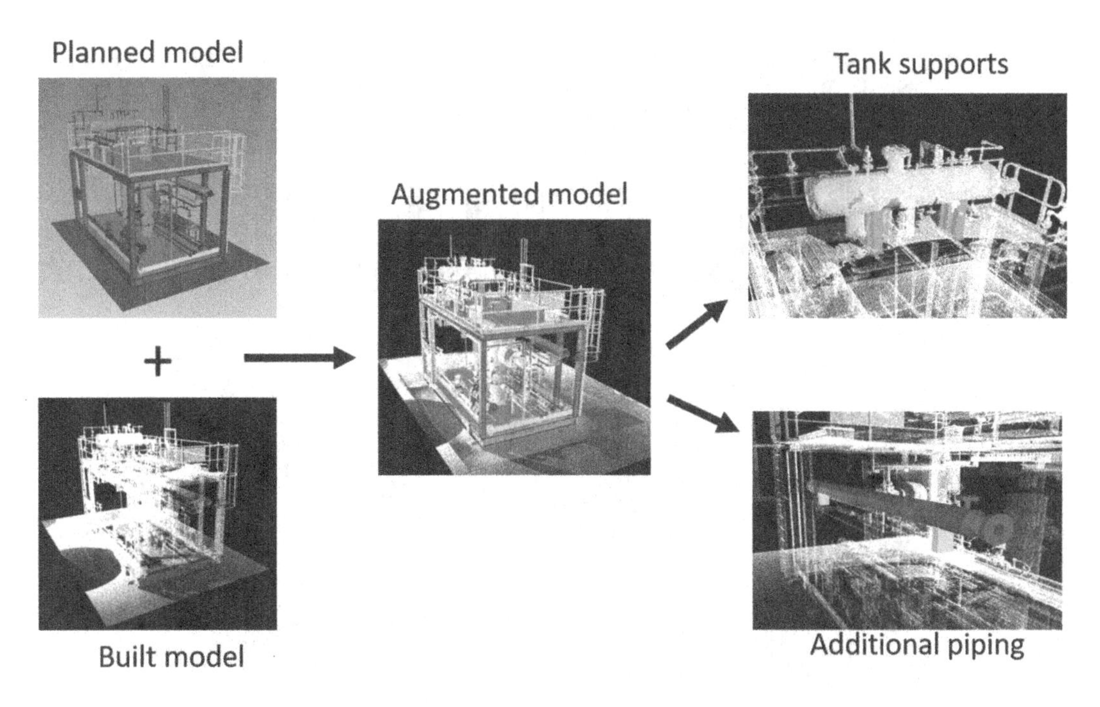

FIGURE 5.26 Built information modelling. (Courtesy: Curtin University.)

Because this is in *AR* (modified on a computer), we can zoom into the structure or out, we can reduce the size of equipment to fit spaces, and in fact, we can see the best method of installing a particular piece of equipment within an existing structure, before we have actually fit it in. This ability of *AR* to test pieces of equipment and modify their design or modify existing operating equipment [(17)] is known in industry as *built information modelling* (BIM). This form of industrial design is now commonplace and has had major use in areas such as improving the scaffold design and efficiency of a structure (e.g. the rebuild of Notre Dame Cathedral in Paris), in order to ensure building stability while maintaining as minimal a cost as possible.

5.5 PRODUCT TRACKING IN INDUSTRY, RFID

An area associated with VR is product tracking. We commonly know of product tracking or product recognition by the use of the product bar code, but industrial product tracking is far more extensive than just the bar code on post office parcels or shopping items. Industry uses far more sophisticated methods mixed with AR to improve understanding the performance of a machine as will be discussed now.

5.5.1 Electronic Signatures of Products – Barcodes

Barcodes were first used to create a method for product identification in US supermarkets during 1973 and were based on the concept of the *Morse* code with short- and long-dashed lines representing numbers and letters. A barcode is a series of lines or patterns that, when read by an optical reader (explained earlier), provide the origins of the product [(18)]. The first barcode had four lines and some were circular. The first product to be packaged in a

FIGURE 5.27 Barcodes. (Courtesy: Shutterstock.)

wrapper showing a barcode was *Wrigley*'s chewing gum which had a *1D barcode* consisting of a row of lines that represented a string of numbers (Figure 5.27). When the wrapper was optically scanned into a computer, a program would send the code information to a product database in which the product's details would be stored. Because this product was being purchased, the database would subtract a single item from the number of items in the store, thereby doing a constant stock-take providing information to the store or grocer's shop of which items need to be ordered to replenish the shelves - and of course the computer provided the product's price to the salesperson. However, only limited information could be stored because the *barcode* method was limited by the number of lines and therefore numbers that could be stored.

In 1994, the *2D square barcode* matrix was introduced as a *quick response* (QR) code by a *Toyota* subsidiary (to track Toyota cars). Because data is read in one direction and then the other as in a matrix of data, more information could be stored so that the database can be far more extensive, providing information on product manufacture (where, when, number, and address of manufacturers, materials used, use-by date, and, in the medical field when used as a patient ID wristband, whether the patient has allergies, medical conditions, and other medical information). These barcodes are also used on business cards or in advertising material because a simple (optical scan) photograph of them allows us to find them easily on the internet.

However, a major issue with barcodes is that the barcode tag needs to be very close to the optical reader, which means only one tag per item can be read at a time, and it requires someone to physically operate the optical reader. In comparison, this is not the requirement with a *Radio Frequency IDentification* (RFID) reader.

5.5.2 Radio Frequency IDentification of Products

From the early 2000s, the *RFID* tag became a preferred method of identifying products in industry. Because barcode reading of each individual product required the optical reader to be close to the tag showing the barcode (usually within a metre), this was not sufficiently efficient for industry that required multiple products to be read and tracked at the same time. Consequently, EM radio frequency metallic tags were developed, which allowed a radio frequency transmitter to read 1000s of tags simultaneously. The radio frequencies used can pass through most objects, so the operator's presence is not required as it is with the barcode. Consequently, production lines can be used with an RFID gate so that products going through the gate can be automatically tagged and thereafter logged in terms of their location. For example, if there is a trolley or production line of products going out of a large warehouse or shopping centre, the trolley or production line contents could all be scanned within seconds (Figure 5.28) whereas that would be impossible with optical barcodes. The RFID data is in the computer database, along with the history of manufacture, where the package may have travelled, cost structure, etc. This improves warehousing and management efficiency.

FIGURE 5.28 RFID scans and tags multiple products at one time. (Courtesy: Shutterstock.)

Another advantage of the RFID tag is that the transmitter/reader can change the information on the tag remotely, in order to update a product's track history of where the product has moved to [19]. This is very helpful in tracking products as they move from one warehouse site to another, as well as for ordering and storing products for many years. One story is that a large offshore valve was needed on an oil and gas platform because the existing valve's seals had deteriorated and the only solution was to find a replacement. The cost of the replacement was expected to be in the order of US$40,000 and it would be a lengthy process of six months to manufacture this old part – but the platform would have to be shut down if the part took that long. However, an unused valve was found in a warehouse within the country (which had been discarded many years before), which saved both the cost of a replacement but also proved the benefit of using RFID tags.

Today, use of the RFID tag extends to VR, when production line equipment fails. It is possible to take a photo of an RFID with a smartphone, and with special software, the equipment history comes up on the screen. This would include maintenance history, origin of the equipment manufacture, and previous failures of the equipment. It is possible to then make a movie of the equipment with an IR lens fitted onto the phone (using a filter *app* in the phone), to then send a photo of that heat map back to the manufacturer, who could then make an immediate interpretation of any possible problem and pass back a photo or movies from *subject matter experts* (SMEs). This is how we now solve equipment problems using the power of the smartphone taking a photo of an RFID tag, and transmitting that to the manufacturer for immediate advice on how to solve the problem.

5.6 EXERCISE

Exercise in tracking a product from New York to London and on to Cairo.

We have recently set up a secure office in Cairo that needs special electronic keys to maintain security. The keys have been manufactured in New York and are the electromagnetic type that allows the user to tap the lock and, if recognised, it opens (like a hotel key). What information would be input to the key manufacturer's database to allow it to:

a. Leave the manufacturing plant.

b. Travel from New York to Cairo.

c. Be installed in the Cairo office.

FURTHER READING

1. How accurate is electronic judging? https://theconversation.com/a-hawk-eye-for-detail-how-accurate-is-electronic-judging-in-sport-8136
2. How does Hawk-Eye work: https://www.bing.com/search?q=how%20does%20hawk-eye%20work%3F&form=SWAUA2
3. Tennis on clay courts: https://en.wikipedia.org/wiki/Clay_court

4. Qaisar, S., Imtiaz, S., Glazier, P., Farooq, F., Jamal, A., Iqbal, W. and Lee, S., 2013, A Method for Cricket Bowling Action Classification and Analysis Using a System of Inertial Sensors: Computational Science and Its Applications. ICCSA, V. 7971, ISBN 978-3-642-39636-6
5. Rowlands, D., James, D. and Thiel D., 2009, Bowler analysis in cricket using centre of mass inertial monitoring: *Sports Technology*, 2(1–2), 39–42, DOI:10.1080/19346182.2009.9648497
6. Soccer dribbling: https://www.soccercoachweekly.net/soccer-drills-and-skills/dribbling/
7. What is Spidercam: https://www.thesun.co.uk/sport/football/2596974/spidercam-video-amazon-prime-premier-league-football/
8. Player Stats Centre comparison: https://www.premierleague.com/stats/player-comparison
9. Objective player recruitment: https://www.statsperform.com/team-performance/football-performance/player-recruitment/
10. Introduction to IoT sensors: https://dzone.com/articles/introduction-to-iot-sensors
11. Monitoring and control systems: https://daniel-gce-al.weebly.com/uploads/1/4/1/2/1412714/361_monitoring_and_control_systems.pdf
12. Predictive maintenance: https://www.onupkeep.com/learning/maintenance-types/predictive-maintenance
13. Computing pedagogy: https://teachcomputing.org/pedagogy
14. Machine Learning: https://www.ibm.com/cloud/learn/machine-learning
15. Stock market predictions: https://www.thestreet.com/markets/2020-stock-market-predictions
16. Avatar (2009): https://www.imdb.com/title/tt0499549/
17. What is Augmented Reality? https://www.fi.edu/what-is-augmented-reality
18. Which barcode do I need? https://barcodesaustralia.com/barcodes/
19. What is RFID and how does it work? https://www.abr.com/what-is-rfid-how-does-rfid-work/

CHAPTER 6

Most Common Active and Passive Sensors

6.1 SENSORS – THE BASICS

When we consider sensors, they can be separated into three obvious types – *natural sensors* which are the sensors that we are naturally born with in our bodies; *passive sensors* are manufactured sensors that, once set, respond to changing conditions around them; *active sensors* which are sensors which have been manufactured and require continued input/output information or supply to respond when particular changes occur in their environment.

6.1.1 All of the Natural Sensors

The three main sensors we naturally respond to result from changes in *heat*, *light*, and *sound*, which operate differently from our sense of *smell* and *taste*. A further sensor we all have is our *nervous system* (sensing physical changes and informing our brain of these changes), which operates similar to a transmission line, passing all sensor data to the *Central Processing Unit* (CPU). However, it is not really understood that most of the main three natural sensors respond to a frequency that results from the changes in air pressure or electromagnetic radiation.

For example, when we sit in the sun, the sun's heat is caused by *electromagnetic* (EM) radiation, otherwise known as sunlight which is dominantly in the *InfraRed* (IR) to *UltraViolet* (UV) range – see Figure 1.22. It is said that, if our body is composed of 60% water, and if sunlight was focussed like the microwave frequencies in a microwave oven, we would be boiled from the inside. Fortunately, it is unfocussed but someone can suffer from sunburn of the skin from excessive sun exposure, even while the body reacts by sweating. So we have a natural body detection of heat because high or low heat can cause body problems – we can tolerate up to about 40°C and down to about 5°C while our own body temperature is typically 37°C.

Normally, we consider that heat is what occurs when a material's atoms vibrate rapidly, and this concept fits in nicely with our natural understanding of heat which occurs in our

DOI: 10.1201/9781003108443-6

body when we exercise or exert ourselves. The natural sensor of external heat to our body would initially be our skin, so that when we cool down too far, we shiver which is a natural reheating of ourselves by vibrating our body's core. Our internal nervous system senses the heat and either causes us to sweat when there is too much, or shiver when there is not enough.

As for *light*, that is within the visible spectrum between UV and IR (Figure 1.22). Our eyes have been developed to naturally use light for external vision which is carried to the brain. Without seeing images around us, the human body would not have developed, and somewhere in history the body decided that the range of vision we presently have was adequate for our survival. Perhaps some humans were born who could see IR or UV wavelengths – UV has a wavelength of just under 1 micron (10^{-6}) while IR has a wavelength of 0.01 micron (10^{-8}) so our eyesight can only see in a tight bandwidth of wavelengths. We see using *rods* and *cones* (1) – the rods see images in black and white (and are used most when it is dark which is why we see at night in tones of grey), while the *cones* see the colours which are provided to the brain (and used during daylight) with the *rod* data (Figure 6.1). This is possible because the 6 million *cones* spaced over an area of 0.3 mm of the retina makes each cone about 0.2 micron wide, so the effect of a wave of light of 0.01 to 1 micron passing our 0.2 micron sensor is like a 2 mm thick leaf quivering in wind that changes in wavelength from a rapid 0.01 mm to 1 cm. That is, the leaf changes from rapidly gusting shivers to a gentle sway in the breeze. The cones will be slightly different in diameter so that they are sensitive to the different coloured frequencies. By contrast, the rods which are spaced over the rest of the retina, are not so sensitive and only register black and white images over the broader peripheral vision that the eye sees.

Considering that the final sensor we have is that of *sound*, we can hear only from about 2 Hz to about 18 kHz (with the higher hearing frequencies reducing as we get older). Light travels at just under 300 million m/s, whereas sound travels in air much slower at around 350 m/s (2). Consequently, the human body has adapted our hearing to have a *drum* in the *Outer Ear* region, which will vibrate in the range of about 2–18 kHz (maximum). When used with the formula: *velocity* = *frequency* × *wavelength*, this gives a wavelength of 350/2 to 350/(18 × 10^3) which is a wavelength range of 175 m to 20 mm (a very wide range),

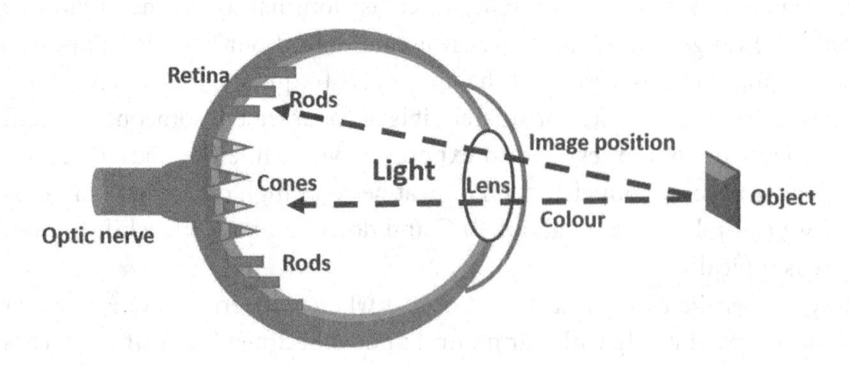

FIGURE 6.1 Light arrives at the retina, with cones determining colour and rods determining shape and position.

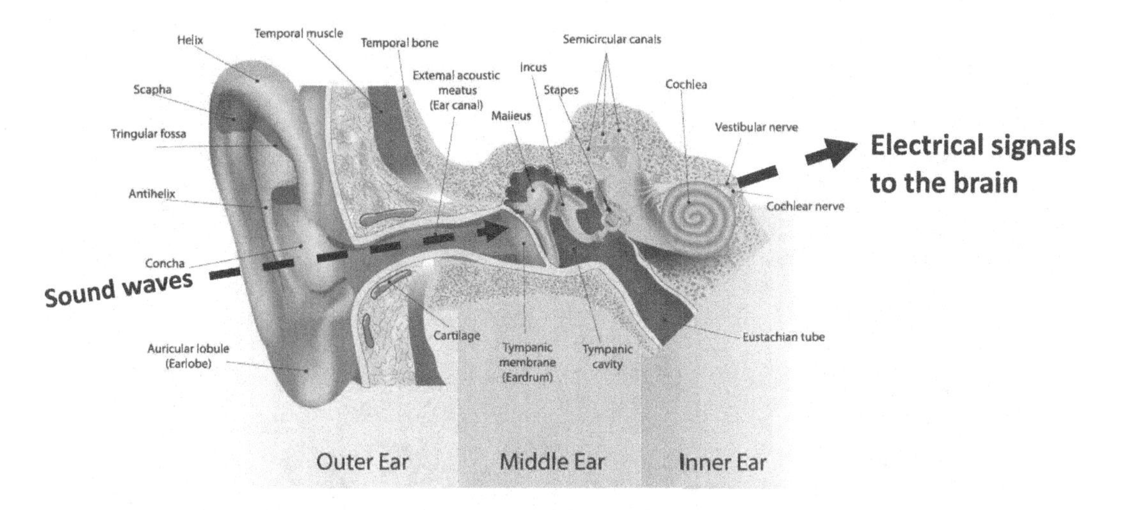

FIGURE 6.2 Sound vibrates the ear drum to produce electrical signals to the brain. (Courtesy: Shutterstock.)

vibrating the ear drum surface which is about 0.1 mm thick (Figure 6.2). The ear drum is attached to the *auditory nerve* in the *Middle Ear* region, which feeds signals to the *cochlea* in the *Inner Ear* region.

The frequency range reduces as we age, because the drum surface becomes stiffer, and vibrates less at the higher frequencies leading to loss of hearing passing less signal to the auditory nerve (leading to the cochlea). Many of the commercial hearing devices either increase the volume of the sound helping the ear drum to vibrate more or they are attached to the outer ear or skin against bone, transmitting the vibrations through the skin and bone back to the *cochlea*. A *cochlea implant* is a wire welded onto the auditory nerve, which bypasses the ear drum and sends signals from the piezoelectric crystal receiver worn behind the ear directly to the cochlea.

Smell is quite a different sensor – we smell when receptors on the surface of inside our nose have an electrochemical reaction to molecules entering the nose, the reaction producing an electrical pulse that goes up our nose along the olfactory nerve to our brain. When we are growing up, we train our brain (develop our personal database) about what smells good and what smells bad. The sense of smell is also linked to our *taste buds*, which are small bulbous receptors on the top of our tongue, which also have an electrochemical reaction in similar manner to our sense of smell, sending a signal from our tongue to say if we like a taste. The signal for *taste* is often linked to the signal for *smell*, since we often smell something before we taste it and the smell tells us what taste to expect (Figure 6.3). Try tasting a Eucalyptus Peppermint leaf without smelling it – the taste is quite bitter with no fragrance, but smell it first and then taste it and you may think it tastes minty. Do it the other way around and it doesn't.

So the three major personal sensors (sight, hearing, temperature) are alerted by a *frequency reaction* whereas the two minor personal sensors (smell, taste) are controlled by a *chemical reaction* [(3)].

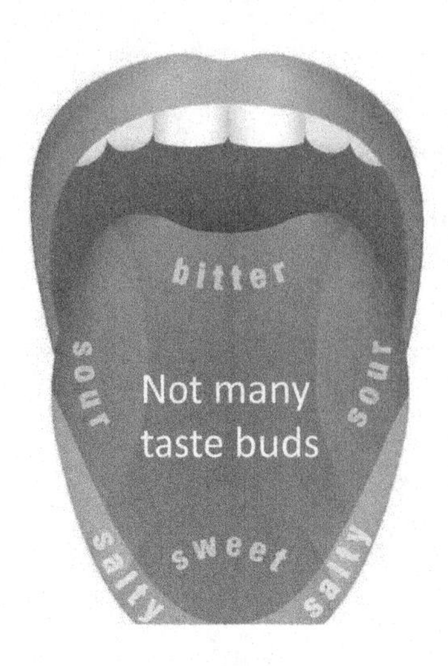

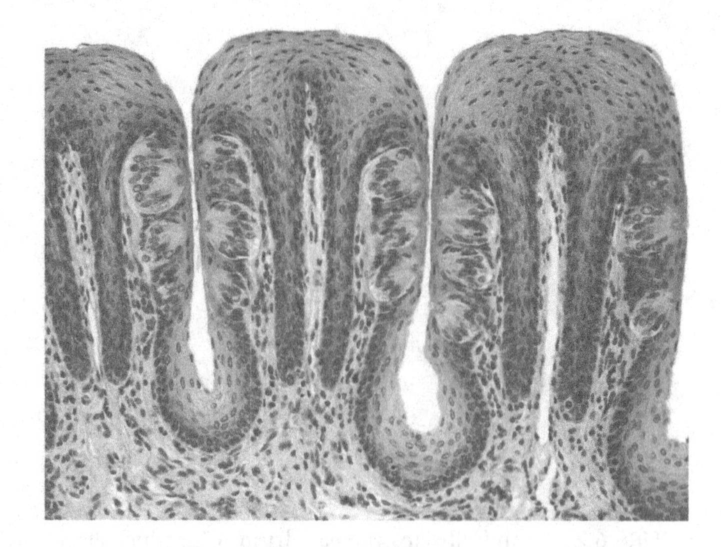

FIGURE 6.3 The tongue has taste buds shown in section. (Courtesy: Shutterstock.)

It was mentioned that the nervous system was a form of sensor, albeit that it transmits data from our three sensors to our brain like a data transmission line. The *taste buds* (which in section look like small plants budding) on the tongue are considered to be at the end of nerves, and the nervous system itself has nerve endings on other parts of the body. If a nerve is pushed or stretched, it provides a signal to the brain that says it is moving due to a physical pressure or strain of some form. These natural personal sensors can also be classed as *personal passive sensors*, whereas in nature, there are other passive sensors that are not involved with the human body.

6.1.2 Most of the Passive Sensors

A *passive sensor* is a sensor that once it is manufactured, it automatically reacts to changes in the environment surrounding it. As it happens, the very first passive sensors manufactured by mankind over the years involve the same properties as those which are *natural sensors* in the human body discussed earlier – that is *heat, light*, and *sound*.

Heat – temperature changes can be detected using a liquid thermometer, in which the liquid expansion or contraction quickly occurs as the applied heat changes. Mercury was most commonly used, with the first useful invention of a reliable temperature scale having been introduced in 1724 by Dutch inventor Daniel Fahrenheit. He put mercury into a bulb from which a sealed glass tube showed the mercury rise and fall with temperature change (mercury was used instead of alcohol, because mercury is consistent in its expansion and has a high coefficient of expansion, which meant that it expanded faster than alcohol). His temperature scale had water freezing at 32°, and boiling at 212°. Swedish inventor Andreas Celsius introduced his thermometer scale in 1742 with 100° as the water freezing point and 0° as the boiling point (which in practice changed to having 100°C as the high and 0°C as the low). He called it the Centigrade scale, because it was 100 points from freezing to boiling – this was much easier to work with than Fahrenheit.

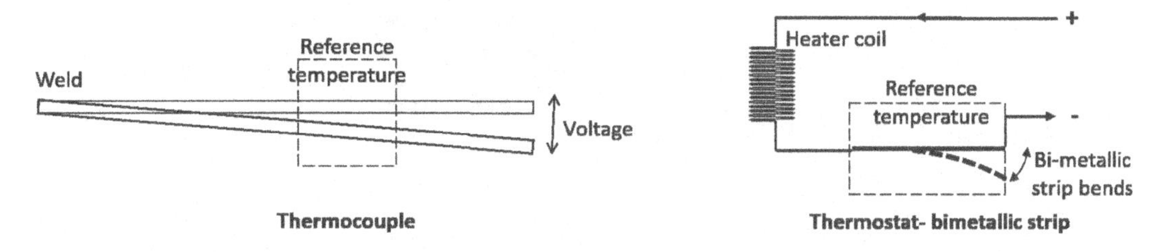

FIGURE 6.4 A thermocouple changes voltage when the weld temperature changes.

To convert Centigrade to Fahrenheit we multiply by 9/5 and then +32. Do this the opposite way around to convert °F to °C i.e. −32 and multiply by 5/9.

Another natural temperature sensor is the *thermocouple*. This consists of two dissimilar metallic conductors bound to each other, which have different coefficients of expansion (responding differently across a range of temperatures). This bimetallic strip expands when heated producing a change in voltage, as shown in Figure 6.4. If we weld a strip of chrome to a strip of aluminium and heat this weld, and then put the middle of the separated two strips into a box that keeps them at the same reference temperature, a small (mV) voltage occurs across the open ends of the two strips. If we heat or cool the weld, the open-end voltage changes in proportion to the heat at the weld. If we know the heat amount applied to the weld and the reference temperature, we can then calibrate the voltage value to a temperature scale in which Sensor Temp = Weld Temp − Ref Temp.

An alternative to this is the *thermostat* which uses a similar principle, but in this case it uses a bound *bimetallic strip* of metal, each having two different coefficients of expansion which when heated together, expand differently causing the strip to bend (Figure 6.4). By bending away from a fixed point, a circuit can be opened – such as when a room is being heated, the current flow through the heating circuit is set to keep the heater on until the set temperature is reached, when the bimetallic strip bends and breaks the circuit – switching the current to the heater off. This is the way heater and room air conditioners are often controlled, with the thermostat bimetallic strip being located in the heater if it is a simple coil heater, or on the wall in the ambient room temperature if it is controlling an air conditioner circuit.

Light – We can make a semiconductor material (usually silicon), which, when light photons impact on its body, the photons have a chemical reaction with the material and they generate an electron flow – which is current. This is known as a *photodiode* – *photo* referring to light and *diode* meaning one-way current, flowing from an anode to a cathode like a typical car battery.

Put a resistor in the line and from *Ohm's Law voltage* $V = \textit{current}\ I \times \textit{resistance}\ R$. The current being produced by a few of these diodes linked together can then be directed away to storage batteries and we have a *solar cell* and lots of these make a *solar panel* (Figure 6.5). The panel is coloured black, which is not reflective so absorbs most of the arriving EM sunlight.

While we are discussing the *solar cell*, consider that 75% of the universe is made of hydrogen. We can make hydrogen energy by taking the solar cell current and passing it through water (called *electrolysis*) which then separates into *oxygen* (O_2) and *hydrogen* (H_2). We can then store the hydrogen and reverse the reaction by later passing the hydrogen

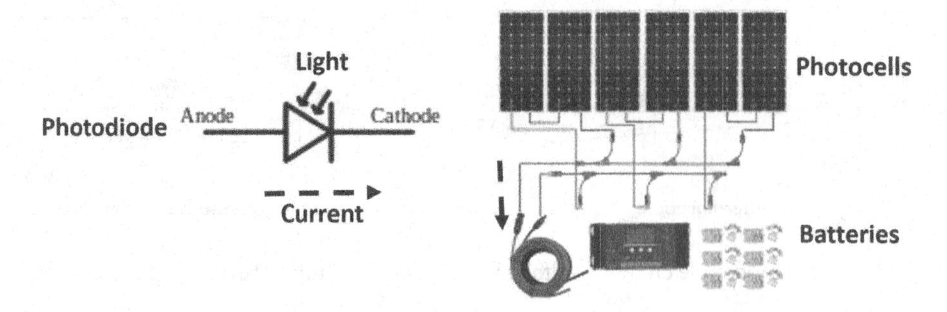

FIGURE 6.5 Lots of photodiodes make a photocell.

mixed with oxygen over the panels (modified to do this) to produce electricity for use whenever needed. These solar cells have a new name referred to as *fuel cells* [4] in Figure 6.6.

> "While we are discussing the solar cell, consider that 75% of the universe is made of hydrogen. We can make hydrogen energy by taking the solar cell current and passing it through water (called *electrolysis*) which then separates into oxygen (O_2) and hydrogen (H_2). We can then store the hydrogen and reverse the reaction by later passing the hydrogen mixed with oxygen over the panels (modified to do this) to produce electricity for use whenever needed. These solar cells have a new name referred to as *fuel cells*."

An area of 10,000 km^2 (100 km × 100 km) of solar panels with an efficiency of only 10%, could produce enough hydrogen energy for 1 billion people with an average consumption of 5kW per person. Suitable open areas are found in deserts around the world, and have the advantage that this method allows the generation of energy on demand, without the production of any form of greenhouse gases (and so it is good for the planet).

Sound - One of the first such sensors commonly used since the 1800s has been the *piezoelectric crystal*, which has been used to passively send or receive sound, a feature that has most commonly been used for the telephone. To make a piezoelectric crystal, we have to heat (200°C for one hour) a mix of baking soda (sodium bicarbonate) with cream of tartar (potassium bitartrate) in distilled water, and keep stirring adding more of the mixture if needed, until the water evaporates leaving crystals. When we squeeze the piezoelectric crystal, a voltage spike is produced as shown in Figure 6.7. When these piezoelectric sensors are inserted into car *airbags*, their output spike ignites a chemical inside the bag which explodes as a pressured gas, rapidly expanding the bag size.

If the crystal is formed as a wafer or thin disc as shown in the figure, we can pass an electrical pulse into it which causes it to expand or contract and, under certain conditions, produce a pressure sound since it also vibrates the air. Because the crystal is small, this produces a sound at high frequencies we can't hear known as *ultrasound* [5]. If we make it thicker using a plastic or a flexible card to hold it, this dampens and reduces the high

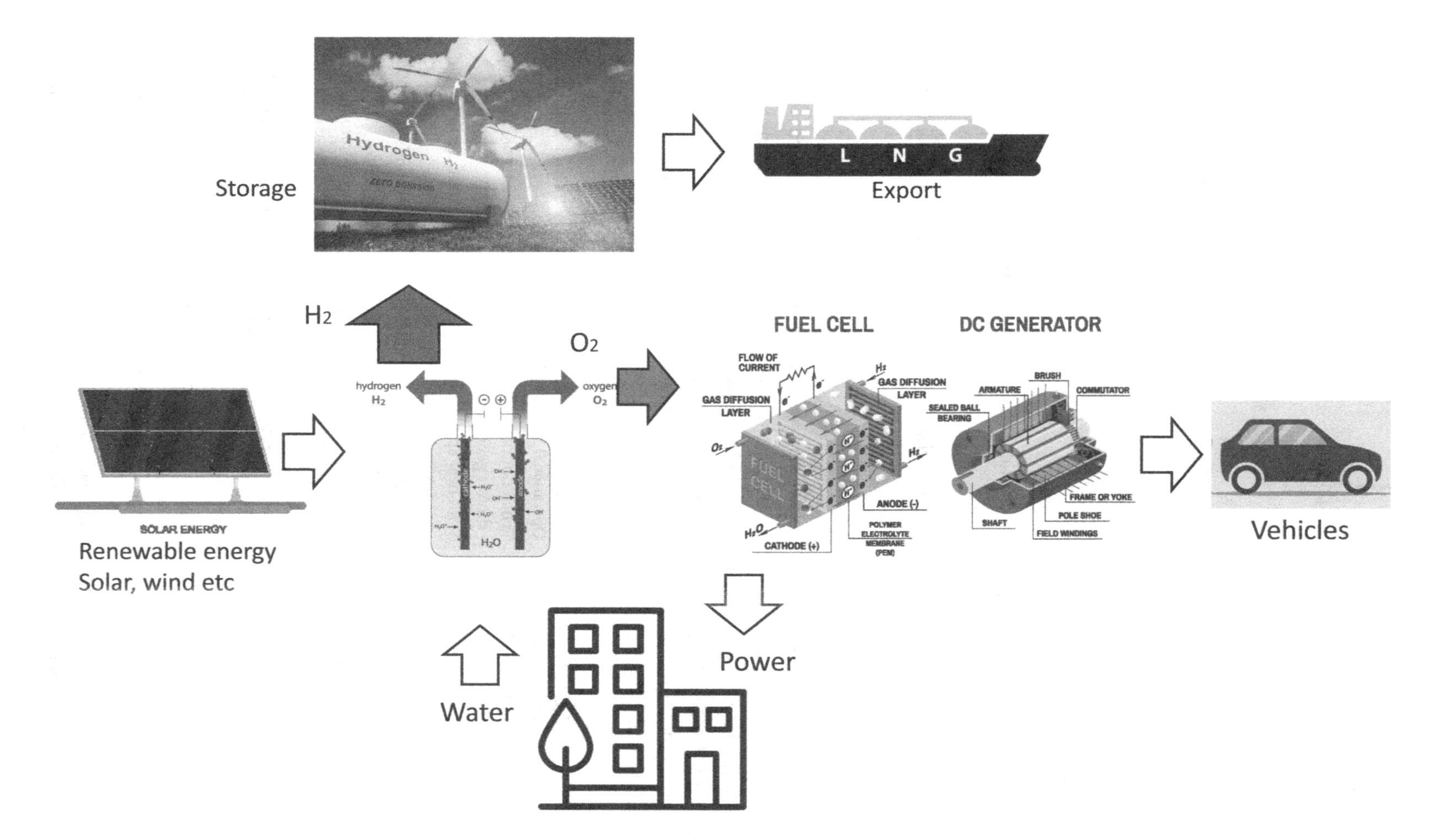

FIGURE 6.6 Solar cell power to electrolysis produces hydrogen and oxygen, which is reversible as fuel cells. (Courtesy: Shutterstock.)

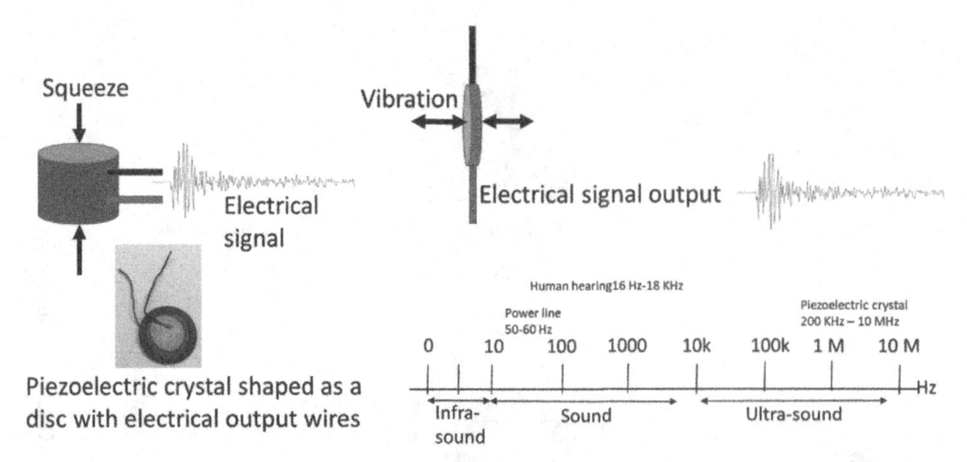

FIGURE 6.7 Electrical signal produced by a crystal, in the ultrasonic frequency range 50 kHz–10 MHz.

frequencies to a lower frequency level and it becomes a phone speaker so that we can hear sound (like a *smartphone* speaker).

6.1.3 Most of the Active Sensors

So far we have dealt with the personal passive sensors of the body, and also the passive sensors that can be manufactured. However, the vast majority of sensors result from a manufacturing process, and they need some form of input to activate them.

General sensor technology uses some form of external input to make them responsive to changes being measured or monitored. Some examples responding to physical property changes would be as follows in Table 6.1.

TABLE 6.1 Most Active Sensors

Physical Property	Sensor Type	Operational Method
Temperature	Human body	When hot the body sweats to cool veins, when cold the body shivers generating heat in veins.
	Thermometer	With heat application, mercury expands/contracts in tube in proportion to temperature changes.
	Thermocouple	Two dissimilar conductors tied together, respond differently at different temperature, producing voltage change.
	Resistive temperature detectors	RTDs use a length of fine wire (copper) wrapped around a body change resistance as temperature changes.
	Fibre optic cable	Imperfections in fibre manufacture allow light to reflect back to source/receiver – as temperature reduces or expands the cable, imperfections move so that reflected light changes in arrival time.
Pressure	Fibre optic cable	Imperfections in fibre manufacture allow light to reflect back to source/receiver – as pressure reduces or expands the cable, imperfections move so that reflected light changes arrival time. This is similar to temp.
	Airflow fan	Fan rotates motor spindle to generate electrical output.
	Pitot tube	Tube with pressure sensor (piezo) at one end to detect changes.

	Airbag	Pressure sensors (piezo) receive spike in pressure, sparks chemical inside bag exploding as pressured gas.
	Diaphragm/piston	Diaphragm or piston movement resulting from mechanical or fluid pressure change closes electrical circuit.
	Strain gauge	Two or more wires which when strained, results in resistance change with power change in current flow.
Sound and vibration	Human ear	Sound along ear canal to drum which vibrates, causing fluid to move and hair cells to vibrate, generating electric pulses up hearing nerve to brain.
	Piezoelectric crystal	When crystal squeezed, gives an electric signal or pulse out. Used in hydrophone, microphone, mobile phone, music pick up.
	Fibre optic cable	Imperfections in fibre manufacture allow light to reflect back to source/receiver – vibration causes reflected light movement.
	MEMS	**M**icro**e**lectro**m**echanical/electromagnetic tiny, mechanical components linked to integrated circuit to produce signal when movement action is applied – used in mobile phones as gyro, senses gravity change.
	SONAR	**So**und **Na**vigation and **R**anging transmits underwater ping and reflection time gives two-way distance. Can be used for mapping seabed or other objects.
	Geophone-seismometer or Accelerometer	Wire suspended in EM field moves with respect to magnet housing due to vibration, gives signal output.
Light	Human eye	Iris (coloured ring) causes pupil to contract reducing light onto retina. Low light causes iris to expand.
	Photodiode	Semiconductor converts light to current.
	Solar cell and LED	Solar cell is lots of photodiodes – LED works the other way, converting electricity to light (in a pixel).
	Fibre optic cable	Light passes up to 17 Km inside glass fibre thread using total internal reflection.
	Movement	Change in light indicates movement, light beam breakage, photoelectric distance to objects.
	Interferometry	Two very thin mirrors spaced so light passes through and used to measure wavelength – used in lasers/telecoms.
	Photoelasticity	Optical (frequency) change indicates stress of material and can give elasticity/stiffness.
Electro-magnetic	EM Induction	EM field developed by current in coil causes induced field – Eddy currents, can be broken by metallic objects – works in wet conditions.
	RADAR	**Ra**dio **D**etection **a**nd **R**anging transmits EM radio waves which are reflected back to receiver.
	GPR	**G**round **P**enetrating **R**adar transmits EM through ground and variations in ground reflected to receiver.
	SAR	**S**ynthetic **A**perture **R**adar joins individual images of (landscape) reflections to make 2D/3D images.
	Hall Effect	Semiconductor crystal gives current output proportional to magnetic field strength, used for wheel speed.
Electrical transmission	Graphene carbon coating	Molecule thin (can't be seen) membrane or layer can change shape with mechanical load, changes resistivity and indicates mechanical changes when electricity passed through.

(*Continued*)

TABLE 6.1 (*Continued*)

Physical Property	Sensor Type	Operational Method
	Capacitance probes	AC current passes through medium, indicates change in capacitance giving permittivity.
	Wheatstone bridge	Four different resistance wires connected together so that change in resistance of one leg causes voltage change.
Chemical reaction or detection	Biosensor	Biological droplets have chemical reaction (e.g. thickens, heats, dissolves) changing properties to light or power.
	Carbon paste	Graphite and liquid, allows electrical transmission in fluids.
	Air-fuel ratio	O_2 sensor gives electrical output when O_2 ions pass over it compared with other side.
	Glycol coolant	Mixed with water, glycol lowers freezing temperature allows faster heat transfer and controls rusting.
	Breathalyser	Breath passing over semiconductor anode of tin oxide, produces acetic acid + water. Current through acid-resistance passes to cathode.
	Fuel cell	Alcohol oxidation of ethanol produces current proportional to alcohol.
	CO/smoke detection	Two electrodes in sulphuric acid converts CO to CO_2 at one electrode and oxygen at other changes current flow. For smoke detector, CO passes over chemical surface, changes colour so IR LED displays colour and current changes.
	CO_2 detection	Analysis of EM frequencies by transmitting/reflecting light on cloud and observe IR level, or passed over MEMS metal oxide semiconductor where change in resistance measured.
	Catalytic beads	Two coils of platinum wire embedded in aluminium bead at 500°C connected to Wheatstone bridge - current passes. One bead oxidises while other bead restricts it. Gas raises temperature, changing resistance, changing current.
	Chemical FET	**F**ield **E**ffect **T**ransistor made of silicon, has current passing through but external chemical changes used as a gate to change current flow.
	Chemiresistor	Resistance of sensor body changes when chemical passes by.
	Dew point	Dew observed using mirror in gas and reduce temp at given pressure by monitoring change in reflected light.
	Hydrogen	Light shone on Palladium surface absorbs H, so change to palladium hydride shown by change in reflected frequencies.
	SAW detector	**S**urface **A**coustic **W**aves run at about 500 MHz - vibration is transmitted through a film on a quartz glass surface - a frequency shift and velocity gives an indication of the chemical film, being lead, mercury, or oxygen.
	Nitrogen oxide	Ceramic metal oxide at high temperature (400°C+) gives different current depending on level of NO.
	Ozone	Transmit light through clean air followed by sample of ozone, absorption of UV shows how much ozone is in air.
	pH	Glass electrodes doped with Si O_2, inserted into medium, voltage between electrodes indicates acidity.
	Potentiometer	Similar approach to pH, used to determine specific solutions - current gives voltage which when removed, remaining voltage indicates electric potential.

6.2 SIMPLE EXPLANATION OF ACTIVE SENSOR TECHNOLOGY

Table 6.1 provides a brief overview of the different sensors, and how they operate. For the most part, the physics of these sensors have simple explanations, except for the chemical sensors which require more in-depth understanding of chemical reactions. However, such in-depth understanding is not within the scope of this book, and so we will keep to the simplest of explanations where possible. It also should be noted that there are lots of specialised sensors which are used in the automobile industry. These are often derivations of the most common sensors, with different applications only associated with maintaining motor vehicle integrity.

6.2.1 Sensors Responding to Temperature

We have already discussed the thermometer and the thermocouple which are both passive sensors (Section 6.1.2). However, the use of heat causing changes in resistance or length has not been discussed.

The resistance thermometer – If we wrap fine wire (copper, nickel, or platinum) around a glass tube or other similar insulating material (Figure 6.8) and check its resistance, we observe the wire changes resistance depending on the temperature applied to the wire (the other way around – in an incandescent light bulb, passing a current through the bulb causes it to heat and this produces photons – aka *light*, and if we were able to measure the resistance, we would find it changes producing increasing temperature). The different wires (copper/nickel/platinum) have different operating temperature ranges through which their resistance is proportional to the applied heat, so the coiled element type is mainly used in industry in which a fine powder is put in the coiled area to allow the wires to move easily and work up to 850°C.

Thermistors – These are resistors that change their resistance with a change in temperature. They can be used as thermometers and *thermostats* (which switch circuits on or off depending on temperature) in the household range of temperatures 0 to 100°C – being used in fridges, ovens, and microwaves. They are also used in cars for monitoring radiator and oil sump temperature.

Fibre optic cable for temperature – Fibre optic cables (made of clear glass fibres) can be used not only for telecommunications data transmission, but also for detecting changes in temperature, pressure, and strain [(6)]. In Figure 6.9(a), an LED light source is fixed at one

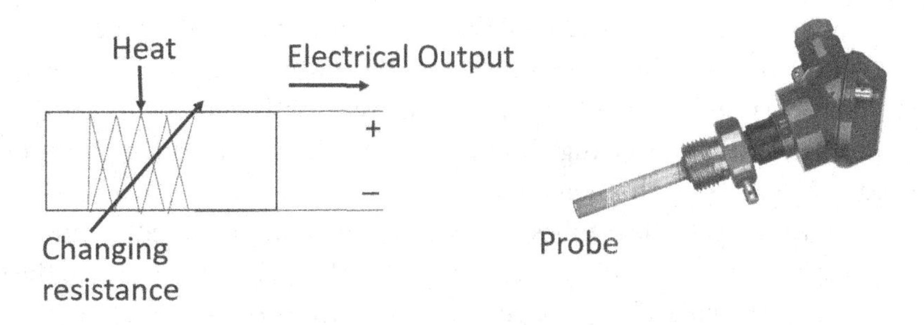

FIGURE 6.8 Resistance thermometer. (Photo courtesy: Shutterstock.)

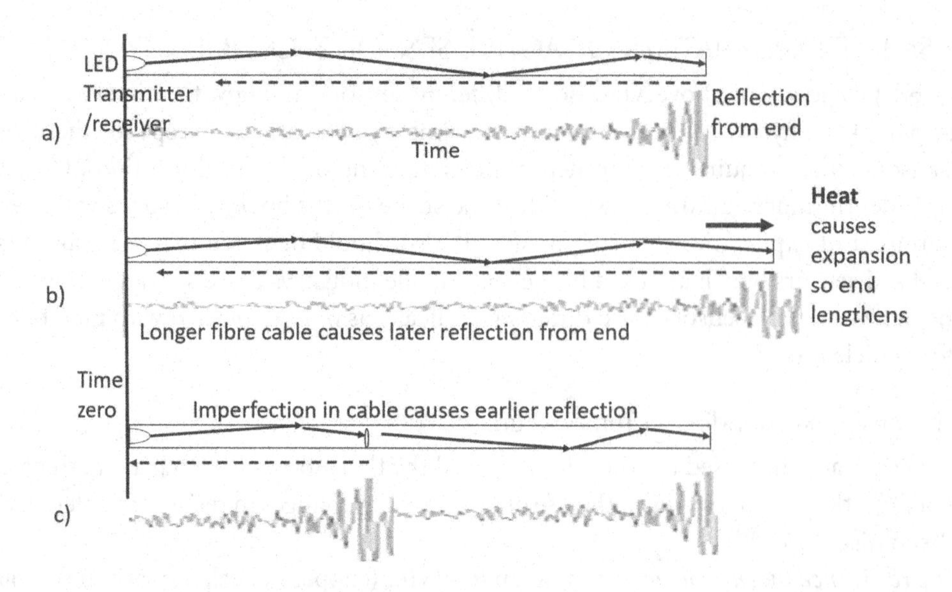

FIGURE 6.9 Fibre optic cable strain gauge and thermometers work by: (a) reflecting a pulse from its end, (b) determining the flight time difference which is proportional to cable length change due to stretch, and (c) a local change in frequency phase caused by a local heat change.

end of a fibre optic cable. This acts as a light transmitter, and when used with a photodiode, they can transmit and receive light pulses.

If we transmit a light pulse from the LED at one end of the cable, the light bounces along the inside of the wall (due to the fibre's refractive index) until the other end, where some of it may transmit out but some will reflect back along the cable towards the photodiode positioned next to the LED (the instrument at the end of the cable is known as an *interrogator*). This reflection will be received as a squiggly line shown in Figure 6.9(a) which is a burst of voltage, and because this takes time from the time zero (aka *zero time*) when the light pulse transmits to the receiving photodiode, then a knowledge of this time taken travelling from the LED to the end reflection and back again, is known as the *travel time* or sometimes it is called the pulse *flight time*. Since we know the two-way travel time and the speed of light (just under 300 million m/s), we can work out how long the cable is the equation *distance = (speed × time)/2*.

If heat is applied anywhere along the fibre cable in Figure 6.9(b), it will expand the cable increasing its length (and if we cool the cable it will contract). If the cable expands or contracts, the travel time (how long it takes for the pulse to travel to the end and back) will change in proportion by a small amount (in microseconds) and the received pulse will arrive earlier or later. This tiny change in time gives us the small amount it has expanded or contracted in length using this equation.

Fibre optic cables are very clear tubes because they are manufactured as pure glass fibre with a loss of light of about 0.2 dB/Km. As explained earlier, a loss of 2 times in the dB scale is 3dB in amplitude, which is about 8% per Km. This means that the maximum length of fibre optic cable there can be without needing some sort of amplifier for the light is about

17 Km in practice. So the telecommunications industry don't like repeater station amplifiers much further apart than 15 Km.

Instead of spending a lot of money on pure fibre cables, if we instead use a cheaper fibre cable with impurities (can be as much as only 20% of the cost of pure cables), this will cause the transmitted light to be reflected at the impurity points along the cable as indicated in Figure 6.9(c). Knowing where these reflection impurities are located will help us, because since we now have more than one reflection in our light, we will have more than one sensor. If we have an impure fibre cable with many impurities and pulse the cable, we will then get many reflections back and, because we know the speed of light, we can work out where physically along the cable, those reflection points are. If the time from a specific impurity reflection to the end reflection is staying the same, we know that only the length of cable from the *interrogator* to the impurity is heating or cooling and not the full cable.

Since the cheap fibre cable has lots of impurities, we have to calibrate it first by pulsing the cable in a room at a known temperature (say 22°C), and then we figure out from all of the reflections where the impurities are. Then we compare those reflection travel times from the impurities in the cable after we have installed the cable onsite, and the difference in travel time between the calibrated point travel time and its site-point travel time will provide an indication of the length differences. So then the cable can be calibrated again at the site and made ready for use.

A fibre optic cable responds more to the higher frequencies than when it is physically vibrated (at lower frequencies). So if we perform a *Fourier Transform* on the reflected pulses, we will find that the high frequency changes (resulting in a phase shift of the received signals) indicate there is a change in temperature somewhere between the reflection points. But if the lower frequencies change, this means the fibre has changed length by a greater amount due to an applied strain somewhere along its length (which could be caused by a seismic vibration). Consequently, depending on how the received data is analysed, a fibre optic cable can be used for telecommunications, and temperature or pressure changes along its length, all at the same time. This is new technology for future use in all manner of communication, as well as pressure and temperature monitoring.

Once installed onsite and the temperatures known along it, we can thereafter compare the reflection travel time and frequency values with the calibrated values. We can use correlation methods (explained in Chapter 3) to establish the travel time difference, and the amount of phase change then provides an indication of temperature change of any section.

Such fibre optic cables can be used in underground tunnels or in tortuous areas where we can run a long cable and so detect heat changes. They are more often being used in wells drilled into the ground, which allows the remote monitoring of stress and temperature conditions in a wellbore. The use of such fibre optic cable variations in the future will have major benefits for monitoring CO_2 injection into underground aquifers, for storage purposes.

A further benefit of fibre optic cables use in temperature monitoring is that they can be run around corners, vertically and horizontally – and can be used for fire detection in tunnels used by vehicle traffic. If the tunnel is not expected to change temperature except when there is a vehicle fire, then a change in length (and hence change in low frequency and/or phase) could be caused by either pressure or a point load of stress being applied to it. Such a fibre cable is therefore used for monitoring earth movement and temperature along its length, with monitoring lengths as short as 5 to 10 m. This includes acting as a seismometer for earth vibrations (to be discussed in Section 6.2.3).

> "*Radiation pyrometer*- This meter (*pyro* means heat or fire) is a fancy name for a meter used frequently for remote checking of people's heat radiation to determine if they have high body temperature potentially caused by fever (COVID-19), as passengers walk through rail or airport terminals."

Radiation pyrometer [7] – This meter (*pyro* means heat or fire) is a fancy name for a meter used frequently for remote checking of people's heat radiation to determine if they have high body temperature potentially caused by fever (COVID-19), as passengers walk through rail or airport terminals (Figure 6.10).

It can have a mirror that allows any arriving light to pass through, and which can also reflect the light onto an IR sensor – which determines the temperature – with an accuracy of about 0.2°C. One version of the instrument allows light to pass through the mirror so that the viewer can see the object being tested, in which case it is possible to put the calculated temperature alongside the picture of the person. The other (cheaper) version has the light simply arriving at the sensor, and the value registering on the screen. These handheld instruments can be used for any form of remote heat sensing, ranging from checking air conditioning blower temperatures to extreme blast furnace heat.

6.2.2 Sensors Responding to Pressure

Some definitions first. *Pressure* is the continuous application of *force* on an object. *Vibrations* are the result of the intermittent application of pressure. *Strain* occurs when two areas of an object are being compressed or extended (*tensile*) due to the application of a force. *Stress* is the application of a force over an area. The general public often therefore uses these terms synonymously, because they do have these linked meanings of the application of force, whether it be continuous or intermittent. We will therefore discuss vibrations in the next Section 6.2.3.

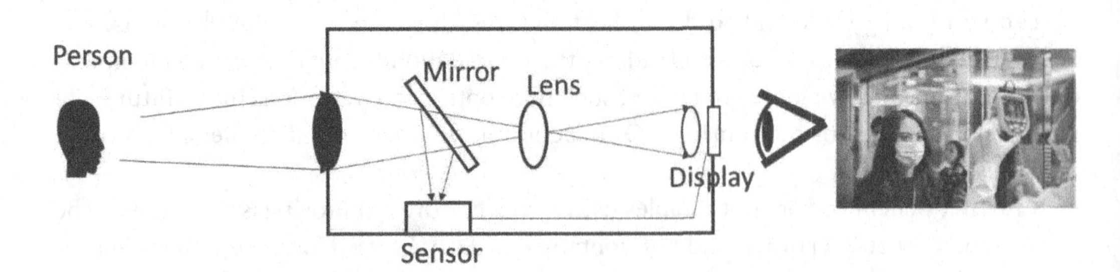

FIGURE 6.10 Remote heat detection – radiation pyrometer. (Photo courtesy: Shutterstock.)

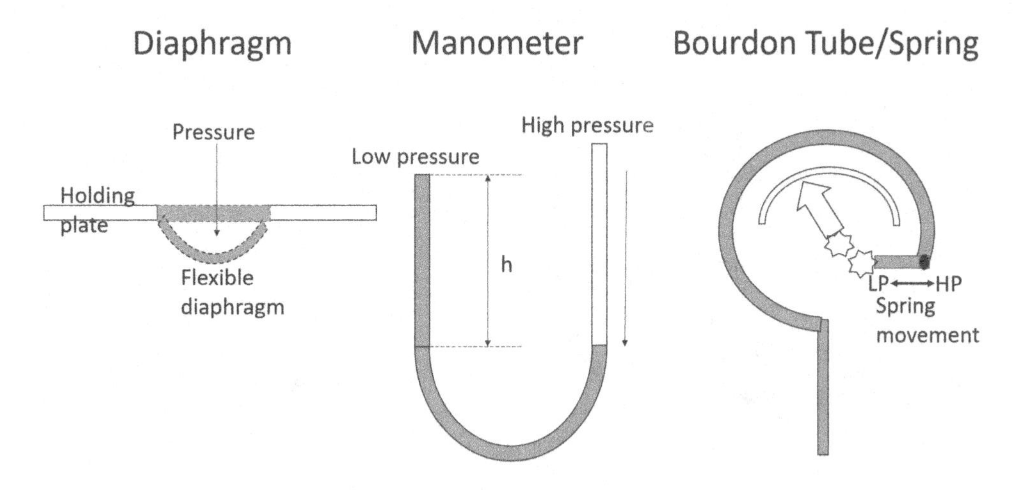

FIGURE 6.11 Diaphragm, manometer, and Bourdon Tube/Spring for measuring pressure.

To measure continuous gas pressure, we need some form of graduated/calibrated device containing a component that is elastic and moves with pressure, whereas to measure continuous pressure on solids we can use some form of spring. In liquids we can use a simple piezoelectric crystal.

Continuous pressure measurement has been with us for centuries, and the most common sensors of continuous pressure have been based on the simple *diaphragm*, *manometer*, and the *Bourdon Tube or Spring* shown in Figure 6.11.

The *flexible diaphragm* is a simple device that flexes when a pressure is applied. The diaphragm can be attached to a graduated gauge which shows the amount of applied pressure due to movement. This is the principle behind a car brake-fluid system, in which mechanical pressure on a diaphragm or piston closes a circuit when the pressure reduces.

The manometer is an extension of the simple thermometer, which uses a U-tube containing liquid to measure pressure on the liquid. By multiplying *height h × liquid density = pressure*, which is actually the difference between atmospheric pressure and the amount seen on the gauge, known as the *gauge pressure*.

The *Bourdon Tube or Spring*, was patented by Frenchman Eugene Bourdon in 1849, and uses the fact that when a circular-shaped tube has internal pressure applied inside of it, the tube tries to straighten out (like a spring straightening, which is why it is sometimes referred to as a *spring*). A metal link attaches a cog with a gauge at the end of the moving spring so that, at high pressure, it moves one direction, then in the other direction when the pressure reduces.

The *diaphragm* method can also be applied in the form of *bellows*. This is most commonly used in the blood pressure machine (aka *sphygmomanometer* - *sphygmos* means *pulse* in Greek, *manoir* means *main* and *meter* means *measure* in French - *main pulse measurement*) or instrument. When we go to a doctor who tests our blood pressure, the doctor wraps an inflatable pressure *cuff* around our arm and uses a *stethoscope* to *listen* to our heartbeat (we will talk about the stethoscope in the next section). He then uses a blood pressure machine to test the maximum blood pressure we have by hand-pumping air from

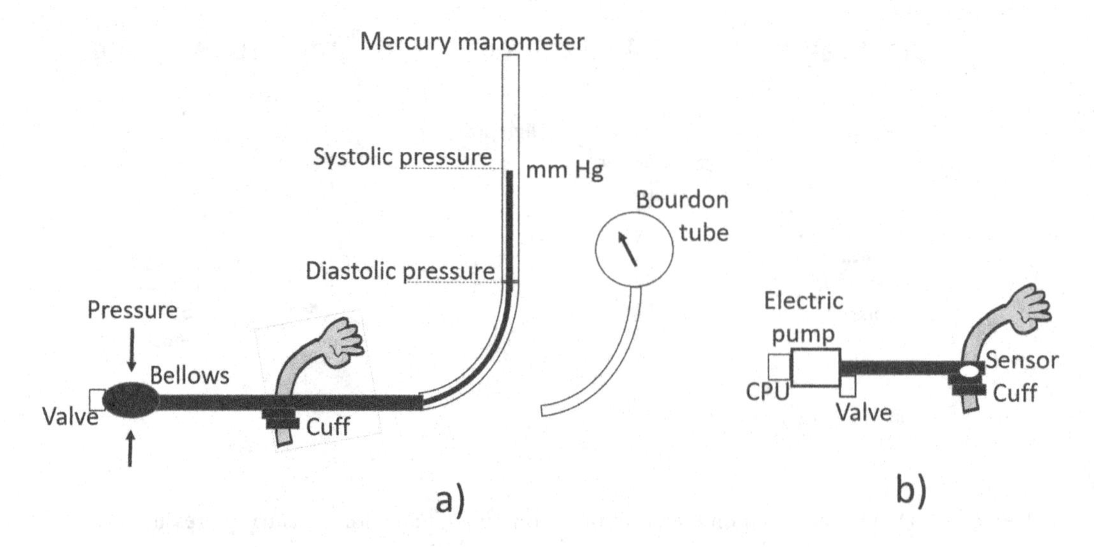

FIGURE 6.12 (a) Sphygmomanometer and (b) blood pressure machine.

a bellows into the cuff to a high pressure – the pump and cuff are in-line with a mercury manometer, so that the pressure being applied in the cuff pumps the mercury up to a high level (Figure 6.12a). The manometer is graduated in millimetres (known as *millimetres of mercury Hg*) so that when the doctor pumps up to a high level of pressure, this temporarily stops the blood flow through the arm's veins. By releasing an air valve in the line, the pressure reduces and just as the heartbeat is heard in the stethoscope, this is the maximum pressure produced by the heart so the doctor takes a note of the reading on the manometer (*mm Hg*) – this is the *Systolic pressure.* The pressure is then gradually released from the valve and at rest pressure is the *Diastolic* blood pressure when the heart is at rest *(systolic* from Greek to 'contract' and *diastolic* from Greek to 'expand'). For the average healthy person the *Systolic pressure* is about 120 mmHg while the *Diastolic pressure* is usually below 80 mmHg. The mercury *manometer* can be replaced by a *Bourdon Tube* gauge instead.

Rather than a manual pressure bellows, we can replace the bellows with a small electric air pump (Figure 6.12b). In this case, the pump is controlled by the electronic CPU, which signals the pump to increase the pressure on the cuff as normal, but the measurements are taken automatically by mounting a piezoelectric sensor inside the cuff. The sensor provides electrical signals to the CPU in proportion to the pressure it receives. When the pressure has reached its maximum according to the sensor, the CPU causes a contactor to open a valve that releases a small quantity of air (reducing the pressure) and the sensor then gives the lower pressure signals to the CPU. On completion, the unit turns off releasing all pressure (so that the cuff can be removed) and provides a digital reading to a display of *Systolic* and *Diastolic* pressures. In this case we don't need a stethoscope because the machine does it all for us.

The wind turbine (aka *wind-farm* when there are many together – *turbine* is the Greek word for *vortex*) receives the blowing wind on its rotor blades, which are angled so that a force is effectively applied at an angle to cause the blade to rotate. The blades are connected through a gear-box to a turbine which rotates, and electricity is produced. Wind turbines are mounted on top of towers to receive as much wind as possible (as high as 30 m) where

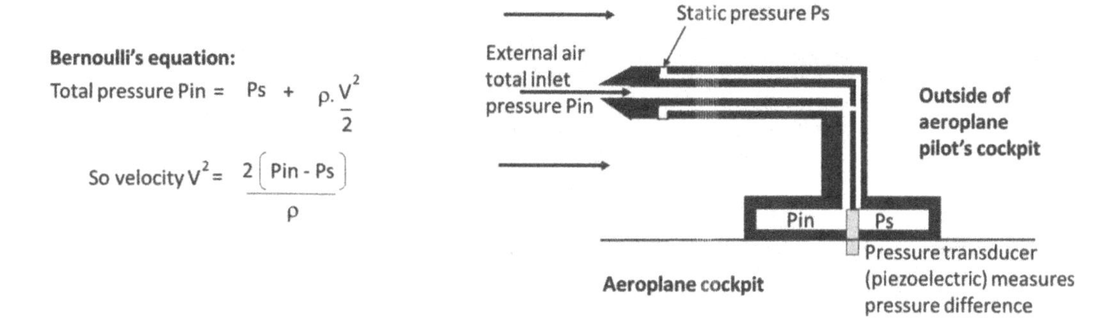

FIGURE 6.13 The Pitot tube measures pressure from which aeroplane velocity is calculated.

the wind is generally faster and more constant than the turbulent breezes at ground level. Most wind turbines have a rudder at the back or tilting blades that move the turbines around to the direction of maximum wind force. In terms of efficiency, the wind turbine is only about 30% efficient in energy conversion, which is greater than the solar panel at about 20% (however, the solar panel is easier to build and cheaper to service) [(8)].

The Pitot tube is often seen near aeroplane cockpit windows. The *Pitot tube* measures the difference in pressure between the outside air flowing past an aeroplane and the outside static pressure (as if the plane was not moving). This difference gives the velocity of the aeroplane through the air, by applying Bernoulli's equation as shown in Figure 6.13. There is also a separate transducer that measures the pressure differential between input total air pressure and static pressure (which is used for the calculation (Pin – Ps), with Pin always being much greater than Ps. The value for the density of air is ρ in the equation.

6.2.3 Sensors Responding to Sound Pressure and Vibration

It has been explained earlier that sound pressure is generally not continuous, but pulses of pressure. Sound is the transmission of vibrating air (like the sound when a bird flies past), so that when we talk to someone the sound is actually the air vibrating from your mouth to the other person's ear drum. Even though we think sometimes that a sound is continuous (like the awful continuous sound of a lawn mower going past), it is actually just an audible set of vibrating frequencies we discussed earlier in Chapter 1. Of course the frequency can be much higher than a lawnmower – like the 'click' produced by bats around 20 to 100 kHz for their in-flight navigation, which we sometimes can hear as they fly past.

In the previous section, we referred to the use of a *stethoscope* for hearing heartbeats. The stethoscope (*Stethos* is another word for chest and *scope* is an instrument) is a centuries-old instrument which was developed by Frenchman Rene Laennec in 1816, who used a short bamboo-type rod to listen to women's heartbeats (rather than put his ear to their breast). If we place a tube over an area of a body, the vibrations travel along the tube and if we put the other end in our ear, we pass the vibrations onto our ear drum. If we reduce the diameter of the tube, the vibrations will still travel along the thin tube and will increase in volume to be a bit louder. So if we attach a small *bell* to the end of a narrow tube, we will then hear a heartbeat louder (neighbours in houses may listen to their neighbours by putting a cup

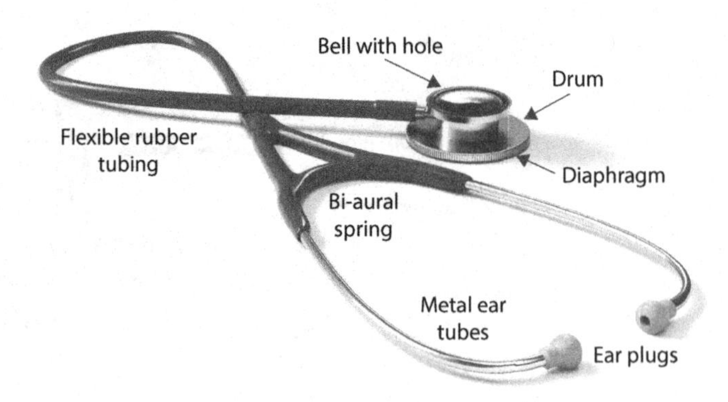

FIGURE 6.14 The stethoscope. (Courtesy: Shutterstock.)

against the wall and their ear against the bottom of the cup). Because the bell gives the low frequencies (due to its diameter being large), we can then put a thin, hard plastic or ceramic diaphragm across the bell's open face to transmit the higher frequencies. Then attach this to flexible rubber tubing (but fairly stiff rubber otherwise the sound won't travel along it so well) to small earplugs connected with metal tubes to the rubber tubing, we now have an operational stethoscope as shown in Figure 6.14.

The use of the *stethoscope* [9] with the *sphygmomanometer* allows a doctor to determine precisely the pulse type *(systolic* or *diastolic*) which can be heard at the same time as the *sphygmomanometer* shows the blood's pressure value.

Mankind over the years has seen the need to use pressure for many industrial purposes. For example, we used *wind* (being a fluctuating pressure wave of air) and flowing water for power many centuries ago, by developing the wind mill and the water wheel. Essentially, both of these active sensors used wind and water power to turn a sail or wheel, which in turn provided basic rotary power to either grind flour or rotate electrical power turbines. Today, we are still trying to perfect using wind power (an issue is how much energy a wind turbine takes to make it turn over and also how easy it is to maintain its moving parts and blades). However, we have got to the point where the largest wind turbines are producing up to 12 Mega Watts of electricity, which is enough to power around 16,000 households per turbine each year – assuming the turbine blades are continually rotating, which they aren't. Much political debate involves the use of wind turbines, but reality states that the wind rarely constantly blows, which is why *wind farm* sites have numerous wind turbines pointed in the dominant wind direction.

By comparison, water or *ocean wave energy* uses the vertical swell motion of water waves passing floating buoys, which are connected to floating turbine platforms to produce *wave power* as the buoy bobs up and down Wave power is different from *tidal power*, a technology that uses any moving water or ocean current to generate electricity (Figure 6.15). While the wind drives the *wind blades* passing this motion on to a turbine in the air, the *ocean waves* cause the *buoy* to move up and down, directly driving an internal turbine or pumping water up a hill for storage and later use. The *tidal turbine* operates from tidal currents which are like the diaphragm discussed earlier, which moves in and out, producing horizontal or vertical motion on a piston-type turbine [10].

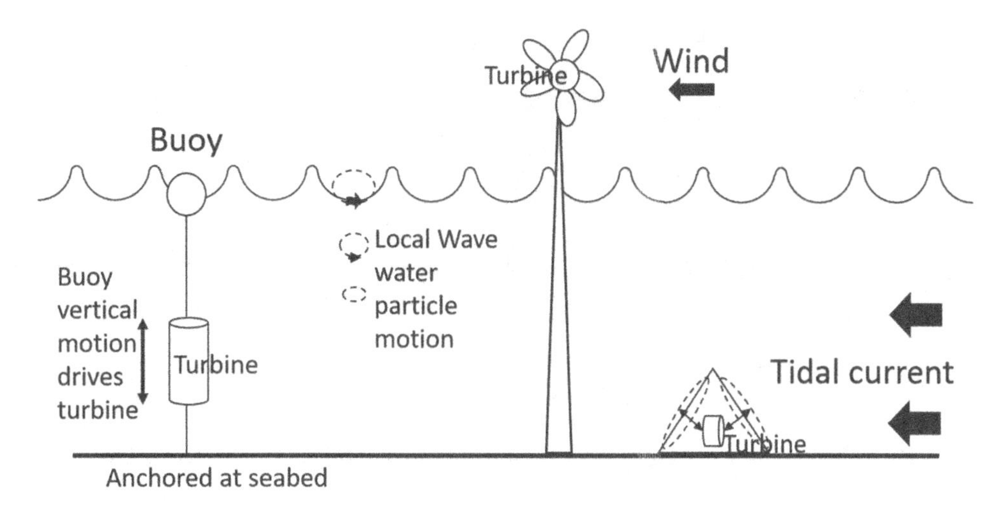

FIGURE 6.15 Wind, wave, and tidal power generators.

Returning to more recent innovations in sound and vibration monitoring technology, the *piezoelectric crystal* is the most common form of active sound sensor which is used in mobile phones, microphones, guitar pick ups and was explained earlier in this chapter and shown in Figure 6.7. This same crystal can also be immersed in liquids (called a *hydrophone* – *hydro* being Greek for water and *phone* being Greek for sound) to give an electrical signal out when vibrated by sound passing through the liquid such as the ocean's water column, and is the method by which marine mammals communicate with each other under water. Small piezoelectric crystals pick up the higher frequencies found in the voice (500 Hz to 5 MHz) which is why they are used for a range of applications in the audible to ultrasonic frequency bandwidth.

In order to sense lower frequency vibrations through the ground, a sensor responsive to lower frequencies than the kHz level of piezoelectric crystal is required since ground vibrations tend to be in the range of 2 – 200 Hz.

The MicroElectroMechancial (aka *MicroElectroMagnetic*) *MEMS Sensor* is now a common instrument used in many applications where a tiny motion sensor is needed. They can be made of interleaved silicon fingers (securely fixed to a base plate) which are separated by an air gap as shown in Figure 6.16. The ability to hold a charge in the tiny gap between any two silicon leaves (known as the *capacitance* to hold a charge) similar to a battery, and with the leaves electrically connected, when they move with respect to the base plate during micro-vibration, the capacitance changes and a pulse is output. Because piezoelectric crystals are also made of silicon, these MEMS sensors are sometimes broadly referred to as *piezoelectric sensors* without explaining whether they are single leaves or solid layers.

The photo in the figure shows a MEMS sitting on a person's finger. This MEMS was developed for insertion within the human body and was used for monitoring blood pressure because when the MEMS body is vibrated in one specific direction, it gives an electrical spike output. This characteristic is useful in any application that requires the direction of movement, with its most popular use being as a tiny gyro in *smartphones*, *smartwatches*

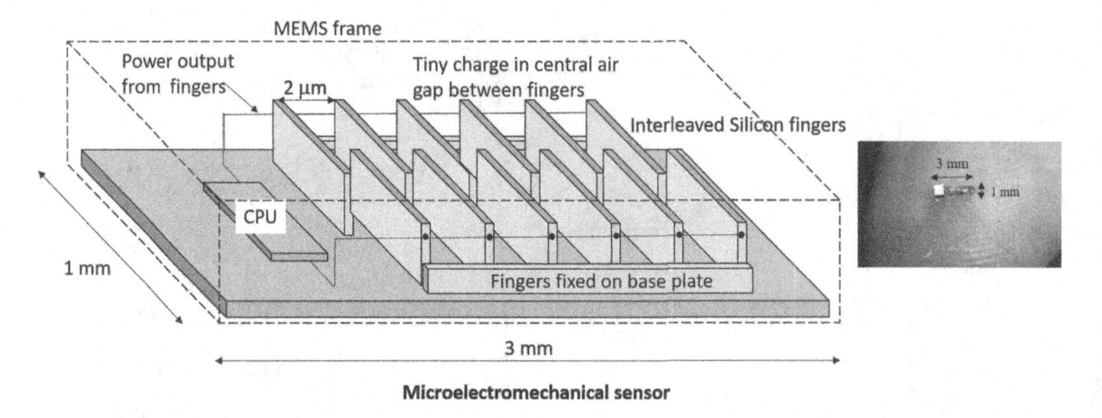

FIGURE 6.16 MEMS sensor made of silicon leaves – photo of bionic piezoelectric MEMS.

(where it can give an output each time we step forward), and sport player's *smart backpacks* among others. A piezoelectric crystal works at limited frequencies and is cheaper to buy than the silicon leaf MEMs but does not have the directionality of the leaf (which is adequate for use in a car airbag for example).

The *geophone* (aka *seismometer*) is such as sensor which makes use of electromagnetic (EM) fields. In this technology, a mass is suspended by a spring at one end and a wire at the other end which itself is suspended within the magnetic field of a permanent magnet as shown in Figure 6.17. If the magnet body physically moves or is shaken vertically, the field moves with respect to the mass and wire, making the wire move across the EM field, breaking the lines of magnetic force. This causes a current to flow in the wire, which when resistors are inserted, gives a voltage output across the wires – as is shown by the voltage signature (aka *wiggle trace*) in the figure. In practice, it typically looks like the geophone body shown on the right of the figure, with a long spike that is pushed into the ground for good coupling with the ground and therefore good response to ground vibrations.

Another form of vibration sensor is the *strain gauge*. As we earlier discussed the origins of strain in which a force is applied to an area causing a change in strain across the area,

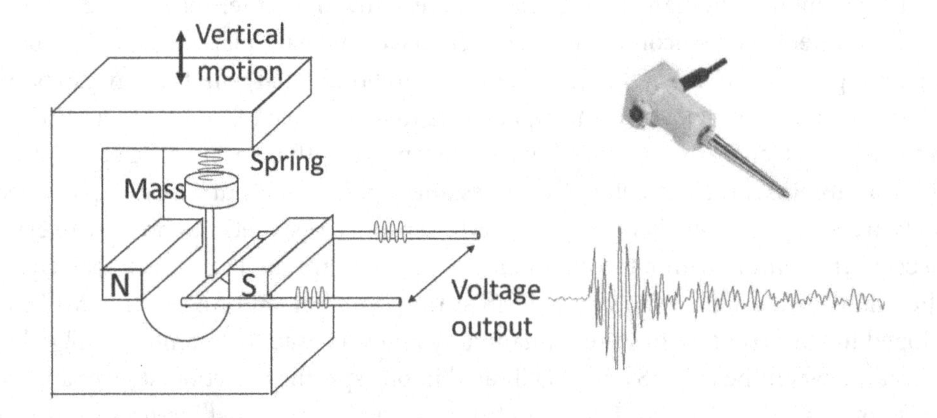

FIGURE 6.17 Geophones (aka *seismometers*) provide electrical output when vibrated.

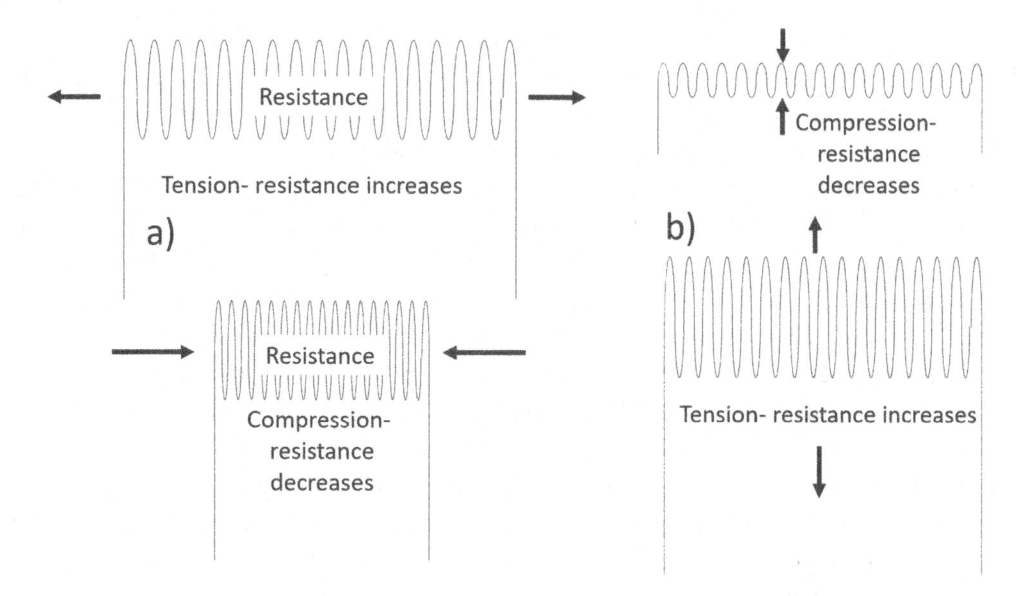

FIGURE 6.18 Strain gauge changes resistance by different amounts with strain direction in: (a) one direction horizontally and (b) at 90 degrees vertically.

then if we consider using the expansion or contraction of a wire which is attached to the ends of a block under strain, we have made a simple *strain gauge* [11]. We can also pluck that wire if it is under extreme strain to make a sound like from a guitar string – its vibrations causing the string to vibrate the air around it at specific frequencies we call *music notes.*

The *strain gauge* wire can be configured over an area, so that any change in strain over the area can be sensed by the wire, which can then expand or contract changing its natural resistance to any applied electric current (as an area is being compressed, it gets denser). But note that the amount of resistance change depends on the direction the wire is strained, since the direction with the greater length of wire will have a larger resistance change than the other (i.e. it is more sensitive in the longer wire direction). If the wire is compressed, it makes the current travel path shorter so the resistance of the wire reduces. If the wire expands under tensile forces and stretches out, it may get longer and therefore causes greater resistance to the current passing through. In Figure 6.18(a) the stress is in one direction whereas it is in the orthogonal (90 degrees) direction in Figure 6.18(b).

6.2.4 Sensors Responding to Light (Visible EM Radiation)

In Chapter 1 we reviewed how sunlight is the visible part of the full field of *EM radiation,* having a specific frequency and wavelength band (see Figure 1.22). Here we review sensors that operate in the EM spectrum that is visible.

Photodiodes were mentioned earlier in Section 6.1.2 as being the most used active sensors which respond to light – these are sometimes also referred to as *photoresistors.* They are made from a semiconductor material and have a chemical reaction when light photons are turned into electrons – they produce current when sunlight (or a reflecting laser light on a CD disc) arrives at their surface – and are often used in security applications with an

LED producing the beam (if the beam is broken, the photodiode will turn-off raising an alarm). A diode only passes electrons in one direction, so these photodiodes produce a tiny direct electric current (DC) when light falls on them [12] and, when many are used, they are called a *solar cell* or *solar panel* (Figure 6.5). Their individual output current can be around 0.1 milliamps, which is very small and if they are to be useful we have to connect many of them to produce the sort of current that would be required to drive a normal light bulb. In areas of abundant sunlight when they are assembled in solar panel form with a panel size of around 2m × 1m, their combined output is over 300 watts which can power five 60 watt light bulbs. Large arrays of solar panels are often referred to as *solar farms*, with the output current going to large banks of batteries. The battery power is a backup fail-safe measure that not only stabilises the current flow providing continuous power to cities irrespective of sunlight conditions, but also provides power typically for home use at night.

Just as we have photodiodes, we also have *photoresistors* (aka *Light-Decreasing Resistor* [LDR] or *photoconductive cell*) which is a passive resistor that decreases in resistance when receiving increasing amounts of light, and useful when needing a circuit that changes in the presence or lack of light.

The Interferometer is a tool that can be used as a light sensor or light filter to separate particular light wavelengths. The main part of this is a *half-silvered mirror* (aka *beam splitter*) which allows some light to pass through but filters out other light. If we take two blocks of acrylic or glass and cut them across diagonally then re-glue them back together, there will be a jointed surface diagonally across them which will allow some light to pass through in one direction but reflect other light of a wavelength which is a function of the jointed surface (Figure 6.19a). This can be thought to be similar in manner to swimming in a pool of water, so that when we are under the water in which the pool's surface is flat,

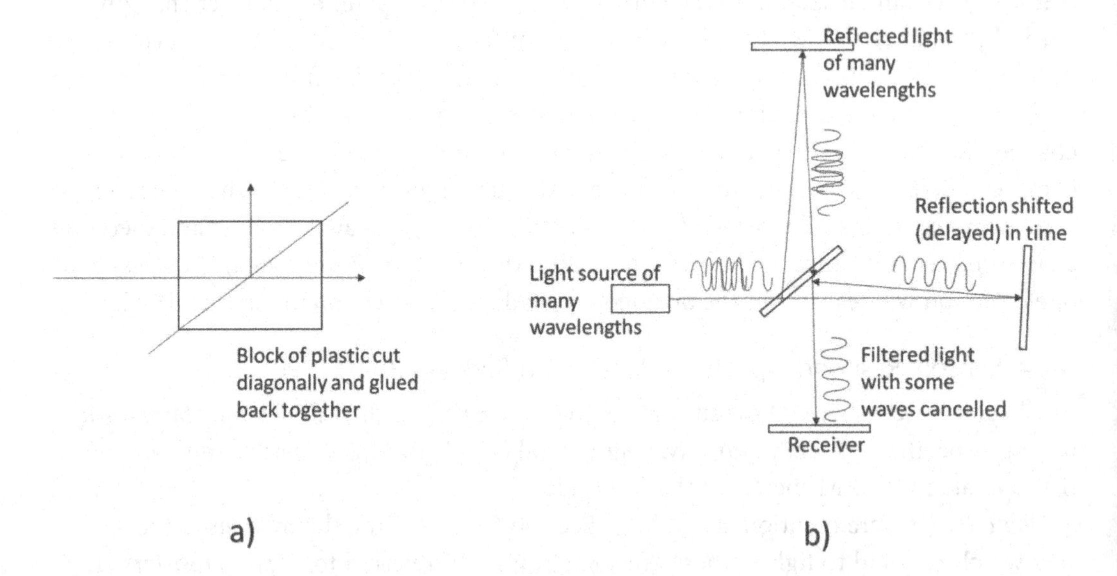

FIGURE 6.19 Basics of interferometry uses (a) block of plastic cut to create an internal discontinuity causing (b) part of a light beam to pass through and another part to diffract along a different pathway.

when we look up at some angle to the sky we can see a reflection of the bottom of the pool, and at other angles we can see the sky.

So the idea here is that we can take a similar half-silvered mirror *beam splitter* and place it so that a beam of light passes through while some light is reflected (Figure 6.19b). Both the passed-through and reflected light beams are then reflected back by two different mirrors, and they are combined together at the beam splitter. The mirrors are placed at appropriate distances from the beam splitter so that their separate reflecting light wavelengths can be either the same or of opposite polarity.

When their light waves meet at the beam splitter surface and individual-specific wave phases are the same, they will combine whereas if their individual specific phase is opposite in polarity, they will cancel. That way, we have now been able to use interference of specific light EM waves to filter out specific wavelengths (or frequencies). This technology is used over a range of instrumentation and data processing, from photographic cameras to fibre optic light transmission, to remote sensing geophysics.

6.2.5 Sensors Responding to Any EM Energy

The presence of a magnetic field was explained in Chapter 1, and how an *EM field* was used in the old analogue TV sets to deflect the travelling electron onto a particular location on the TV screen.

EM fields can be used to detect cracks or metallic objects embedded in them which both change the field's magnetic flow (and so cause their presence to be felt). Figure 6.20 shows an alternating current supply to a coil of wire which is suspended over a conductive material. The current flow through the coil will generate its own alternating magnetic field which will then set up a separate field in the conducting material and other currents which circulate in the material known as *Eddy currents* [13]. This principle of *EM induction* with induced *Eddy currents* is used in many applications.

Crack detection – If the *impedance* of a coil (which is like *resistance* when an alternative current is flowing) changes, what has happened is that the crack has stopped the *Eddy currents* circulating, and this affects the flow of the coil's *EM field*, showing up as a change

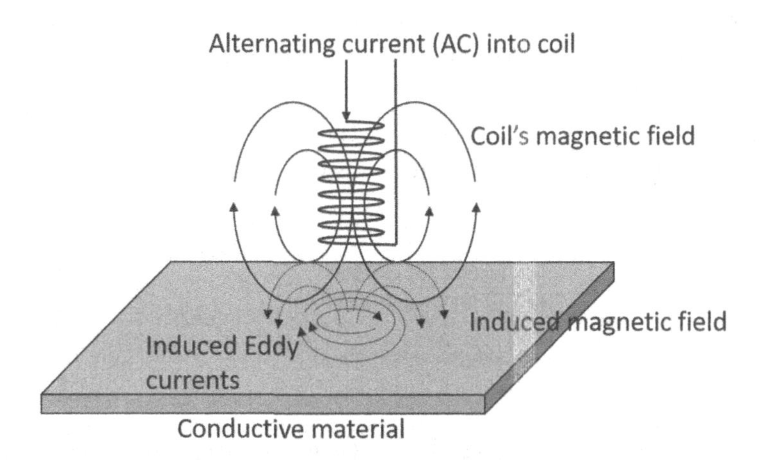

FIGURE 6.20 Input an AC current into a coil, and its magnetic field induces Eddy currents.

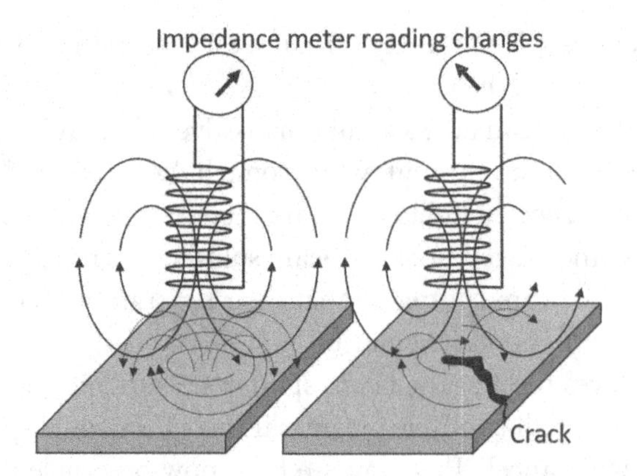

FIGURE 6.21 In a conductive plate, a crack causes changes in coil impedance.

in the coil's impedance (Figure 6.21). So if I have a meter that can read the coil's impedance, this value will change as I run the coil over a crack. This principle is used in non-destructive testing for cracks in pipes or other metallic surfaces where you don't need contact with the conductive plate being tested.

Magnetic card detection – If, however, a magnetic card is placed within the alternating electromagnetic field, this field will cause *Eddy currents* to be generated, which develop their own small EM fields in opposition to the main inducing field (Figure 6.22). This change in the EM field indicates a magnetic presence. If the magnetic card has a magnetic code embedded in it, this changes the main magnetic field in such a manner that the outgoing alternating current from the coil changes and is digitised to represent the number on the card.

Alternatively, a person standing or walking through an airport security detector gate before going onto a plane, will produce an alarm in the receiver coil if the person has metal coins or other metal products on their body. This is also how metal detectors work (Figure 6.23), but in this case a steady very low frequency AC current is transmitted from a transmitter coil, and a co-located receiver coil's field changes if there is a metal object interrupting its receiver field. In this case, the metal has *Eddy currents* induced in it, to affect the receiver's field. Alternatively, an EM pulse maybe transmitted which is reflected back to the receiver by the metal object.

EM Induction sensing – a transformer – There are two types of current – *Direct Current* (DC) as used in car and solar batteries as well as many household appliances, and

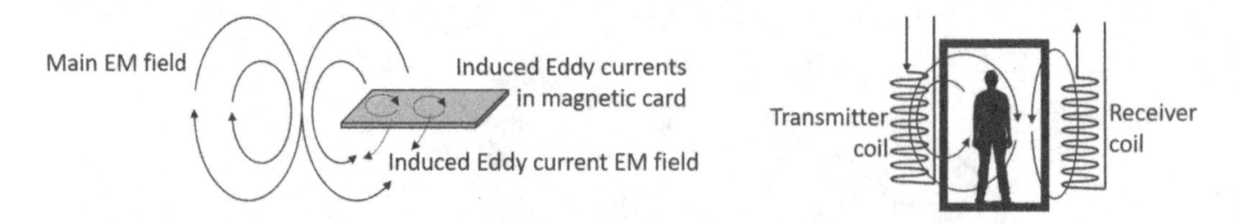

FIGURE 6.22 A magnetic card placed in an EM field will change the field, while a person in a metal security walk through metal detectors will produce an alarm if metal is present to change the field.

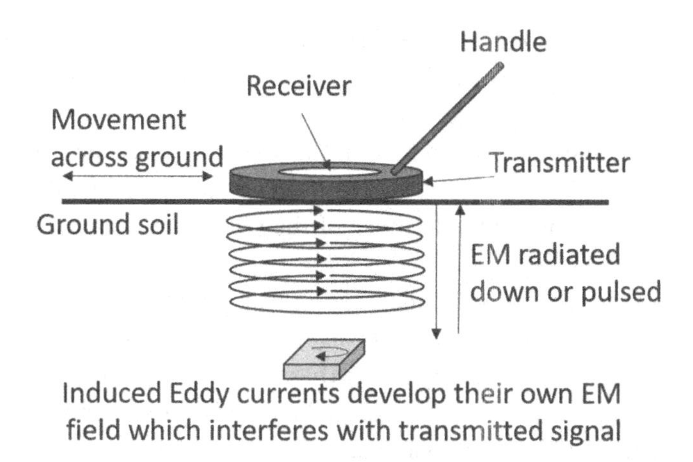

FIGURE 6.23 Metal detectors either transmit a steady low frequency whose wave shape is changed by submerged metal or they transmit a pulse which is reflected back by the metal object.

Alternating Current (AC) as used in household mains electricity. DC has simple current flowing from anode to cathode while AC has current operating at a frequency typically 50 or 60 Hz. Batteries tend to have no higher voltage than 12 volts (each battery cell can be charged to about 2 Volts) and can be used with an *inverter* to convert back to AC. When we see large overhead power lines, they can be any voltage value from 3 kV (3000 volts) to 32 kV AC and often have *step-down* transformers which reduce the voltage level down to the typical 110 volts or 220 volts used in houses and offices around the world.

These work by taking a simple iron ring and wrapping a set of wire turns around each side (Figure 6.24). The voltage input to the coil of wire (we call the *primary winding*) at one end causes a magnetic field around it, and this field generates a *magnetic flux* (which is a type of current called an *Eddy current*), travelling around the iron ring *core*. The *Eddy* current develops its own magnetic field then generating a *secondary voltage* in the *secondary winding*, and the amount that is caused depends on the amount of wire (number of wire turns) that is wrapped around the secondary side of the core.

In 1831, *Michael Faraday* demonstrated this first type of electromagnetic sensor when he showed that there was a relationship between input voltage at the primary side (*Vp*) and

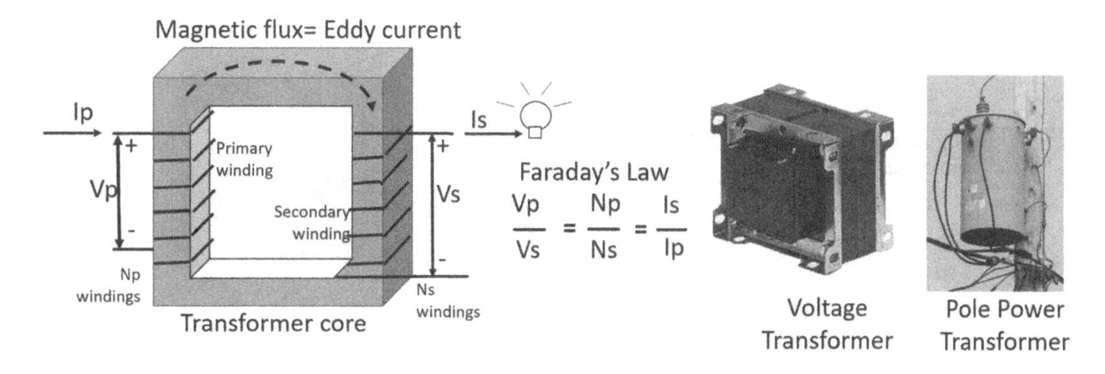

FIGURE 6.24 Schematic of simple transformers used for low and high power supply.

the output voltage at the secondary side (*Vs*), which was in proportion to the number of wire turns that were wrapped around the two sides. This transformed the voltage to a new value, so he called it a voltage *transformer.*

By reducing the number of turns to half on the secondary side, he could transform the voltage from 220 Volts at the primary to 110 Volts on the secondary side. The number of turns in the primary was *Np* while those of the secondary was *Ns*, while the current amounts were called *Ip* and *Is.*

Using this he showed what is now called *Faraday's Law*, where $Vp/Vs = Np/Ns = Is/Ip$.

This type of transformer is a *Voltage Transformer* as shown in Figure 6.24, but you maybe familiar with *Pole Power Transformers* that are on the tops of poles in many streets around the world. In this type of transformer, the core is immersed in oil to keep it cool and to avoid any sparks (oil being a good insulator).

We also use this principle in *current transformers* (CT), where we use the magnetic field of a conductor passing through a *toroidal core* or *ring*, to generate an *Eddy current* which produces a secondary current in wire wrapped around the ring (proportional to the primary current passing through its centre – Figure 6.25). The secondary current can be used to drive a current meter (aka *ammeter*), so that this type of current sensor tends to be used in determining the current flowing in *busbars* found in power distribution stations and large power consuming factories.

These current transformers still conform to Faraday's Law in terms of their secondary current being a function of the primary current, and the inverse of the number of coil turns. Typically a CT will produce about 5% of the primary current since it only has to drive an ammeter. Note that this time the primary is a single conductor while the secondary is the main monitoring instrument, and that some electricians can use a portable handheld clamping CT for testing any wiring.

Household Electricity Meters – also work on this principle. The next time you look at your household meter, you will see that it has an aluminium disc turning around whenever the power is turned on in the house, and the turning disc turns the electricity dials so that your meter tells the power company how much electricity you have consumed [14]. This

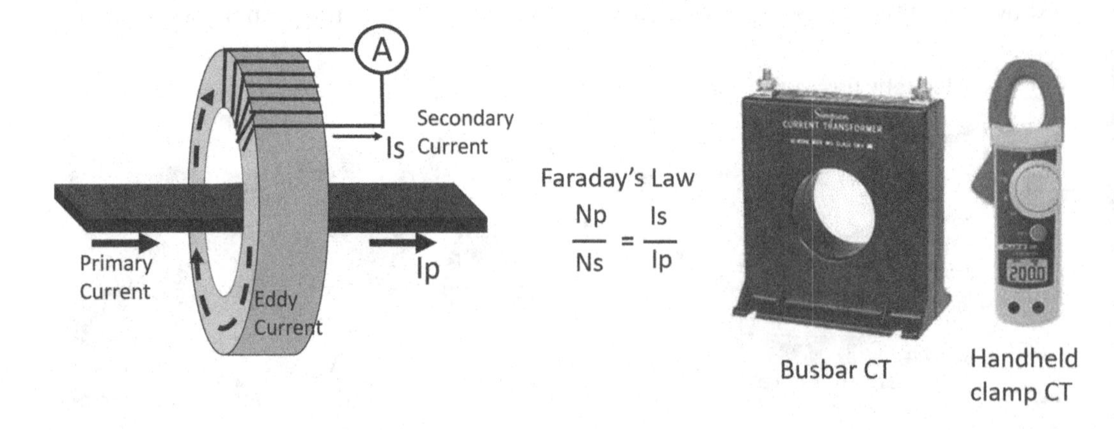

FIGURE 6.25 A simple current transformer. (Photos courtesy: www.electronics-tutorials.ws)

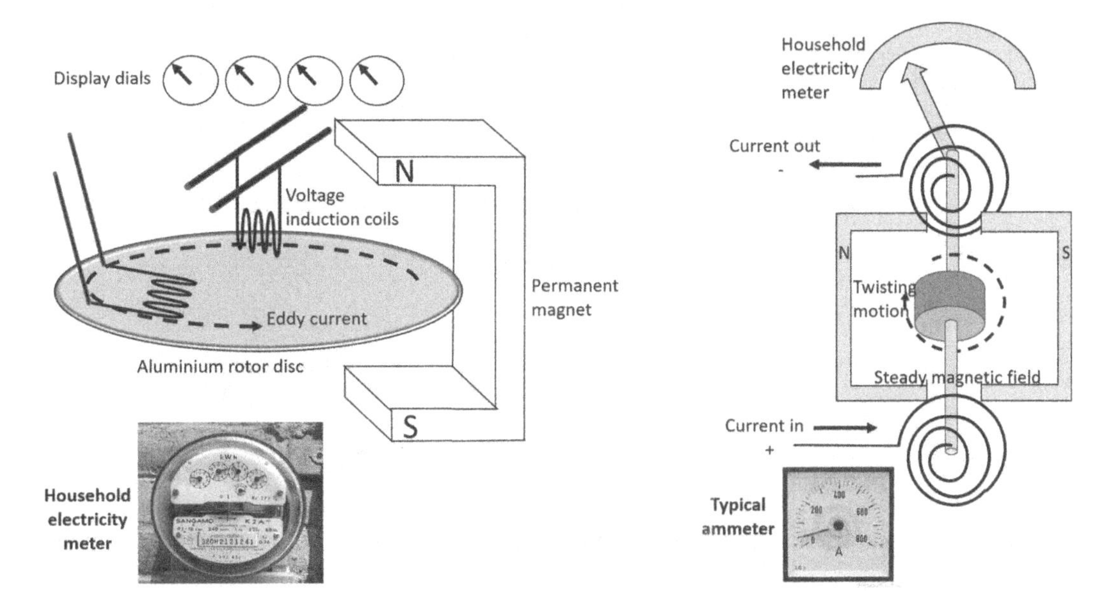

FIGURE 6.26 Household electricity meter and ammeter operate on similar principles.

works by having the main power wires connected to two induction coils – one for voltage and the other for current (Figure 6.26). The induced flux produced by the two-coil EM fields causes a physical force on the disc to make it rotate, with the induced Eddy currents running in a circular manner in the disc. The disc is linked to the display dials which show how much current is being used, and a permanent magnet is put on one side of the disc to slow it down so that the meter readings are accurate in terms of how much current is going through the main power wires.

The concept that the electromagnetic field produces a force that is circular was used in the development of the first current meter (aka *ammeter* or *galvanometer* named after its inventor *Galvani* in 1820) and of course was mentioned in Chapter 1. Instead of making a rotor disc go around as done with electricity meters, current flowing through a thin metal strip across a permanent magnetic field caused it to develop its own magnetic field, which made the strip rotate and twist a needle around a graduated scale to produce the everyday ammeter.

Elevated high-speed trains (aka *bullet trains*) such as the Japanese *Shinkansen* trains (meaning in Japanese 'new trunk-line') are trains that almost fly along rails with hardly any friction [15]. That is because their carriages are levitated up and have motion along the rail line using a linear magnetic field (Figure 6.27). A normal motor has a magnetic field developed by a cylindrical stator containing electromagnets producing flux force which rotates a metallic rotor in the middle. However, in the case of a linear motor, the stator is laid out along the ground, and the rotor becomes the carriage base.

Originally a concept developed by Wheatstone in the 1840s, a working model was demonstrated as possible by Dr Laithwaite of Imperial College London in the late 1940s, and a fully operational commercial train called the *maglev* was opened in 1984 operating on a monorail between Birmingham-airport, UK and the city. Only recently did the United Kingdom seriously consider this method for their new high-speed rail system.

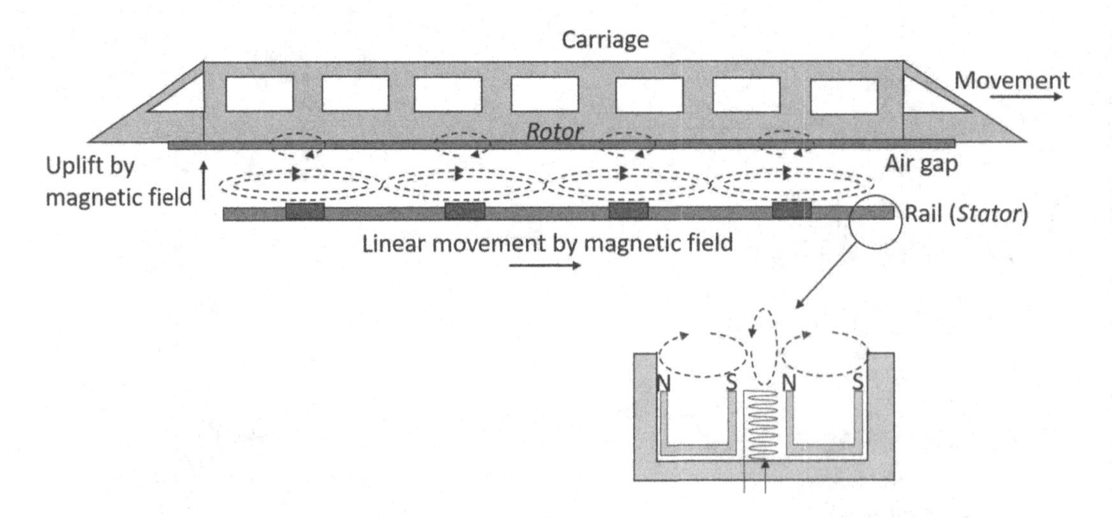

FIGURE 6.27 Linear friction-free train movement using linear magnets with magnetic field uplift.

In Figure 6.27, the rail is actually a U-shape steel rail with electromagnets in its base either side of a central coil. When alternating current is applied to the magnets and the coil, a magnetic field is set-up in the air gap above the U-rail. Just as with the household meter explained in the previous section, a rotating magnetic field is established in the carriage base (which can be a thick aluminium plate), in which Eddy *currents* circulate (as in the meter) and their own *Eddy current* EM field is in opposition to the U-rail field. This opposition causes the magnetic fields to push against each other, so that the force is along the track and the train starts to be pushed by the EM force. The carriages now move smoothly without touching the rail, except that there is a steel strip that keeps the carriage straight and on track should it move away such as when the carriages go around corners.

Radio Detection and Ranging RADAR has been discussed earlier in Chapter 1 (Figure 1.9), in which EM radio waves are transmitted by a radar transmitter that constantly rotates 360 degrees. When the EM waves reach a steel object (ship or bridge), a rain cloud or coastal mountain range all of which reflect the EM energy back to receiver, the radar receiver records the reflected *ping* onto the radar screen which has a bar representing the position of the rotating radar transmitter/receiver. When mounted on the mast of a ship, this allows the ship's navigator to steer a course past reflecting hard objects in the night.

When mounted on an airport tower, it helps the aircraft controllers see where the airborne aircraft are flying at any point in time, so that the controllers can advise the pilots by radio which way to steer the aircraft to make a successful landing at the airport. The circular rings in Figure 1.9 are equidistant so that the aircraft can be seen and tracked as they move across the screen, and their relative distances can be measured. The rough line in the figure could be either the coastline for a ship at the centre moving towards the coastline (the circles' lines maybe one mile apart) or the rough line could indicate the radar is located on an island in the ocean.

Ground Penetrating Radar (GPR, aka a *metal detector*) has adapted the transmission and reflection of EM waves through shallow ground, so that when variations in reflected EM

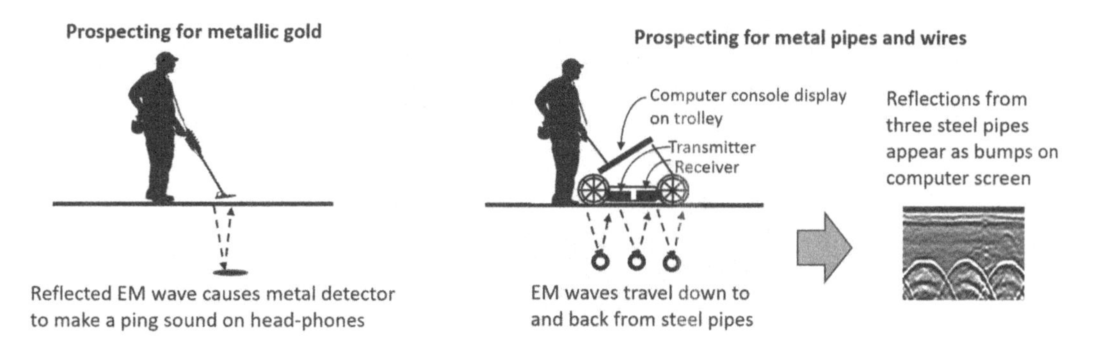

FIGURE 6.28 Using a metal detector to sense gold and a GPR trolley to detect steel water pipes.

energy occur at the receiver, the reflections' arriving pulse makes a *ping* noise in a hand-held metal detector, and may produce a profile of the shallow ground on an industrial GPR trolley (Figure 6.28). The reflections indicate the presence of a solid metal (such as gold or a steel water pipe) or changes in soil constitution (such as when using a GPR over an old grave site). The image that is produced for the operator to see on the GPR console [(16)] is the reflection off the top of the pipe and appears as a curved image due to the diffracted EM waves being received as the operator walks along – they do not represent the shape of the object, only its presence.

Synthetic Aperture Radar (SAR) uses individual reflections of radar EM energy transmitted and received from the earth by planes or satellites (or transmitter/receivers on tripods in mine sites), to reconstruct images of landscape (or open-cut mine site) views. With high frequency radar, the EM reflection gives us a distance to the reflection (like it does with normal radar detection) and if we continuously record these reflections as they would appear on a flat plane, we can view fine details of the surface down to a few centimetres and eventually produce a flat map of the surface (Figure 6.29) which maybe a strip of the earth about 100 Km long.

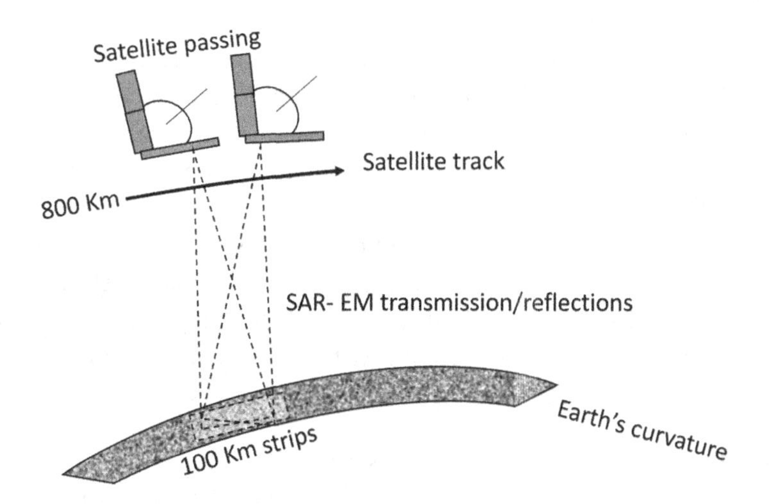

FIGURE 6.29 SAR mapping the earth's surface with radar.

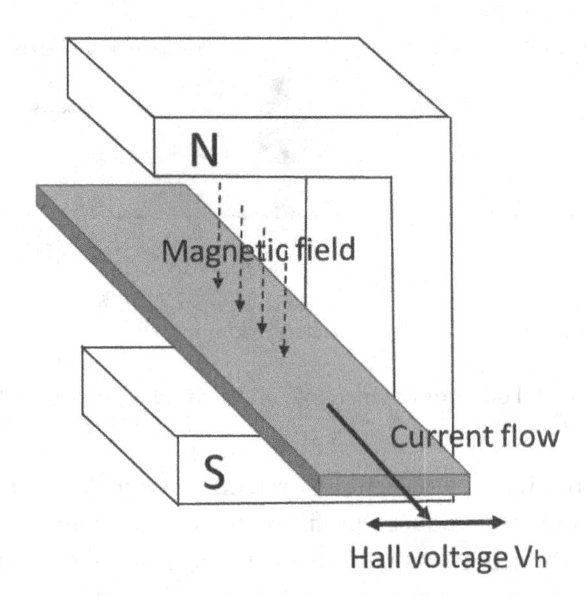

FIGURE 6.30 Hall Effect is current/voltage produced when semiconductor passes through magnetic field.

Just as we see one image with one eye and the same image with the other eye, we can construct 2D and 3D images by overlaying these pictures on each other (as we discussed earlier when we reviewed 3D glasses). Consequently, SAR produces 3D images of landscapes and topography.

This makes a view of any 3D surface (such as the moon) appear in depth with interesting landscape relief, while, at open-cut mine sites, one 3D picture subtracted from another taken three months earlier, will provide the amount of rock that has been removed from the open-cut sides of the mine site. SAR pictures are well established in the media where they show satellite 'pictures' of objects such as buildings on the ground, and are commonly used by *Google Earth* to show maps of cities. They are also used during times of warfare between countries to detect troop and vehicle movements.

The Hall Effect is the use of a semiconductor material that gives a current output proportional to an applied magnetic field strength and was discovered by American Edwin Hall in 1879 (Figure 6.30). When he placed a semiconductor crystal (a material that conducts electricity when other physical parameters are changed) in a magnetic field he found that a voltage (the Hall voltage) occurred across two sides and that current flowed. If the magnetic field moved, this changed the amount of current flow, so this Hall Effect is used as a sensor at times when an output motion needs to be translated into an electric current, such as is used for measuring a vehicle's wheel speed in anti-locking braking systems. Because it has no moving parts but can be placed next to moving parts, it is also used in limit switches, DC motors, and joysticks to control hydraulic valves.

6.2.6 Sensors Responding to Electrical Transmission

Graphene carbon coating is a very thin layer of carbon molecules, so thin that it can't be easily seen without a microscope. It has electrical properties in which, if it changes shape, its resistance changes so that when current is passed through it, we can see the resistivity

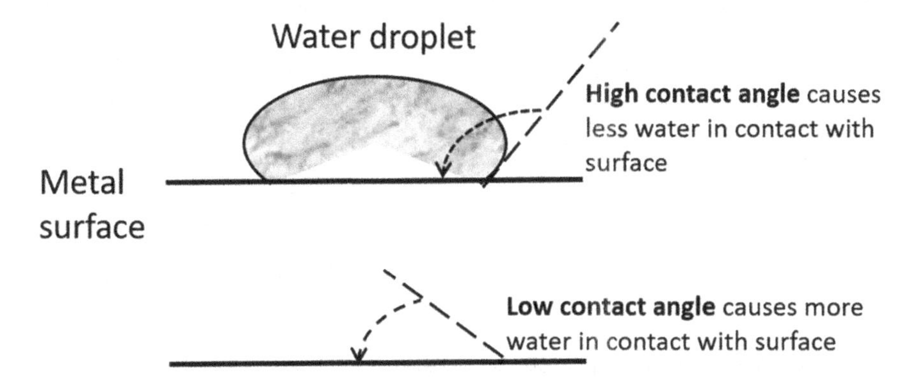

FIGURE 6.31 High contact angle reduces amount of liquid in contact with surface, so less water is retained on surface.

change. This is useful if we want to detect mechanical load (we can cover an area of ground with graphene so that, when a truck runs over it, the area changes resistance and effectively becomes a truck weighing machine), and also this material is very strong with a high contact angle (so it wears for a long time and dispels water – the reason it is used for coating cars instead of ceramic paint – see Figure 6.31). Its reasonable level of heat conduction can make it useful for conducting heat while its strength is useful in building as a composite material or as a tough layer over moving parts (thereby reducing friction wear).

Capacitance probes shown in Figure 6.32 are used in metal tanks as a method to determine liquid levels in which two conductor rods provide a value of *capacitance* when an AC current is input (in practice, the rod and probe are a single instrument with the current running down one probe, through the liquid and up the other). Another application is in personal identification (such as at an immigration counter when they ask for two fingers to be put on a screen) in which a finger on a clear conductive screen completes a circuit with low level AC current being applied through a circuit board to a copper pad – a small camera can also be mounted under the screen so that a photo can be taken for correlation with any stored fingerprint in the database.

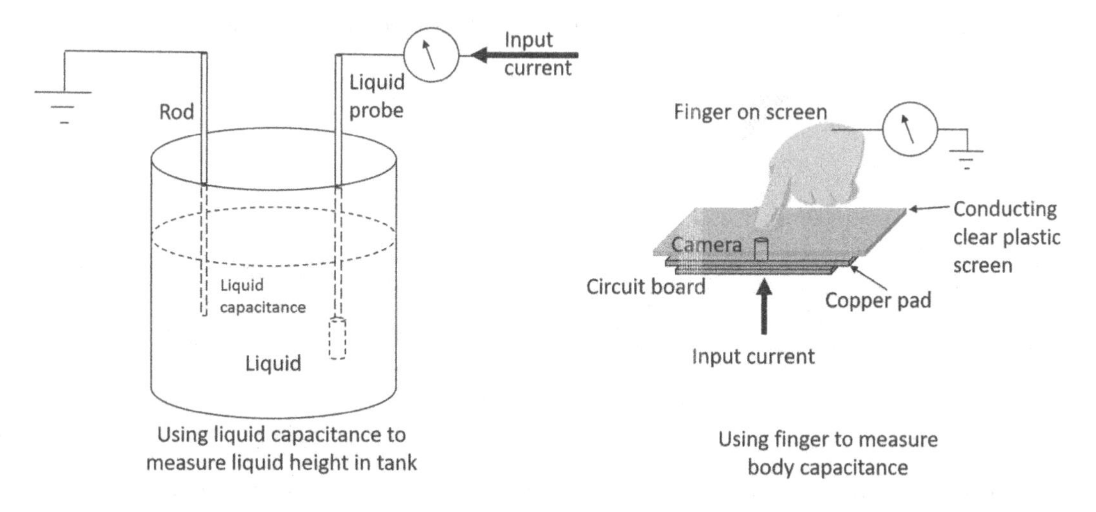

FIGURE 6.32 Measuring liquid level in tanks and body capacitance for identification.

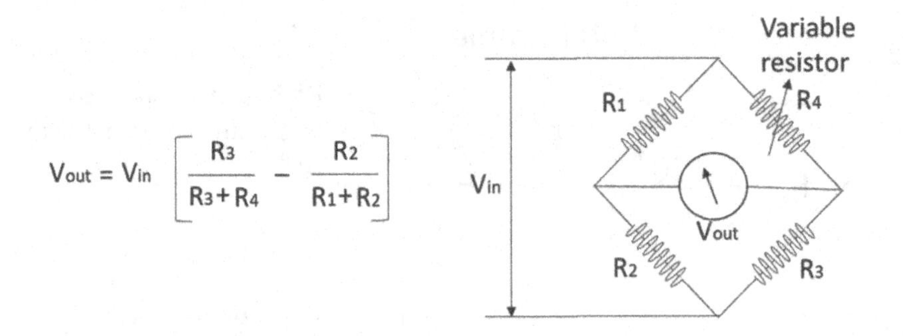

FIGURE 6.33 Wheatstone Bridge equation and network.

The Wheatstone Bridge is explained here because it is the most often used circuit for applications in sensing, where the resistance of a sensor varies with movement, or temperature etc. The *Wheatstone* bridge circuit was originally invented by Sam Christie in 1833 but became popularly named after Sir Charles Wheatstone 10 years later when he used it for analysing soils. Three different resistors are connected together as *legs* in a *bridge* circuit as shown in Figure 6.33 with the variable resistor connected to them as shown by the arrow on it. When a steady voltage is applied across the bridge (*Vin*) and the sensor (R_4) changes its resistance, a voltage change occurs across the middle of the bridge (*Vout*). The amount of voltage change is given by the *Wheatstone Bridge* equation, so that a change in our sensor resistance output then causes a proportional change in output voltage in millivolts, which will drive a voltmeter display or give an output to any feedback loop control circuit. Such an output is useful in remote control of equipment, which is the reason the *Wheatstone Bridge* has numerous applications in automation.

6.2.7 Sensors Responding to Chemical Flow

Biosensors are used mainly in the health industry for detecting changes in biological fluid composition. As the fluid passes over a biosensor, it has a bio-receptor surface sitting on a semiconductor material (nanomaterial), which has a chemical reaction with the fluid (e.g. thickens, heats, or dissolves) to cause the semiconductor to pass a current (Figure 6.34).

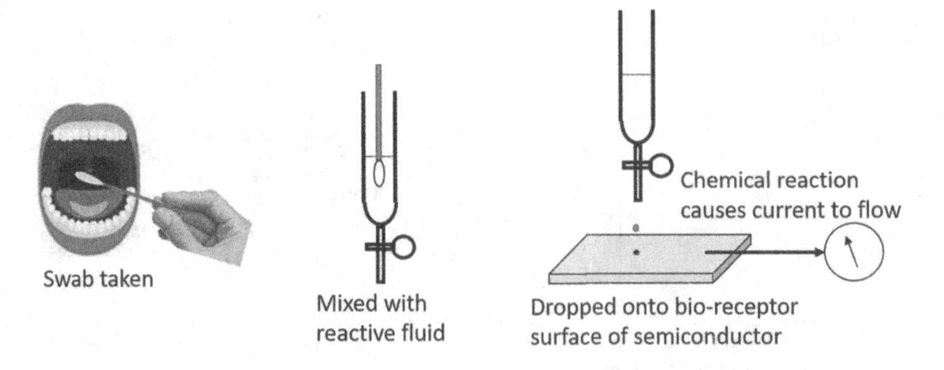

FIGURE 6.34 Bio-receptor surface chemical reaction causes semiconductor current flow.

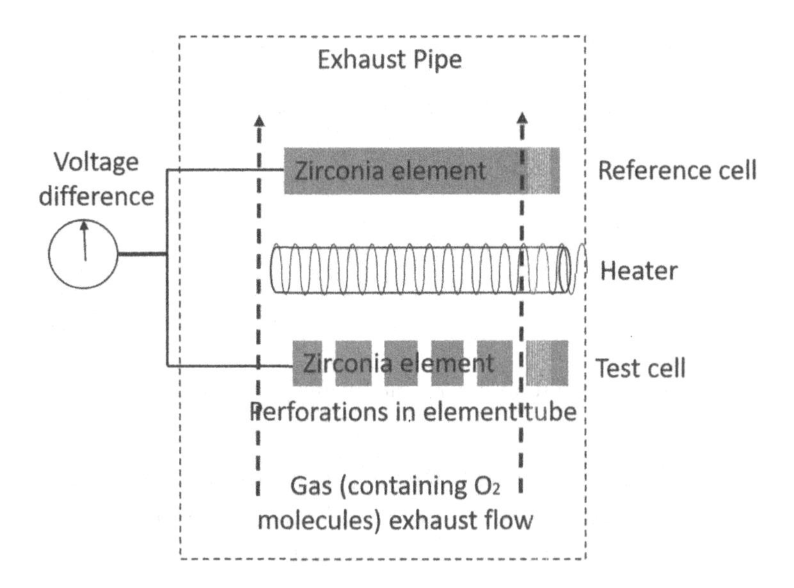

FIGURE 6.35 Air-fuel (O_2) sensor.

This current is then amplified and can be fed to a meter for display. An example of its use is in DNA detection or to detect onset of flu.

Carbon paste electrodes have a gel paste of graphite and a pasting liquid applied on the probes or electrodes, allowing electrical transmission from one probe to another. When the paste is spread in a liquid filled tank, it can be used as a conductor in the fluid – having the ability to be used for measuring its liquid level.

Air-fuel ratio sensors (gauges) are used in vehicle engines or industrial furnaces to provide data on the exhaust oxygen mixture leaving the fuel combustion process, or the amount of oxygen that remains after burning when used in a vehicle's exhaust system. Getting the fuel mix is important to make an engine run economically so that fuel costs are minimised for the best engine performance. The sensor (Figure 6.35) is based on the fact that zirconium produces low level current when hot air or specifically oxygen ions pass over it. It is made up of two zirconia cells – one reference cell and one test cell which is perforated to allow gas to pass through it. The cells are built as a single unit so that when placed in an exhaust pipe, the difference in voltage between the reference cell and the perforated test cell due to the hot gas passing, is proportional to the amount of oxygen in the gas stream.

Glycol coolant, often referred to as *antifreeze* coolant for vehicle radiators, is a mix of glycol with water (17). The glycol lowers the freezing temperature of water as a result of faster heat transfer (thereby reducing the chance of vehicle radiator water freezing) and reduces rusting of metal surfaces.

Breathalysers, originally invented in 1958 by Robert Borkenstein, are used to test the level of alcohol in your breath since this is proportional to the amount in your blood. When blowing into the breathalyser mouth piece, the breath passes over a semiconductor anode of tin oxide or platinum (with an electrolytic material between its electrodes), which produces acetic acid and water. Current is passed through this anode (which changes

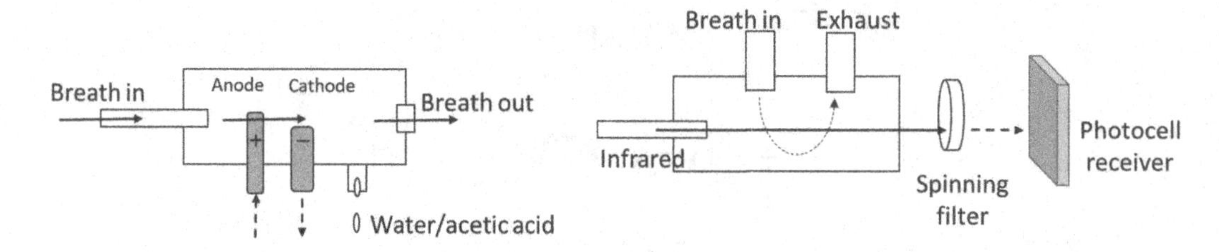

FIGURE 6.36 Two types of breathalyser – acetic acid type and infrared type.

resistance based on the amount of acid that is produced) to the cathode. The resulting voltage change across the electrodes is fed to an *LED* display so that the greater the acetic acid produced, the greater the blood alcohol level (Figure 6.36). Another type uses infrared frequency transmission through the breath cloud (ethanol in the breath stream filters some transmitted frequencies), and another is based on a fuel cell in which oxidation of ethanol to acetaldehyde on the electrode, changes the resistance of the electrode in similar manner to how it operates with tin oxide.

Smoke and CO alarms – There are two common types of smoke alarm – one is the ionisation (aka electrochemical) type which is useful for fierce fires where not much smoke is produced, and the other relies on smoke breaking an infrared LED light beam (Figure 6.37). The ionisation (electrochemical) sensor has two electrodes in an electrolyte (sulphuric acid or other reactive agent) passing current from one electrode to the other. When CO arrives, one terminal converts CO to CO_2 and the other absorbs the oxygen – but this process changes the circuit resistance. The *IR-LED* light beam type is good with CO, since specific frequencies are absorbed, and is often used in large under-cover spaces such as warehouses. For other LED type smoke detectors, CO passes over a chemically treated pad which changes colour when smoke passes over it and a photodiode detects the colour change.

CO_2 detection – The same *IR-LED* type detector in Figure 6.37 can be used for detecting CO_2 since CO_2 is in the 4.26 micrometre adsorption bandwidth. Alternatively, a light beam can be passed over a *microelectromagnetic sensor (MEMS)* in which a metal oxide semiconductor's change in resistance is measured.

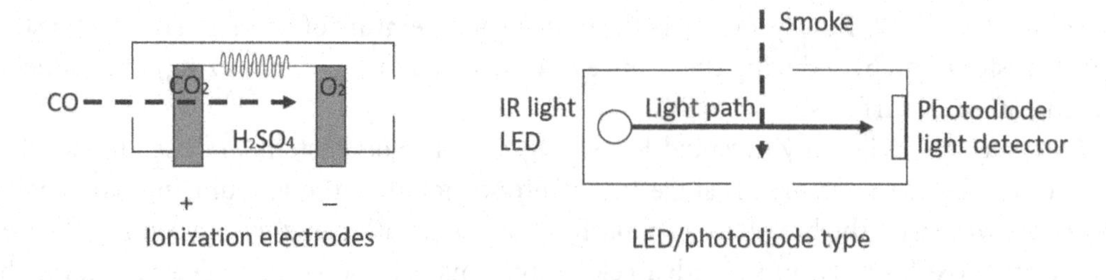

FIGURE 6.37 Smoke/CO alarms are mainly ionisation or LED type.

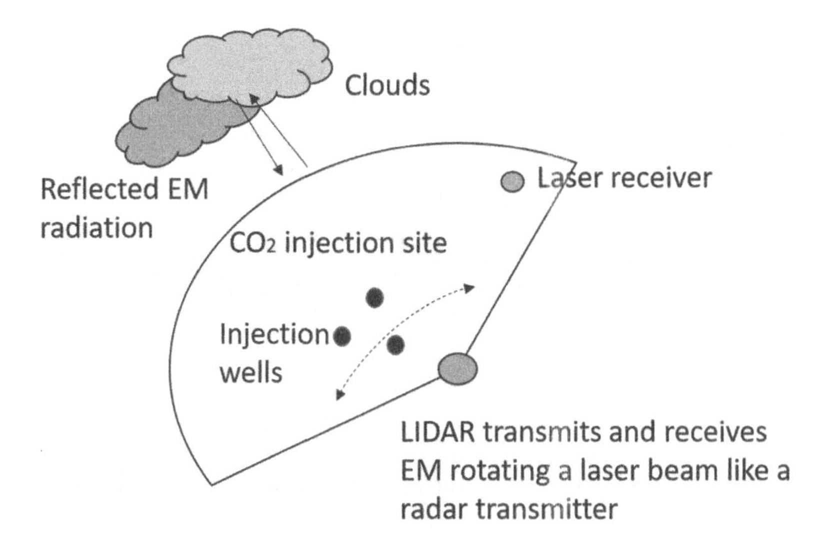

FIGURE 6.38 LIDAR transmits/receives EM laser light to monitor atmospheric CO_2 levels.

Light (Laser) Detection and Ranging (LIDAR) *for* CO_2 or atmospheric pollution is used over very large outdoor areas, in which a rotating laser beam can be used as a flash light in light house manner around an area, and analysis of the reflected or transmitted EM frequencies, either through the atmosphere or reflected from a cloud, allows observation of CO_2 or any other pollution levels (used in large outdoor areas such as at CO_2 storage sites and in cities for checking pollution levels – Figure 6.38).

Catalytic beads were introduced as a sensor in underground coal mines to sense the presence of hot coal gas (Figure 6.39). Current is passed through two coils of platinum wire embedded in an alumina bead, with the bead suspended in gas temperatures up to 500°C.

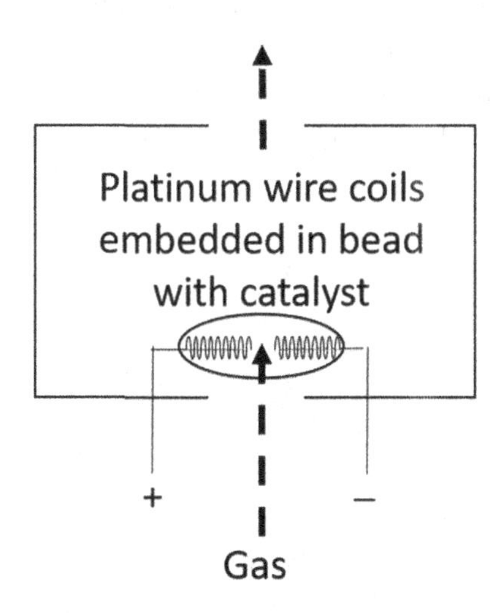

FIGURE 6.39 Catalytic bead oxidises as hot gas passes over it, changing resistance.

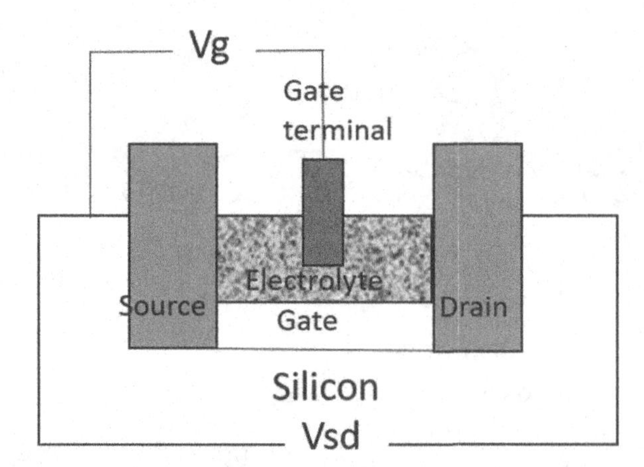

FIGURE 6.40 Chemical Field Effect Transistor (FET) uses gate terminal in electrolyte to control current.

The bead oxidises as hot gas passes over it, the alumina changing resistance and therefore changing current – this is fed to a *Wheatstone bridge* connected to external monitoring meters.

Chemical Field Effect Transistor (ChemFET) made of silicon, is a semiconductor that has three terminals – two to pass current from one terminal (source) to its output (drain), with the third used to control the current flow referred to as gate. The gate's electrode can be in contact with a chemical fluid (Figure 6.40), so that changes in chemistry controls the gate which changes the current passing through by varying the resistance (similar to an ion-sensitive ISFET) with the ChemFET being used to detect if specific chemicals are present.

Chemiresistor is a resistor that changes value when a chemical is placed in the circuit containing an analyte (Figure 6.41) and is sometimes referred to as a *chemical tongue* because when the chemical is present, the circuit resistance changes and the current changes just like the tongue transmits an electrical signal to the brain when it experiences a particular taste (the difference between an analyte and a catalyst is that there is no reaction with a catalyst – it only helps a reaction along – whereas there is a reaction with an analyte and it can change form/weight in the reaction). This chemical maybe a metal oxide in contact with a vapour of nitrogen oxide, hydrogen sulphide, and oxygen among others while graphene

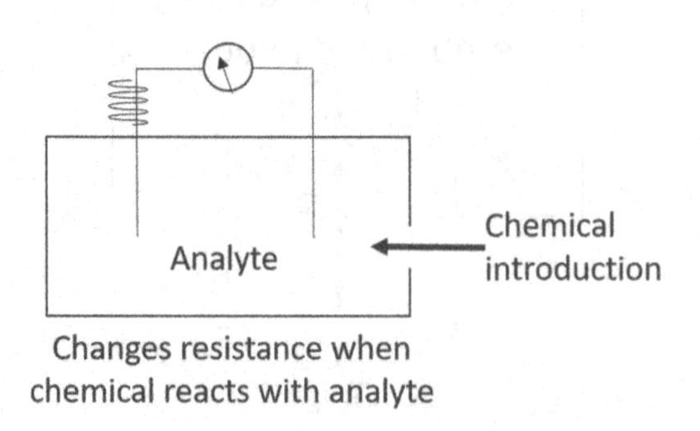

FIGURE 6.41 Resistance of circuit changes when chemical reaction with analyte.

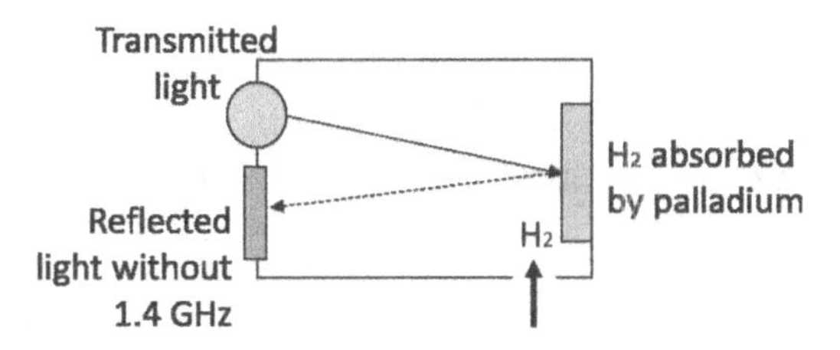

FIGURE 6.42 Reflected light from palladium contains zero 1.42 GHz if H_2 present.

is used for carbon sensing (see also *ChemFET*). The *Chemiresistor* may also take the form of a fine film over an electrode.

Hydrogen gas sensor uses the property that Palladium absorbs hydrogen gas to form palladium hydride. The palladium surface reflects when a light is shone on it and, because hydrogen has a frequency of 1,420,405,751 Hz (1.42 GHz), if that specific frequency is not present in reflected frequencies, then H_2 is not present (Figure 6.42). These can be used at the end of optical fibre cables, so that a simple FFT performed on reflected light by the CPU can detect the presence or not of H_2 (no 1.42 GHz in reflected signal).

Surface Acoustic Wave (SAW) detector is used to sense the presence of lead, mercury, or oxygen in a material (Figure 6.43). This is done by vibrating a SAW sound wave at about 500 MHz along a thin film of piezoelectric material overlaying a surface in which a frequency/phase shift or velocity change in the received signal gives an indication of the chemical fill along the surface. *Microelectromechanical sensors* (MEMS) can do this with a CPU which analyses the transmitted signal and compares it with the received signal, and gives an output signal proportional to the difference. The signal value is related to the presence of the chosen property in the material's surface.

Ozone (O3) and the ozone layer is important because the ozone layer absorbs harmful UV EM radiated energy from the sun – a hole in the ozone layer can allow excessive UV through leading humans to experience sunburn and in the worst case skin cancer. To detect ozone, we can transmit light through the clean air followed by a sample of ozone, and any absorption of UV shows how much ozone is in air [18]. We can make ozone by making an electrical spark in the air, and some of the smoke observed (and smell) is ozone – which is blue in colour. This is part of the reason the sky looks blue – because there is ozone in it (but the major part is that our eyes see the blue-frequency of the spectrum better at 700 THz than the lower red end at 400 THz, so the sky appears blue).

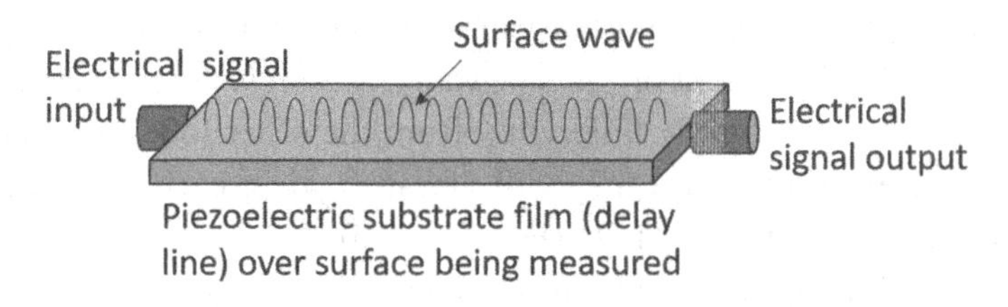

FIGURE 6.43 Surface wave vibration at about 500 MHz changes frequency if specific properties in the medium's surface being measured are present.

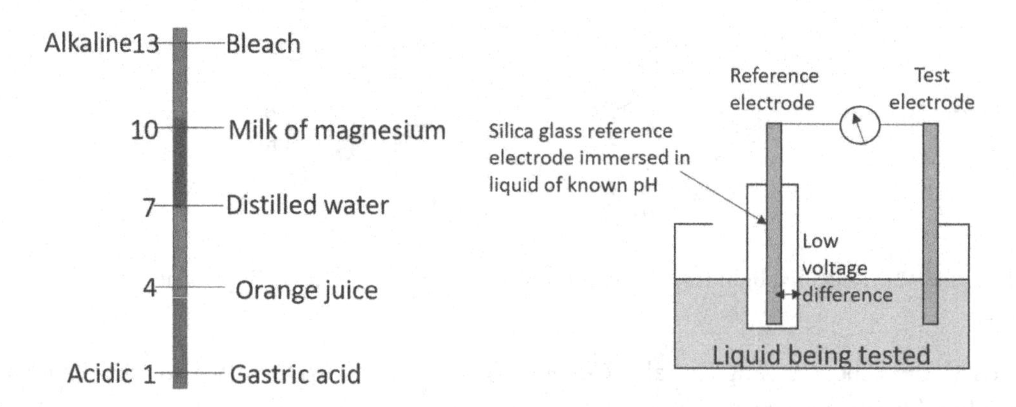

FIGURE 6.44 pH scale for acidity uses comparison of hydrogen ions.

pH (which stands for *power of hydrogen*), is the figure of the *inverse* of how many *Hydrogen* ions are in a liquid using a log scale – we call this *acidity.* The scale is a bit confusing because the more hydrogen ions a liquid has, the lower it is on the scale of acidity – while it is a $\log_{10}$ scale because it is so broad a set of values, by using a $\log_{10}$ scale we can to reduce it to something more manageable.

If it is highly acidic like lemon juice, it is a low number around 2. With distilled water it is neutral about 7 and soapy water or bleach is alkaline at about 13 (Figure 6.44). To measure pH, we can put some litmus paper in the liquid, and this changes colour so it is easy to read off the colour to compare with a colour chart. However if we want to be more analytic, because an acid has more hydrogen ions than alkaline ions, this means it can produce more current between two terminals immersed in it, and so produces a greater voltage the more acidic it is.

To test this, we put a thin glass reference electrode (containing an internal test probe immersed in a liquid whose pH we know – say neutral potassium chloride) into the liquid to be tested (Figure 6.44). A low level voltage will occur between the glass electrode outside and the liquid of known pH. Put a test electrode in the liquid being tested and compare it with the reference electrode, and a voltage difference will be shown. There are more hydrogen ions in the test liquid than inside the thin glass reference tube liquid, making hydrogen ions exchange places with the neutral potassium chloride inside the thin glass electrode – developing a voltage difference.

Coriolis Flow meters (aka *water pipe noise*) are used when a fluid flows through a pipe, and its mass against an abrasive pipe wall causes a twisting motion of the fluid. If we flow water through two pipes together alongside each other and put a bend in the pipes, because either, their path length is different or the pipe wall abrasion is different, the water flowing through one pipe will twist slightly differently from the other (due to different pipe wall abrasion or resistance). In the bend they will vibrate against each other making a noise with frequencies around 80 Hz to 1 kHz (Figure 6.45a). This is the noise heard in household pipes. Either turbine or ultrasonic measurements can be applied to obtain the mass flow rate in which the frequency of turbine rotation or arriving noise can indicate mass flow.

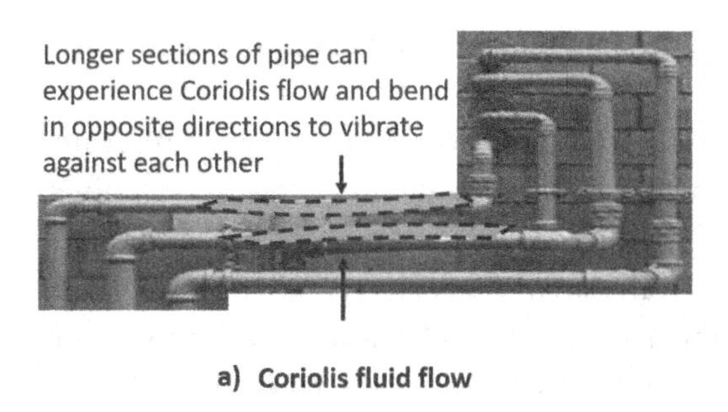

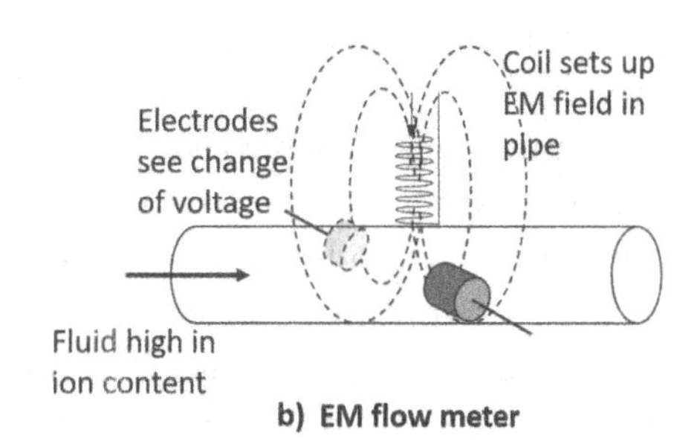

FIGURE 6.45 (a) Coriolis flow noise caused by fluid flow twisting pipes; while (b) EM field voltage changes when liquid high in ions passes through.

Equally, an *EM flow meter* is a coil fixed externally to a pipe, generating an EM field so that when a liquid containing ions passes through the pipe, it changes the magnetic field causing a change in voltage across a pair of electrodes (Figure 6.45b).

Venturi flow meters are used to determine flow velocity using pressure changes during flow according to Bernoulli's equation (Figure 6.13). Basically, when an incompressible fluid (such as water) flows through a constriction or *choke*, a reduction in pressure occurs with an increase in flow velocity (Figure 6.46) and a float in a tube rides on the flow rate. Turbine flow meters can be used which are based on a turbine blade spinning in the flow, just like a *bicycle goniometer* which is geared to a counting machine.

6.2.8 Haptic (Tactile) Sensors and Their Future Potential

Haptic is the Greek word for *touch* so any sensor which is haptic is tactile and considered one of the most unused technologies of our natural human senses This is not referring to the use of touch-sensitive screens but to the human ability to receive a touch from something or somebody external in order to experience some tactile sense which is meaningful.

The most obvious tactile sense which a human can experience is when an explosion occurs, and the loud noise, carried through the air by its pressure wave, can impact on our ears and we hear the explosion, as well as feel the impact on our bodies and physically shake the body as the pressure wave moves through and past us. We have already seen how transducers can be used to vibrate at different frequencies, an adaptation of this being the vibration a smartphone makes when it rings in your pocket – which is the first common use of haptics on the body. In this case a tiny vibration linear motor (*Linear Resonant Actuator* – LRA) develops a tiny magnetic field which causes a sprung metallic disc to vibrate at frequencies around 150 to 250 Hz (which you can feel and hear). The *LRA*

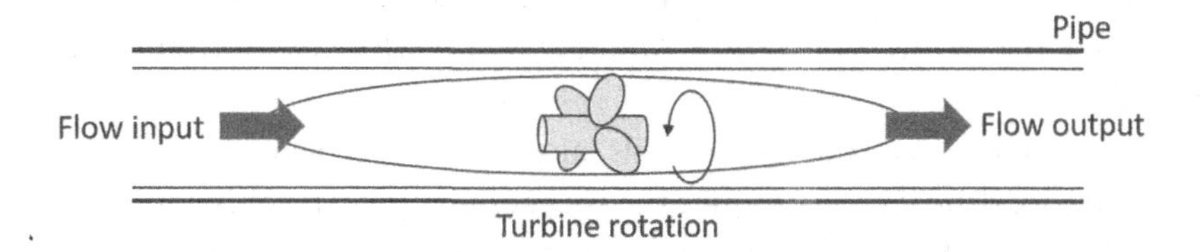

FIGURE 6.46 Venturi flow meters often use turbines to compute velocity of flow.

with a spring allows the vibrations to last longer than a piezoelectric crystal at these lower frequencies.

However, when coupled with a force feedback loop system (which is explained in greater detail in the next chapter) this tactile sense can be used for a number of other applications. For example, consider wearing 3D glasses and looking at a 3D picture of a piece of rough rock on a screen while wearing gloves that contain small piezoelectric (MEMS or LRA) transducers on the fingertips. It is then possible to have an electrical feedback loop (connected to software running a 3D screen image) to allow you to feel the tactile sensation of running fingers over the 3D rock's surface. The feeling of haptic vibrations gives a sense as if running fingers over the rock's rough surface.

This was how tactile sensing was first commercially developed in the 1990s to demonstrate haptic technologies. In the future, it could be used to remotely sense rock properties if future spacecraft has vehicles that drive around a planet and transmit their robotic arm finger sensor data to observers wearing tactile gloves monitoring the vehicle on earth.

3D sensing haptic games and VR were developed using this technology during the 1990s, which were mainly joystick driven. 3D image software drove a pressured joystick, so that by moving the joystick around using a finger, an *emoji* of your finger on the screen (an *emoji* is a Japanese word for a small cartoon image) would show your finger on a 3D image. You could then sense how dense or movable a 3D image might be. This was a fun way of getting a tactile sense of the solidity of a 3D object.

However, its industrial introduction was limited to sensing rocks. One application was in 3D imaging of a core of geological rocks in which a geologist could run a finger over that geological core and sense which rocks were stiff (well cemented) and which rocks were more elastic (and porous). Unfortunately, this was treated as a game and the technology failed to be developed further as a commercial application – perhaps interplanetary exploration may see its reintroduction.

The most use that can be made of these haptic sensing devices is when they are used with *Virtual Reality* (VR) headsets (see Figure 5.22). In this case for example, in the game of tennis, the racquet can have haptic sensors along its length which when pulsed with an electrical spike, causes the whole racket to shake. So when used with *3D VR* on a tennis court, it is possible to play tennis with the headset on and hit a ball with the racquet making the same vibration transmitted to the hand as is felt when a ball hits it. If a sound of a racket hitting the ball is produced at the same time, it can be very lifelike and appear real. This would also be the case when playing baseball or using cricket bats.

The Resonant actuator (aka *resonant motor)* is discussed at the start of this section citing the *LRA* in which the motion is *linear* (along a single line or axis as in a smartphone). The motion of such *motors* can also be radial/eccentric (the force acting around a circle) and these motors are referred to as *Eccentric Rotating Motors* (ERMs). Both motors are shown in Figure 6.47.

In the LRA, two tiny electromagnetic coils (aka *voice coils* because their action is similar to coils used in a loudspeaker to produce the speaker vibration movement) are mounted in a housing with a spindle running through, which generate a rotating magnetic field current (like the *Eddy currents* in the meter rotor disc of Figure 6.26). However in this case,

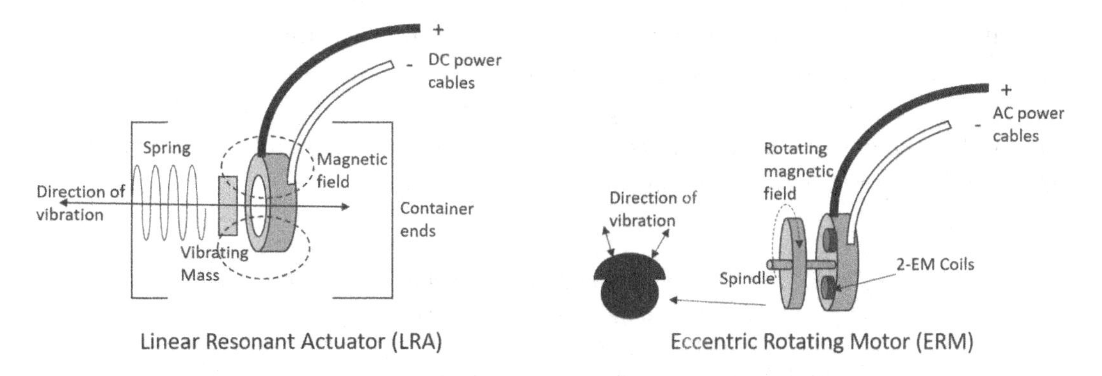

FIGURE 6.47 LRA vibrates along a single axis, with an ERM which vibrates eccentrically.

the rotating disc is shaped with more mass on one side which makes it vibrate laterally (eccentrically). This action is often used in the food canning and pharmaceutical industries when a lateral action is needed to move products across a production line.

Future use of haptics can be imagined in the games area. If football (soccer) boots were built with internal *haptic* vibration sensors (such as *piezoelectric* discs for short bursts or *LRAs* for longer bursts of energy) they could be controlled by the electronics backpack football players wear. This would set up the potential for *virtual* running around a field wearing the *VR* headset, playing football with an *avatar* like Messi or other world-class footballers.

If we broadened the use to a bodysuit into which multiple *haptic sensors* (19) and electronic receivers had been sewn, we could then play *VR* games with anyone, for example slide-tackling other players while sensing their body force against us (to a limited degree). Of course this all depends on the ability to run anywhere wearing a light-weight *3D VR* set of eyeglasses with a microprocessor that knows every move via GPS, and computing capability that is incredibly fast (which likely won't happen any time soon).

Another possibility is in learning a trade or manual movement – for example, it would be possible to learn how to saw wood, hammer nails, or solder wires together by doing these activities wearing *VR glasses* and *haptic* devices. Such haptics are already employed by the education industry for teaching everyday events such as learning to drive cars and as extreme as learning to drill an oil well (as is taught by Aberdeen University).

As discussed earlier, the success of using these sensors requires an understanding of how the sensors work, how their data is processed and how the data feedback operates, if they are going to work at all. We will cover how data is processed in the next section (and feedback loops in Chapter 7), but for now we have a basic understanding of how the majority of common sensors work.

6.3 DIGITISING SENSOR DATA, USING CONVENTIONAL COMPUTER INTERFACES AND QUANTUM COMPUTING

It is one thing to have sensor data output as a continuous stream of varying current or voltage (which is what basic data is), but that is inadequate – the data needs to be in digital form if we are going to use it. As was explained in Chapter 3 (Figure 3.18), data needs to be translated from a continuously varying value (of current/voltage) into a set of digital values that accurately represent the analogue form [20]. This must be recorded and made into computer words, so that we can use computer languages to understand the data and then act. We call this *digitising data* (taking it from analogue continuous values to digital values). This chapter will discuss how this is done in its simplest form, and then go on to explain how it is used by industry and in our everyday life.

> "data needs to be translated from a continuously varying value (of current/voltage) into a set of digital values which accurately represent the analogue form. This must be recorded and made into computer words, so that we can use computer languages to understand the data and then act. We call this *digitising data* (taking it from analogue continuous values to digital values)."

6.3.1 Digitising Sensor Data

When an analogue signal is received, it appears as a continuous stream of current or voltage values. At the time of recording, these can be displayed on paper charts similar to a continuous line on graph paper, which we refer to as being an *analogue recording.*

In such a continuous set of data, let's say the price of gold over many years, the values go up or down as determined by the gold market but if we want to make sense of why it has gone up or down, we need to look at the past 5 or 10 years of price activity. The gold market price is influenced by world stability, i.e. whether there are political or health issues around the world (COVID-19 or a poor harvest resulting in less product and therefore a higher price being demanded by the consumer).

If we do want to study segments of data, we must be able to convert the data into something separated in time. Consequently, we have to take the data and convert it into an output that has a value for each chosen time of interest. This time is known as the *sample rate*, and while it maybe months in terms of the price of gold, it is more interesting in milliseconds when it comes to sensor data. The process of converting analogue data to something more useful is called *analogue-to-digital conversion* aka *A to D* (and A/D or ADC) and a piece of hardware that does this is an *A/D Converter.* Every piece of electronic equipment we operate in life today that has a computer process in it has some form of *ADC.*

In Chapter 4, we discussed how on TV, a cricket (base) ball's trajectory position once it is released from the bowler's (pitcher's) hand can be determined at a steady time rate, and with the values of height/length/speed, then we can determine its average flight and make possible predictions on where it may travel over the next split-second in time. In order to make this prediction we must change the analogue position value to a digital value by digitising it at a set sample rate.

Typically in industry, this sample rate depends on the frequency of operation (frequency was explained in Chapter 1). If we sample once per cycle, our digitised output will be a flat

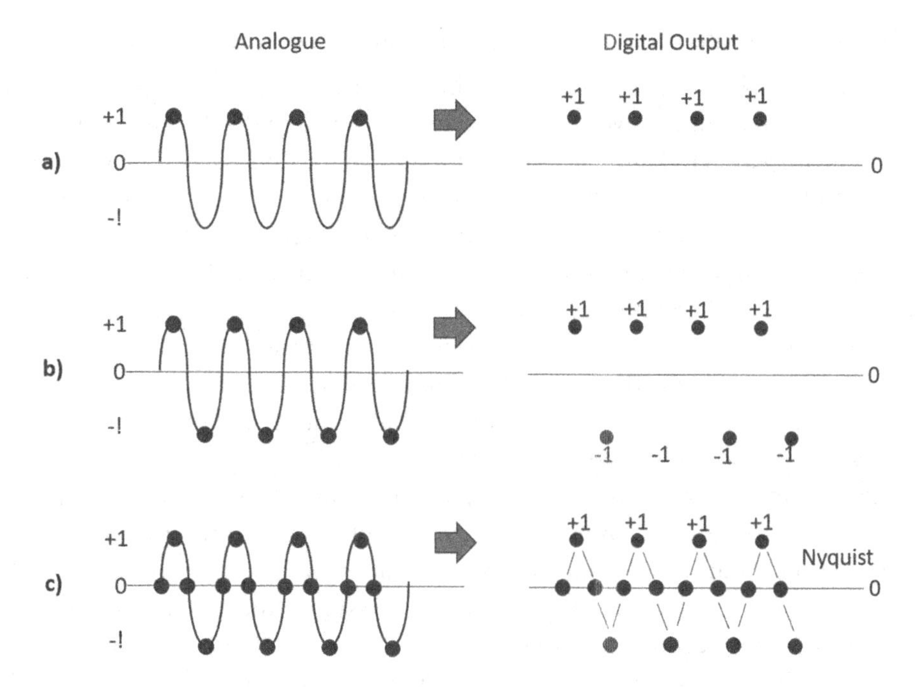

FIGURE 6.48 (a) Digitising once per cycle, (b) digitising twice per cycle, and (c) digitising four times per cycle.

line (Figure 6.48a) at a value of +1, so that clearly is wrong. If we sample twice per cycle (Figure 6.48b), that only gets the high and low (*peak and trough*) values of the cycle (in this case +1 and –1), but although it is a better representation of the analogue wave, again it is likely to give a meaningless output.

To sample four times per cycle (Figure 6.48c), would give a first reasonable representation of this frequency after we have joined the sample point dots, and avoid the potential for *data aliasing* as previously explained in Chapter 3 and displayed in Figure 3.3.

Whatever frequency is to be digitised, the sample rate of its ADC must not be longer than this in time, and is a concept known as the *Nyquist* point or *Nyquist Criterion.* In other words, any wave that is to be converted from analogue to digital must be sampled at a rate of at least ¼ of its wavelength. So if we have a frequency of 1000 Hz (i.e. 1000 cycles per second), we must sample it at a rate of 4000 Hz or 1/4000 seconds = 0.00025 seconds = 0.25 milliseconds.

The speed at which an ADC must operate is also dependent on the size of the computer word it is putting together and how fast the recording, memory, and data writing is when writing the data onto a computer disc or USB. Consequently, the *Nyquist* sets the minimum rate for digitisation, while the other system recording requirements such as the degree of resolution (how fine the sampling should be) sets the maximum rate of digitisation (within the hardware capabilities). It is worthwhile noting that the speed of a computer is generally the speed required to run the CPU, so as it samples, manipulates, and converts data to a digital format as needed, this is known commonly as the *clock speed* [(21)].

6.3.2 Sending the Output to a Recording and Computing Control Point

As in Figure 6.48, taking more samples will improve the data resolution and we then may require a code set to form data words required for recording on the database as we learnt in Chapter 3 and displayed in Figure 3.5. As a reminder, this output set of data words contain each data point value as well as the conditions under which recording has occurred (and being written in word form on magnetic media shown in Figure 3.5) if needed. Today, data is recorded using today's technology, so in the future when we look back at this old data, we must be able to read and understand it, and link this with the more recently recorded data. This is sometimes referred to as *backwardly compatible* data formatting, which all data formats must comply with. This is the same as having an old *Word* file written with an old computer operating system and trying to read it with the latest *Word* software (the latest software must be *backwardly compatible*).

The next step in the process is to pass the recorded data onto a database that may either be close to the recording location or in the case of multiple recording systems, at a central recording station. In any continuous process operation where a product is either being manufactured (such as a food canning process) is passing through stages (such as in the parcel delivery sorting centre for postal or courier delivery distribution) where *barcode identification* (ID) is required (see Chapter 5 Figures 5.27 and 5.28), the *ADC* is being employed to pass the product data onto the company database. If data is transmitted to a digital receiver (like an FM radio channel), the data has been digitised by the ADC. However, if it has to be analogue (at its original continuous frequency form like radio AM channels), we would need to convert it back to analogue using a *D to A converter*, which used to be called a *MODEM*.

Before the smart *3G/4G/5G* networks offered smart digital phone messaging access (before 2000), to send a message to someone we would scan the message from paper, and then dial up to send the scan or *facsimile* of the message (aka *fax*) to an analogue phone line connection. The *MODEM* made a loud raucous warbling noise as it scanned all digital inputs and acted as a *D to A converter* so that digitised data could be transmitted along the analogue phone line in a format that the receiving station could understand (the *MODEM's* raucous warble noise indicated that the telephone line had connected). The *MODEM* provided the analogue signal to the phone line for transmission to a computer somewhere else, which had a *MODEM* acting as an *ADC* converting the signal back to digital. Then the computer's CPU data processor recorded the data and ordered it into a format that could thereafter be easily found, read, and ready for display. The *MODEM* was no longer necessary once we adopted the new generation (*3G/4G/5G*) digital networks (see Chapter 1) as our data was already digital, and therefore of course, we no longer hear the piercing warble of the *MODEM* [22].

When there are multiple data sets for recording (often referred to as data *channels*), they each have their own formats, computer word size (aka *number of bits*), values, and sample rates. These all arrive at the same time (known as *parallel* inputs) as shown in Figure 6.49 in which six data *channels* are to be recorded, apart from other local and remote data. Before recording commences we must feed these data channels into the computer one after another (which is known as *serial format)* so we can record them on disc (see Figure 3.5

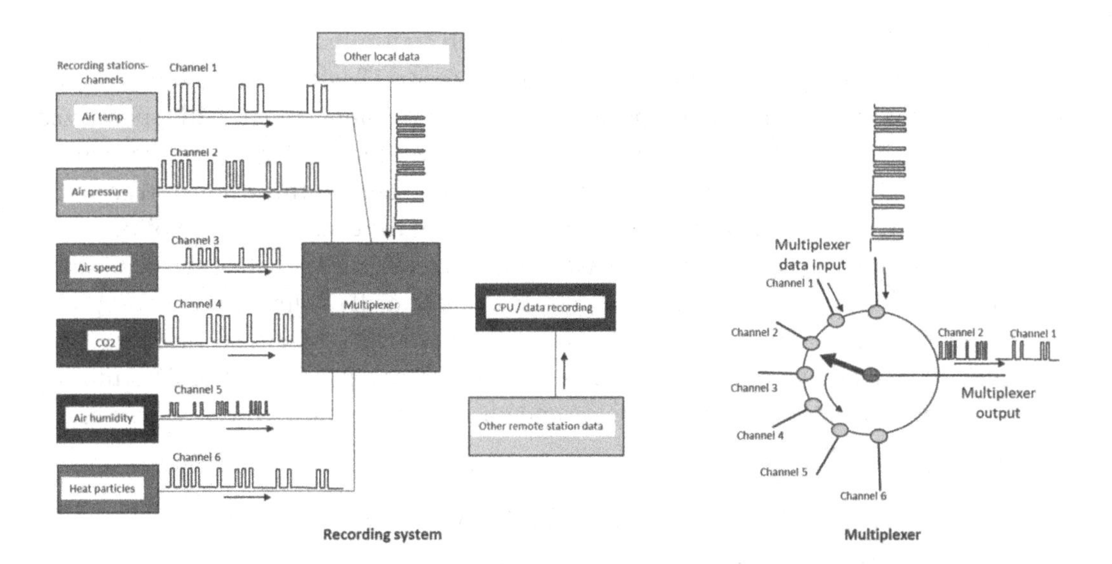

FIGURE 6.49 Recording system with parallel input to multiplexer, and serial output.

which is typical of an environmental monitoring station), and therefore we must change the data from *parallel to serial* format. The device which does this is known as the *multiplexer*. Basically the *multiplexer* is like a clock's hand going around (Figure 6.49), making contact with each channel one after another [(23)], and giving an output of each channel one after another in *serial format*. This is the reason we refer to the computer speed as the *clock speed*, because data is clocked through from the input to the output.

The data is now either transferred via the internet to a computer somewhere in the world (which we refer to as being on *the cloud*), or it is left on a local CPU disc database ready for further local use. Around the world, trillions upon trillions of data sets are being transmitted and stored 24/7 in similar manner to this, and it only stops when an operator interrupts the process or a local power/line connection failure occurs.

So far we have discussed recording large amounts of data at a control centre, which is fine when we want to monitor (observe) the data, but if we want to compute average values and analyse data, which is similar to the predictive analysis discussed in the sense of sports analytics in Chapter 5, we have to have fast computer recording, display and computation of data.

6.3.3 Speed of Modern Control Operations

When we record serial format data, we tend to only need a single PC level of computer. If we want to see a graphical display of the data, our computer needs a special graphics display card, which really is an extra CPU that operates faster than the computer normally needs to be and is used to display only graphics. Consequently, our computer *clock speed* must be sufficiently fast to perform all of these operations at the same time. Typically, today's computer may range from 20 Mbps to 450 Mbps, which means that the slower 20 Mbps would work on any network 3G, 4G, or 5G, whereas the faster 450 Mbps computer becomes limited to using only the 4G and 5G networks, as discussed in Chapter 1 (Table 1.1).

When building an industrial control system which transmits sensor data to a control centre, these computer *clock* factors are taken into account. Basically all components are built into a box that is similar in operation to a normal computer but with specialised software, and then called a *recording/digitiser control unit* (or similar). It will have all of the data acquisition/ADC and multiplexer hardware discussed in this chapter, transmit a stream of data either by optical cable (land line) or *wifi* to the central control point.

The main control centre in any industrial facility therefore receives all of this data at the same time, and is constantly monitoring and checking that all equipment is working well and efficiently, with continuous displays of data and specialised software controlling most reactions. *Data analytics* computer programming becomes a major asset in determining whether a component may fail in the future, and how to optimise any process for the greatest possible efficiency [(24)].

6.3.4 Computer Programs and Toolboxes for Analytics, Python, and *R*

It was explained that in order to analyse sensor data that was being sent to the central control point for recording and analysis, there needed to be software capable of receiving and analysing the data. The computer programming language of choice to perform simple but rapid predictive analysis is called *R* which was released for public use around 1997 at the University of Auckland (New Zealand) by Ross Ihaka and Robert Gentleman (so *R* came from their first names). *R* does most of the basic mathematical manipulations required for data analysis, has good analytical graphics, and its code is used in many smarter languages which provide other software attributes. Sometimes, the simple ability to call up a set of computer lines can make the computations much faster, and, when multiple programs have this capability, they are often referred to as a *toolbox*. *R* has many toolboxes and stock sets of lines that can be used to produce publication quality graphics. It is also one of the easiest data analytical computer packages to learn.

Another common analytical language used by industry is *Python* developed after 2000. It provides a more general approach to analytics and can incorporate much of the *R* code. It is often used for machine learning (see Chapter 5, Figure 5.20) and can combine the data patterns with predictions for display purposes and go some way towards artificial intelligence for future predictions. This software is part of the world of *big data*, which really just means a lot of data flowing to computers (as discussed in the previous section).

In order to manage these programs in industry, there are specialised software packages such as *Anaconda* developed around 2012. This simplifies the management and deployment of *R* and *Python*, and links in with *Windows* and other common software like *Linux* and the *Macintosh Operating System (OS)*. Of particular benefit is the *Graphical User Interface (GUI)* software which allows navigation and applications to be opened (similar to using icons on a smartphone).

6.3.5 How Quantum Computing Will Change the World

We have learnt how to use data analytics to understand how equipment works and to predict how well it will operate in the future. As previously explained, the speed of electronics limits how fast we write data to disc. However, electronic speed is steadily increasing, but

FIGURE 6.50 Supercomputers are simply laptop CPUs stacked together – here they are immersed in oil and stacked on their sides. (Courtesy: DownUnder Geosolutions.)

at a reduced rate. To make the PC run even faster, a simple approach is to stack them one on top of another (without their screens), and immerse them in cold oil circulating around them to reduce generating too much heat (resulting in temperature overload). An example of such a computer is shown in Figure 6.50, in which the left photo is that of a super-computer room at *DownUnder Geosolutions* in Perth Australia, and the stacked laptops (on their sides) are shown immersed in oil in the right photo. Once you have the parallel compute power, the software has to be written to operate in parallel so that segments of computer programs would be fed into these *parallel* computers, and we call it a *supercomputer.* We need to develop something faster than this which does not involve stacking lots of laptops alongside each other.

Physicists have been working on using existing knowledge of electron flow to speed up the process. Consequently, we are now finally entering the stage of what physicists call the *quantum computer*, which I will try to explain as well as I can in simple English.

The word *quantum computing* as opposed to *parallel* or *supercomputing* derives from the use of *quantum* mechanics (*quantum* being Latin for *sum* or *amount*). In Chapter 1, we discussed how during World War II, Dr Turing developed the *Enigma* computer which was used to crack the German military coded messages sent from headquarters to field staff. Subsequently, we developed computers and computer programming (software) to logically solve complex mathematical problems (which we call *classic computers*).

A basic understanding of *quantum mechanics* provides an explanation of how matter is constructed at the submicron (subatomic) level. Using that theory we can adapt our approach to practical applications. We know that an atom is made of positively charged protons and neutrons at its core with negatively charged electrons orbiting around it (Figure 6.51a). The atom is also made of *quarks* which are held together by *Gluons*, which also holds the protons and neutrons together. The electrons and *quarks* create matter particles whereas the *gluons* create force. There are considered to be four forces – EM, gravity, and other strong and weak forces. The photons referred to earlier carry the *EM* force

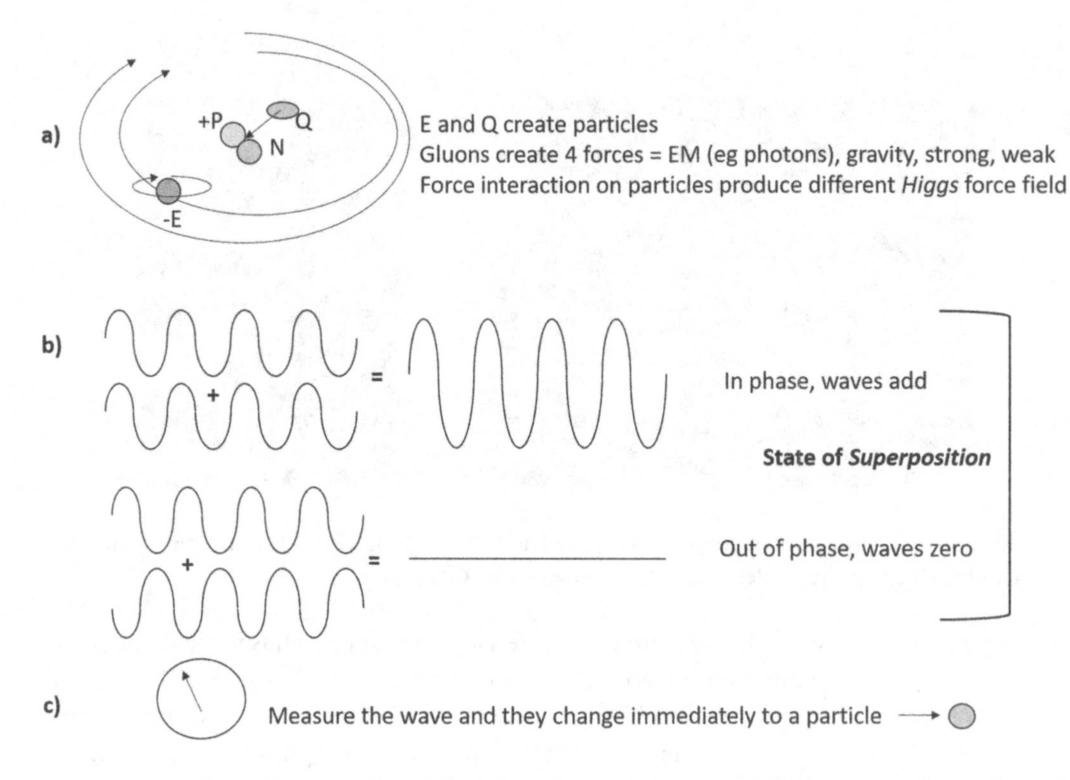

FIGURE 6.51 (a) Quantum mechanics looks at subatomic structures which consist of particles and forces, (b) which have a state of *superposition*, and (c) reverting to a particle when measured during *entanglement*.

because they are produced at *EM* frequencies (as mentioned in Chapter 1). So they can travel as an *EM* wave field.

Electrons store or release energy dependent upon the field they are in. When they change state they can move from one orbit to another around the atom's nucleus [(25)]. This is called a *quantum leap*. Also, two protons can fuse together to produce heat. This theory has already been proven by the development of lasers and integrated circuits.

Particles (like *photons*) can behave like waves (Figure 6.51b). When they meet and are positive (in phase) they can sum or cancel (if out of phase) each other, which is known as *superposition*. In measuring this difference, we may change the values of energy causing the electrons to reduce to particle-like behaviour rather than waves (Figure 6.51c).

This links with Heisenberg theory which says that when we measure something we actually change what we are measuring. For example, when a goniometer measures the speed of a bike wheel, it actually slows the wheel slightly due to its cog using some energy to turn the speed dial.

Electrons can *spin* when acting as a particle. When two particles are together one has a *spin* 'up' while the other has a *spin* 'down' (known as *entanglement*), and can communicate with each other at speeds faster than light (Figure 6.52).

As one particle is entangled with another, and a third particle arrives it links with the first particle. This may change the state of the first particle which automatically changes

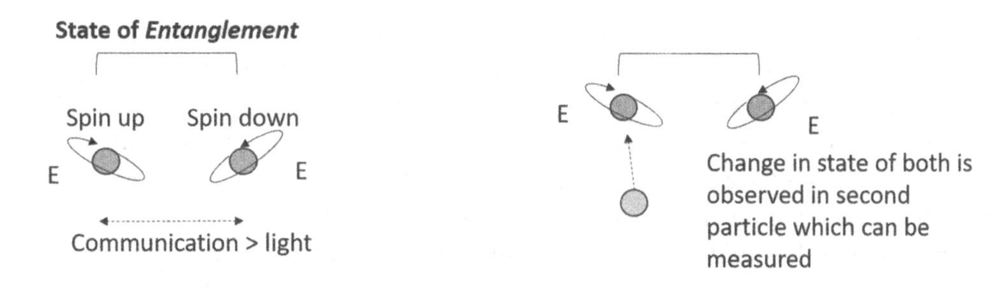

To equate to classic bits, it is easiest to think that the classic bit is the electrical current passing through the ADC over a fixed time at a fixed speed, whereas the qubit is passing itself through the ADC equivalent to a single electron at the speed of light

FIGURE 6.52 Changing state if another particle communicates during *entanglement.*

the state of the second particle. This change can be observed and has an application in teletransportation of particles – by generating a copy (*important note*: this concept is unproven as yet and many physicists argue that this transportation is nonsense).

A *quantum* computing bit ('qubit') is similar to a normal computer (aka *classical computer) bit* we discussed in earlier chapters. With classic computer *bits* as discussed earlier, if a *bit* is present, it is considered to represent a '1' but if not present, it is '0'. With a *qubit,* the bit can be either '1' or '0' since it is *entangled* (linked) with an adjacent particle like an electron which is *spinning up* while the other electron is *spinning down.* So it exists as the value being measured and its *entangled* value. This means that instead of a classical *bit* represented by 1 or 0, a *qubit* is represented by two values in *entangled* form, which can change value if put in a state of *superposition.* A classic *bit* has a form of 1 or 0 while a *qubit* form can change 4 times over.

As soon as we use two *qubits,* the value squares and with the addition of more *qubits,* we keep squaring our speed and computation ability. This means we are able to run numerous *quantum* computations in parallel at the same speed as a single computation each time. The manufacture of such computers currently being produced have around 12 *qubits* with claims that 1000 *qubit* computers are being developed. This compares with our 18 or 24-bit processor which our classic ADC operates today. If 1 *qubit* has 4 states, 2 *qubits* = 16, 3 *qubits* = 16^2 = 256, 4 *qubits* = 2562 = 65k, 5 *qubits* = 4 billion and so on. All this fast speed and computation requires specialist software to be developed, and since *quantum* computing is in its infancy, we don't expect there to be many applications yet since little software has been written.

One problem with *quantum computers* is that they produce small errors in computing results, which are negligible with simple data sets but can build with complex data sets, so we have to run many different computations to reduce errors to as small as possible. This is not an issue when each single computation can be done billions of times faster than classical computers. Consider that we have replaced the microsecond (10^{-6} seconds) computer sample rate *clock speed* in a classic computer with attoseconds (10^{-18} seconds) in a *quantum computer.*

So, instead of the classic computer digitising analogue data into *bits* which are input from six processors (each having a clock speed of microseconds) and then passing them on to a *multiplexer* as we saw earlier in Figure 6.49, we are now passing more than 20,000 processor output *qubits* (with a clock speed of attoseconds) to our *quantum multiplexer.*

A simpler non-complicated way of thinking about it is to consider instead of electrical current (which is a high density group of electrons) passing through the *ADC*, instead a *quantum computer* has individual electrons replacing a current flow of many electrons, passing the electrons individually through the multiplexer at the speed of light. This makes the *quantum computer* unbelievably faster than the classic computer of today – and a reason why observers refuse to make a speed comparison between *quantum* and *classic* speed.

The *quantum computer* output could be to a graphics monitor or TV, and in sports viewing it would be possible to run an animation at what appears to be the same time as the live play, with statistical predictions being displayed during the action. Of course, these animation and statistic data displays happen after live data collection and processing [26], and it is simply that computations are so fast that they appear to the viewer's brain to be displayed in real time.

This opens up the ability to play games and learn a skill with *avatars* as was suggested in the previous section. All manner of real-time computations including screen displays in automobiles and trains, driverless cars, and public transport could be performed. The power requirements of this form of computing would likely be far less per computation than for existing computers. However, a solution will have to be found for the other *quantum computing* problem that is to reduce the heat developed during computation. Maybe cold oil immersion will be adequate.

6.4 EXERCISE

Exercise on using sensors.

We operate a shopping centre which has its own multi-level car-park. Because the car parking lot can only take a limited number of vehicles each with a defined maximum weight, suggest smart methods to detect the number and weight of vehicles as they drive into the parking entry – we don't want to dig up any road or pathways.

FURTHER READING

1. What are rods and cones? https://www.cis.rit.edu/people/faculty/montag/vandplite/pages/chap_9/ch9p1.html
2. The speed of sound in different materials: https://soundproofpanda.com/speed-sound-changes-different-materials/
3. A chemical reaction: https://www.britannica.com/science/chemical-reaction
4. A fuel cell: https://www.britannica.com/technology/fuel-cell
5. What is ultrasound? https://www.healthline.com/health/ultrasound
6. Mechanics of strain: http://www.bu.edu/moss/mechanics-of-materials-strain/
7. The origins of radiation pyrometer: https://www.sciencedirect.com/topics/engineering/radiation-pyrometer

8. Wind versus solar energy: https://alternativeenergysourcesv.com/wind-turbines-versus-solar-panels/
9. How stethoscopes work: https://science.howstuffworks.com/innovation/everyday-innovations/stethoscopes.htm
10. What is tidal energy? https://www.nationalgeographic.org/encyclopedia/tidal-energy/
11. What is a strain gauge? https://www.michsci.com/what-is-a-strain-gauge/
12. A photodiode and its applications: https://www.electronicsforu.com/resources/photodiode-working-applications
13. What are Eddy currents? https://www.magcraft.com/blog/what-are-eddy-currents
14. How to read an electricity meter: https://www.thespruce.com/how-electric-meters-read-power-1152754
15. Three secrets behind the Shinkansen: https://www.tsunagujapan.com/3-secrets-behind-the-wonders-of-the-shinkansen-japanese-bullet-train/
16. GPR and its operation: https://www.sciencedirect.com/topics/materials-science/ground-penetrating-radar
17. How a glycol cooling system works: https://www.towerwater.com/how-does-a-glycol-cooling-system-work/
18. What is ozone? https://www.livescience.com/ozone.html
19. Haptics in the auto world: https://www.precisionmicrodrives.com/vibration-motors/automotive-applications-vibration-motors/
20. Analog to digital conversion: https://www.sciencedirect.com/topics/engineering/analog-to-digital-conversion
21. Computer clock speed: https://www.intel.com/content/www/us/en/gaming/resources/cpu-clock-speed.html
22. What is a modem? https://www.easytechjunkie.com/what-is-a-modem.htm
23. How a multiplexer works: https://www.circuitbasics.com/what-is-a-multiplexer/
24. Optimisation using data analytics:https://journal.hep.com.cn/fem/EN/10.1007/s42524-020-0126-0
25. Properties of the nucleus of an atom: https://nuclear-energy.net/what-is-nuclear-energy/atom/atomic-nucleus
26. Data collection: https://www.qubit.com/data-collection/

CHAPTER 7

Automation and Simulations

Having discussed sensors, how they work and their applications, the next step is to have an understanding of how their output is managed, used, manipulated, and displayed. When we drive a car, we accelerate or decelerate depending on the conditions, and we are *monitoring* the speed instruments (aka *the speedo*) to ensure we stay within speed limits. When an instrument is providing data on a steady process (in this case, *speed*), we can take the data and record it (in our brain), and determine its high and low threshold levels which are not to be exceeded (Figure 7.1). If we do exceed them, we take an appropriate action (hit the brake or press down on the accelerator) thereby applying some form of *gain*. In this system, we have a *feedback loop* (our eyes telling our brain the speed) in order to adjust our speed through the application of our brake or accelerator pedal (to provide *positive* or *negative gain*). In industry when a steady operation is to be performed, we use sensors to monitor the performance and *feedback* the data which provides us with decisions to action. Here, the speed data comes from a gauge we *monitor* – if the speed is not right we change it using a feedback loop applying a feedback gain [(1)].

Sensors normally provide initial data in analogue form. In its simplest form, when we write a word or sentence we can write it in *continuous* flow form (in which letters are joined-up to make the word) often referred to as *running* or *cursive writing* or we write it as *individual letters* – just like this typing. When we read the running writing, our brain interprets this as a word, but our brain also interprets the individual letters of *continuous* as the same word. So when we have a sensor, it gives a set of output data that is most often continuous but if it were provided in separate bits of data it would be easier to handle. In fact can be handled with a computer as we discussed in Chapter 3 when we digitised continuous data to make a prediction, and in Chapters 4 and 5 when we were digitising the vision on a TV screen of a ball bouncing in a cricket or football match.

In this chapter, we take this digitised output and massage it into not only something we can use (like football statistics) but also use it to control an operation. This is what we consider to be the first step of *automation* [(2)].

DOI: 10.1201/9781003108443-7

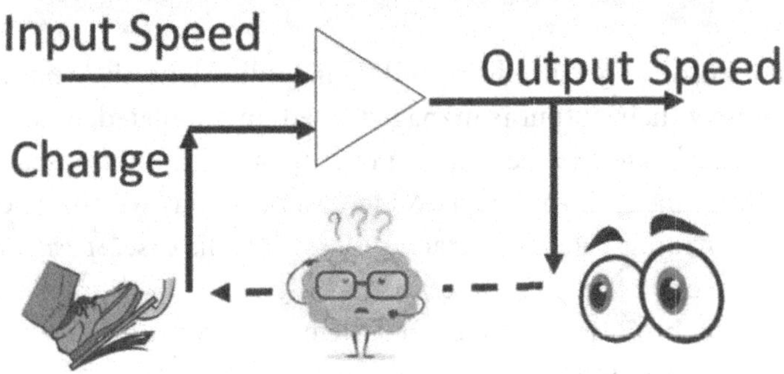

FIGURE 7.1 The simplest feedback control system.

7.1 MONITORING, FEEDBACK LOOPS, AND REMOTE CONTROL

When we have an output of a sensor that is in digital format (see analogue to digital conversion in Chapter 3), we have to make sure the data is unambiguous meaning that it has been sampled and converted from analogue to digital form correctly (see Section 3.1.5) without being aliased (Chapter 1). When receiving or *streaming* data (aka *downloading data* via the internet), after arranging it for display on a screen, we can either use it to control and optimise an operation or process, or we can use it to predict what is going to happen in the near future (Chapter 3).

If we want to use it to control future operations, we need to take the output data and compare it with what is coming in so that we can see what differences there may be and adjust them – this is known as applying a *feedback loop* (aka *recursive filtering*) [3]. Imagine we have a 1.5 volt rechargeable battery and it has gone flat – and we want to recharge it. In Figure 7.2(a), we have the rechargeable battery but it is only giving an output of 0.5 volts. We fit the battery into a recharger which contains a comparator circuit that compares the battery charge with a 1.5 reference charge – which is what we want the battery to recharge to. Figure 7.2(b) shows the comparator as a triangle having two inputs from the battery and its reference. The comparator here subtracts the battery voltage from the reference voltage so that the difference is output.

Then we want to charge the battery so we can input this 1 volt difference (recharging) into the battery in Figure 7.2(c) and after it is charged up it is giving zero volts output from

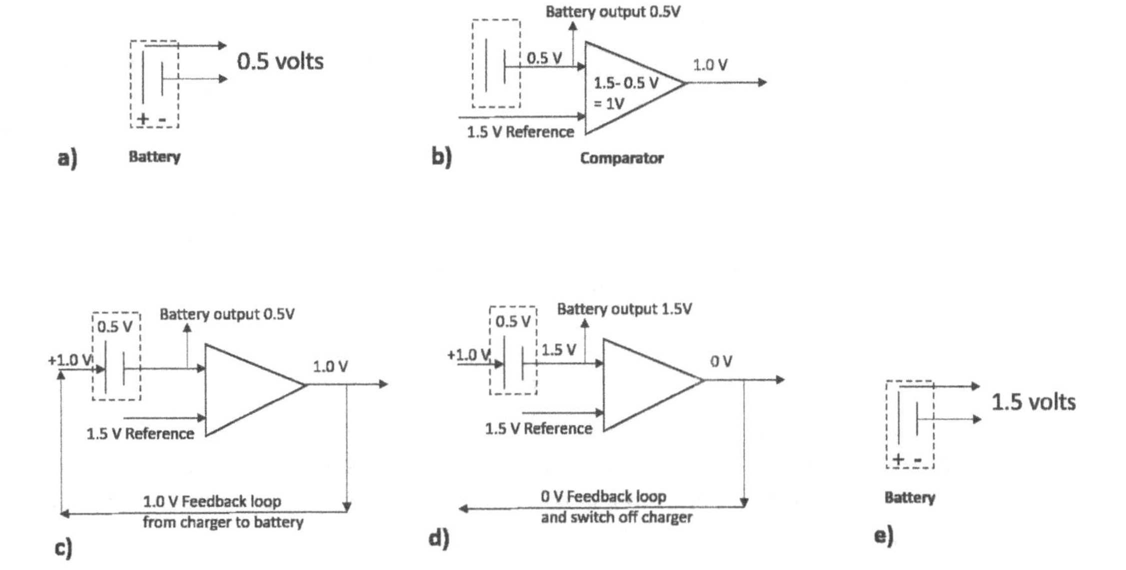

FIGURE 7.2 A rechargeable battery (a) is put into a charger (b), which has a reference voltage in (c) and is charged using a feedback comparator loop (d) to be fully charged again at (e).

the comparator in Figure 7.2(d). Now that the comparator is giving zero output to the input of the battery it is effectively turned off, leaving the battery with an output of 1.5 volts in Figure 7.2(e) and the battery is fully charged, ready to be removed from the charger. Note that this is a form of automation, since the comparator giving a zero output which is feed-back to the input, is effectively automatically turning the charger off, which is just what we want to do when our battery is fully charged (some of a battery's life deteriorates when it is continually charged).

This is a simple explanation of how a battery charger works (i.e. the *operation* of charging a battery), and it demonstrates how we can use output data, compare it with how we want it to be, and then modify it by the use of a *feedback loop.*

The *operation* of the *comparator* in this case is to compare two inputs, and determine the difference, which is effectively subtracting one from the other. This operation is known as the comparator's *transfer function.* The comparator can perform any mathematical operation we want, the most common being to change input signal voltage to be able to read it with some form of measuring device or gauge. Consequently, the triangle here represents an amplifier but could be any other possible circuit that may change the input.

In Figure 7.3, we have an amplifier that takes the *data streamed input signal* and either adds to it to, or multiples it by, a given amount. The amount that is determined at any one time may be set by an external authority. Say for example, that we work in a factory which is canning fruit into aluminium cans. It is very important that the food is poured into the cans at a set temperature, otherwise it may not last very long and deteriorate before it is sold to the food purveyor industry. The data stream is providing us with the temperature in the canning room, and if the temperature reduces (because the food boilers are sometimes susceptible to changes in power) then our *transfer function* is comparing our temperature

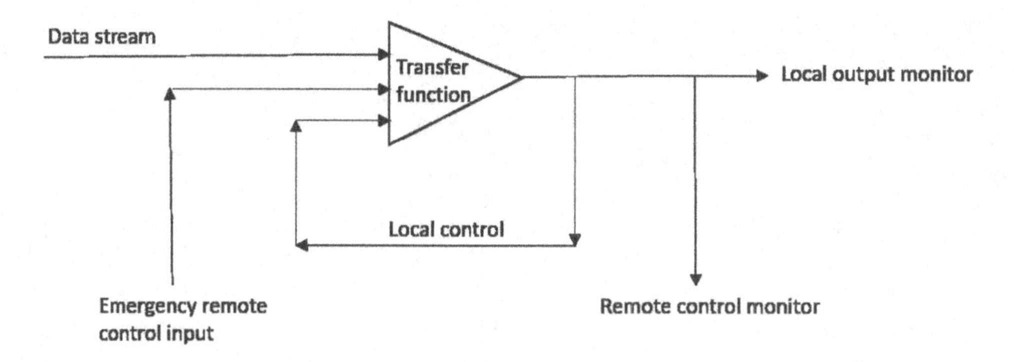

FIGURE 7.3 Local and remote control of operating systems.

data stream with a set temperature we want the food. Then calamity occurs, when we find that one of the power plant air conditioners has frozen up and must go *offline* so we need external information to tell us what to do about temperature.

In this case, the *transfer function* is a simple difference between output and input, and also the transfer function has to accept additional input information to say how to change in an emergency when an air conditioner is about to go offline. This will be the remote control input of additional data into the *transfer function*. Now we have both local control and remote control, operating through a system *transfer function* which caters for the controlling system needs, and we have both local and remote monitoring of the system with a *transfer function* which we use to remotely control the operations, apart from its own abilities to locally control the production operations. These operations can become much more complex than this simple temperature case history, with the *transfer function* including the ability to perform *prediction* of data, and include scheduling for equipment *maintenance*, alarm setting, and so on. However, this is the fundamental operation of *local* versus *remote control* and monitoring.

7.1.1 Sensing Data and Local Control Systems

The operation of a sensor is normally analogue (that is, continuous operation) without a break to record changes in say a voltage level. If we are to use the data as discussed above, it must be digitised which was briefly mentioned in Chapter 3 and discussed in terms of parallel inputs in Section 6.3.2 (Figure 6.49). In that section, we discussed the ability of having parallel sensors each with their own digitising system giving different outputs, requiring a *multiplexer* to give a serial output, perhaps to a central controller.

Consider now the individual sensor and its local control system, which will later be linked with a central control unit. The sensor will have power inputs and an analogue output which is then digitised. There will be an input from a reference point (not necessarily the central control unit), there will be a feedback loop with a *transfer function*, and the output will pass to a *multiplexer*. In Figure 7.4 there are just three sensors working in parallel representing the outputs of temperature, pressure and speed of a process. Each sensor has its own stand-alone unit, with *analogue-to-digital converter* (ADC) [(4)], its own independent feedback loop with *transfer function*, a local digital short-term recording system, and *wifi*

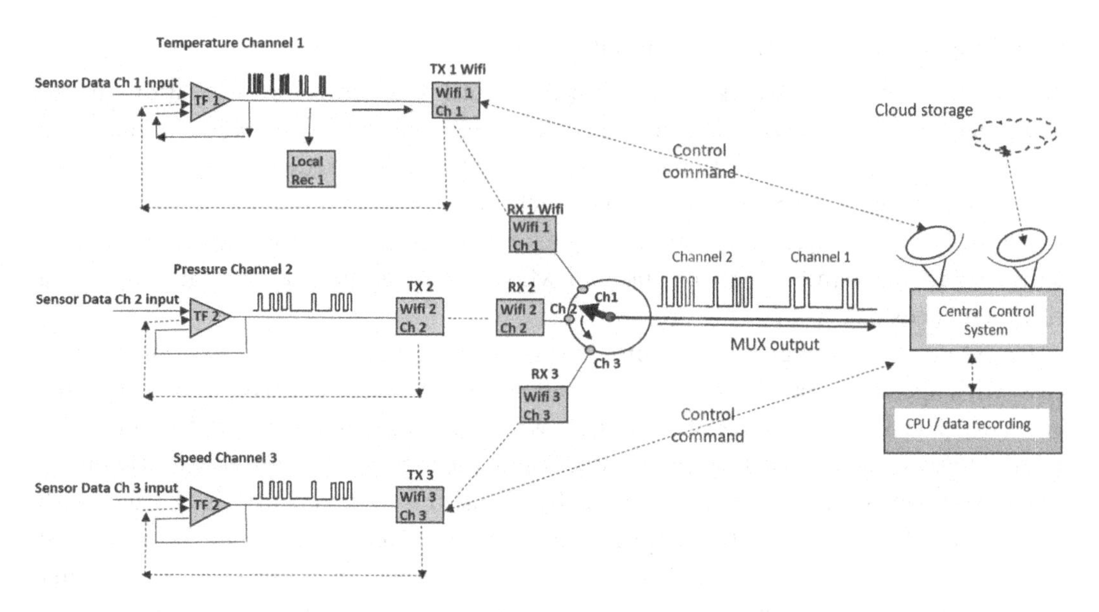

FIGURE 7.4 Local control also requires central control.

to *multiplexer* at the central recording system which can also receive commands from central control. The local system operates independently from all other systems, and transmits data to the remote recorder.

This arrangement allows each independent local system to control its own programmed actions, and also to record both actions and data locally. When required, these actions can be provided to the local *wifi* network, so that central control is aware of actions taken and the history of actions taken locally.

7.1.2 Transmitting and Multiplexing Local Control to Remote Control

The local data will be transmitted to the remote control centre for recording and analysis. Since each sensor may record data locally and be manufactured by different sensor providers, the sensors will have their own data sampling formats and sensitivities. All data must be transmitted via either *Bluetooth* (see Chapter 1) or *wifi*, and must be in a format (that is, in a coded form) that is able to be transmitted and read by a receiver that understands the code. Consequently, after each local *ADC* has digitised their individual data package, it will be reformatted into a coded series ready for *Bluetooth* or *wifi* transmission, and given a channel number (e.g. '1'). The data of first sensor number 1 will then be sent from *wifi* transmitter TX1 and received by *wifi* receiver RX1 somewhere else at the site, probably at the control centre.

This transmission would be sent along with the transmissions from the other two sensors (TX2 and TX3 in this case) by *wifi* to receiver channels RX1, RX2, and RX3 which are then fed into a *multiplexer*. The *multiplexer* reorganises the data changing it from parallel to serial format (see Chapter 6) feeding out Channel 1 data, followed by Channel 2 data and then Channel 3 into the Central Control System. This then passes the data through to the *Central Processer Unit* (CPU) and may be recorded locally.

7.1.3 The Remote Control Centre and Its Operation

The remote control centre exists to monitor, record and control all operations by all equipment 24/7. It consists of computers that are programmed to take the data and examine it to determine if it is of adequate quality as well as predictability. It also feeds the data into real-time monitor screens which provide observers (aka *operators*) with the ability to view active data changes in sensor conditions. The monitors are located in a manner to allow an operator or observer whose job is to interpret the data, to act if a situation needs a decision.

Consequently, some groups of monitors may show data in a specific area of the process, with operators trained in analysing and acting on that data arriving from specific areas of the centre. A number of computers will be used in an advisory role, constantly monitoring the data and displaying trends. While the local sensors are constantly modifying their sensor data to control the local process, the control centre is charged with monitoring all areas, and acting appropriately where action is needed. In the event that one area has a problem that could very easily cascade down into a problem for another area (as was discussed previously where an air conditioner was to be shut down), a decision needs to be taken on what to do with the current problem and what to do if it cascades into other areas. Operators will have been trained in what to do about this – an area known as *Hazard and Operability* (aka *HAZOP*). A *HAZOP* is a systematic examination of a process or operation in order to identify potential problems which may occur during normal operations. Simulations of hazardous problems will have been tested both in a simulation lab and also with the field equipment to ensure that operators are well versed in their response to process problems. Typically operators work in teams, so that if a problem does cascade into another area covered by another team, the other team is ready to take affirmative problem-solving action.

Consequently, when one visits a typical process control centre (which may take the form of a chemical plant control room, a food canning control centre, an airport flight control tower, or a railway monitoring centre), all operators appear cool and calm with little going on according to the casual observer. However, these operators are constantly waiting for signs of a problem and ready to intervene and take the appropriate action.

7.1.4 Recording, Analysing, and Predicting Future Data

Data arrives from the multiplexer ready for display. There may be more than one multiplexer, and typically there may be some 5000 data sets arriving at a chemical process monitoring room at any one time. In Figure 7.4, the data is channelled into the CPU, which separates the data and effectively makes it parallel again, so that individual local sites can be more closely monitored. The CPU will direct some data to one team and other data to another team. It may also record the data on a local computer disc as well as transmit it via *satellite transmitter/receiver dishes* to the *cloud*.

The *cloud* is really just a name given for data that is transmitted via satellite overseas to countries where other satellite receivers accept the data and record it in their local large computer storage centre [5]. So a data set may be transmitted to the *cloud* from Mumbai in India, and actually be recorded and stored from the cloud in Singapore (a well-established site for commercial data storage). In fact, consider where there is a lot of data (aka *big data* as discussed in Chapter 5) that needs heavy computation. Imagine having 5000 data sets being transmitted 24/7, each constantly requiring some form of prediction that only a supercomputer can do.

> "The *cloud* is really just a name given for data which is transmitted via satellite overseas to countries where other satellite receivers accept the data and record it in their local large computer storage centre."

In such a case the data may actually be split up into segments and the computing performed by *cloud computing*, in which case one data set may be computed in Singapore while another data set may be computed in Kuala Lumpur. After computation, the data could be transmitted via the *cloud* to London where it would be recombined back into a form that could be retransmitted back to Mumbai. This is typical of the world of *cloud computing*, and typical of *big data* type data sets handled by these control centres.

The data analysis may include typical predictions of the type discussed in Chapter 4. The ideal model of a Central Control System is to monitor individual sites, combine their operations to ensure individual issues stay with the individual sites and do not cascade into major issues, to record all data whether local or via the cloud, to predict whether any local sites may have issues in the future and to take action to control these sites. This may include transmitting commanding signals to the individual sites to change their operation remotely. For example, Central Control to individual *wifi* receivers, modifying their *feedback loop* to upgrade the reference voltage. This ensures hazards are being avoided to secure a smooth operation of the process being controlled.

7.1.5 Effect on Operating, Repairs, and Maintenance Scheduling

The smooth operation of a process occurs when action is taken to maintain the supply of materials and the maintenance of equipment. Any process that requires the constant mechanical or electrical movement of components is open to normal fatigue through wear and tear of components. A smooth process cannot rely on equipment failure – a motor vehicle break down often occurs due to the lack of preventative maintenance rather than randomly picking up a nail in the street which eventually punctured a tyre. In fact, in some recently manufactured vehicles, a reduction in tyre pressure due to a small air leakage can be detected and the vehicle driver informed of the impending issue.

The use of data for equipment performance prediction has resulted in a major step forward in using the data to establish preventative maintenance programs in many forms of industrial process systems [(6)]. For example, on an offshore production platform, there may be as many as 10,000 production valves operating under different temperatures, pressures and stress states. In the past, it has been hard to perform preventative maintenance, other than replacing valves after a number of years of operation, even though the valves may be operating perfectly well beyond their manufacturer's shelf-life statement.

Because we know the signs of failure of a valve, that is, a valve tends to vibrate or increase in heat, we can position a vibration or thermal sensor alongside the valve. As soon as unexpected vibration (aka *outliers*) or temperature occurs, these sensors will transmit their data to the control centre, which will be able to record the data and commence prediction computing. In this way, instead of replacement of 10,000 valves (most of which may be in fine working order), only those valves which are predicted to fail are automatically replaced. Both conventional and predictive maintenance activities are generally a part of the maintenance operation scheduling of any process control system today. This makes the automation of maintenance activities far more economic than previously [7].

Future predictive maintenance programs will include web searches for the cheapest replacement, equipment availability, warehouse control and all of the attributes of storing and replacement of goods and service technologies previously discussed in Chapter 5. The next step on our automation journey is to understand how process systems operate so that we can link them with suitable control systems.

7.2 AUTOMATED PROCESS CONTROL SYSTEMS, PFDs TO P&IDs

We have seen how a simple feedback loop is used to control the output of any sensor. Each sensor system has its own *transfer function*, *ADC converter*, and digital data output format which is ready for transmission and subsequent monitoring. Each sensor system can also take commands from other external sources, and putting these together produces a complex diagram as has been presented in Figure 7.4.

When we construct an industrial *Process Flow Diagram* (PFD), it can be so complex that we have to minimise the local control details and just have the important components as part of the flow process. Consequently, we construct a *Piping and Instrumentation Diagram* (P&ID) which is a detailed diagram showing the piping and process equipment together with the instrumentation and control devices. This compares with the PFD which indicates the more general flow of plant processes and the relationship between major equipment of an industrial process.

7.2.1 Explaining a Simple Automated System as a PFD

Often, we use two controllers in-line, one after the other, which is known as *cascade control*, so that the *supervisory control* sits alongside the *data acquisition system*, which is given the term a *SCADA system*. In *cascade control*, we are controlling a main process, but we are monitoring its performance as other variables change. For example, a steam-fed water tank has a thermometer at the outlet measuring temperature (Figure 7.5). We need a signal to say 'add more heat' and the steam is turned on or off. Another controller decides when to turn the steam valve on and off as extra water is added or some is removed. This is done at the outlet using a tank temperature secondary controller. Each controller may vary its gain using its *transfer function*, so that the set temperature maintains the steam *valve* or *regulator*. Note the **'X'** symbol for a valve, the **'T'** for valve adjustment control, while the main controller in this figure is an 'oblong-box' rather than a 'triangle' (a triangle most often indicates some form of electrical *gain amplifier*). However, this figure would be the basis of a PFD diagram for the system.

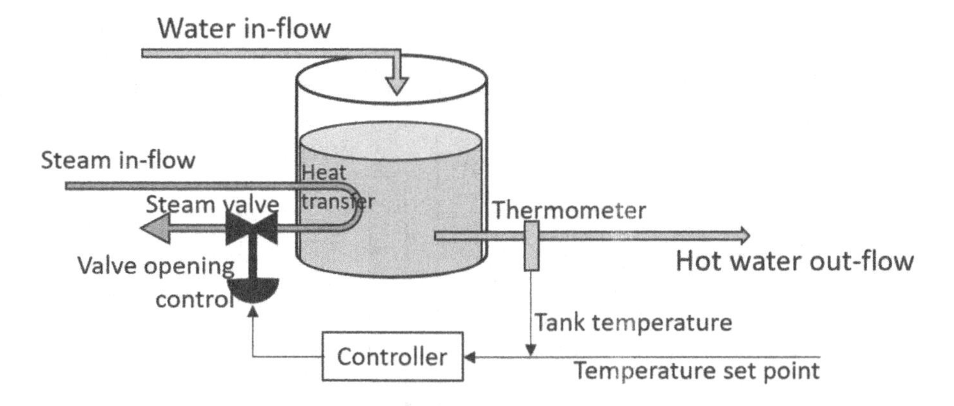

FIGURE 7.5 Cascading controllers (tank temperature controlling steam supply).

7.2.2 P&ID Schematics

In a variation of this example, let's say we want to maintain a water tank (such as a toilet cistern tank) at a fixed water level, with the feed water coming in at the top. A water *level transducer* has a *level transmitter* (LT) which sends a signal to a *Level Controller* (LC) which compares this value with a *reference set point* value of water height. When this is exceeded, the LC sends a signal to the *level control valve* (LCV) which turns on a water pump to drain the tank. If we pump too quickly, the tank may not fill up adequately but if we pump too slowly, the tank might over-fill. This causes oscillation of the system (aka *hunting*) and the water level goes up and down constantly switching the pump on and off. When steady, this system is *tuned*. If we make the input water supply steady (constant), and output is required to change, we have to change the tank level to allow a change in output. So we have to graph these against each other to determine the ideal controller value to allow for all possibilities (including disruption to the water supply).

From Figure 7.6, we can see that there are specialised symbols that are used to represent many different types of valve, controllers, tanks, heat exchangers and so on.

Figure 7.7 provides a simple display of some of these representations, which are typically used by process control engineers around the world. There are a large number of valves each with its own form of operation – as explained previously, a typical offshore gas

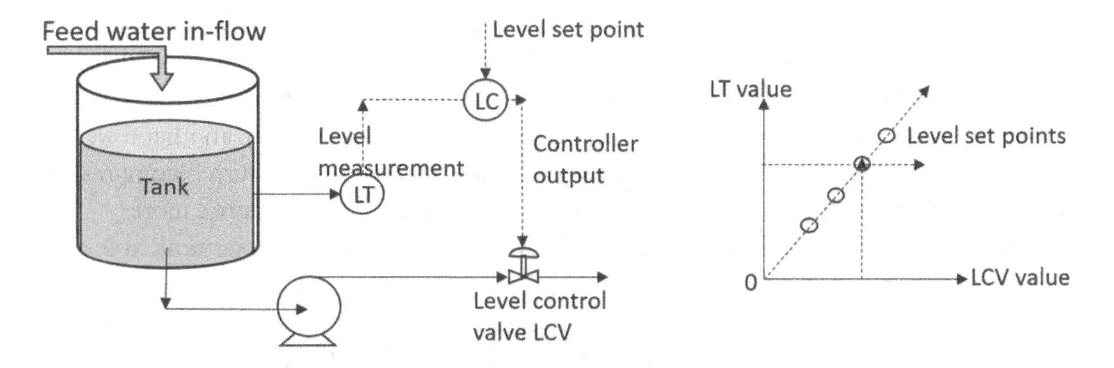

FIGURE 7.6 By adjusting the level control valve, we can maintain a steady tank water level.

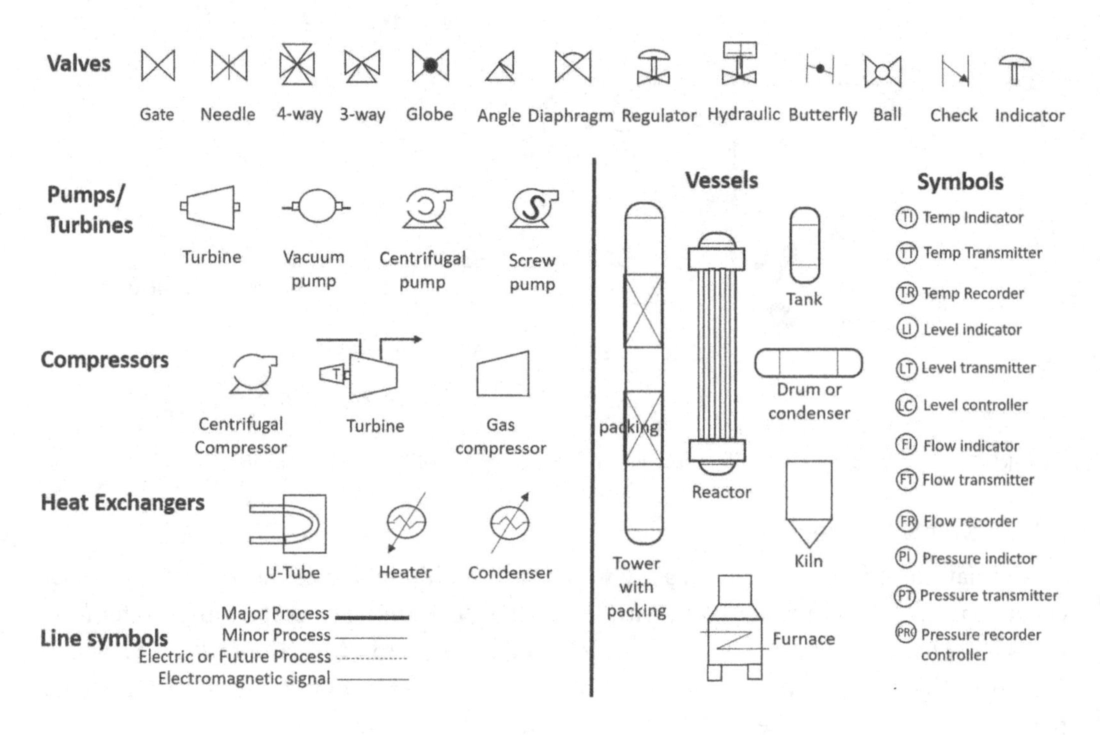

FIGURE 7.7 Typical symbols used in PFDs.

production platform would have around 10,000 valves, and it is very difficult to constantly maintain them without the use of monitoring equipment. That is why we have all of the indicators, data recorders, and transmitters seen on PFDs. The figure also shows some of the more common pumps and turbines, compressors, heat exchangers, and line symbols (which are often used to connect the processes together).

Putting these together, we can produce a PFD for a typical crude oil distillation plant shown in Figure 7.8. Although this appears complex, using the flow diagram it is easily explained by observing that the crude oil is fed into the process on the left, passes through a valve at 1 and is pumped into a heater. The hot crude is now fed into second-from-top of the tall separation column at 2.

Vapour from the top of the column passes through a condenser at 3 (Figure 7.8). The compressed vapour passes into the top of the cooling tower where cooling tower water is pumped and circulated over the internal piping containing the vapour, cooling it into condensate. This condensate then falls due to gravity, and is collected on arrival at the condenser, where it is then fed into a reflux drum, after which can either be pumped as reflux to another tower or to a storage tank (Kerosene tank 1). The uncondensed vapour from the drum is compressed and sent away through piping back to the column at 4 to start the process once more.

Intermediate products from the crude column are pumped into a mixer tank at 5, and the mix is then pumped through a heat exchange reactor at 6 where it cools into a refined oil and exits the process into Oil Tank 2. The heavier oils at the base of the separation column may be split with one part reheated in a boiler and returned to the separator column to recommence the process. The other part will be pumped to Bottoms Tank 3.

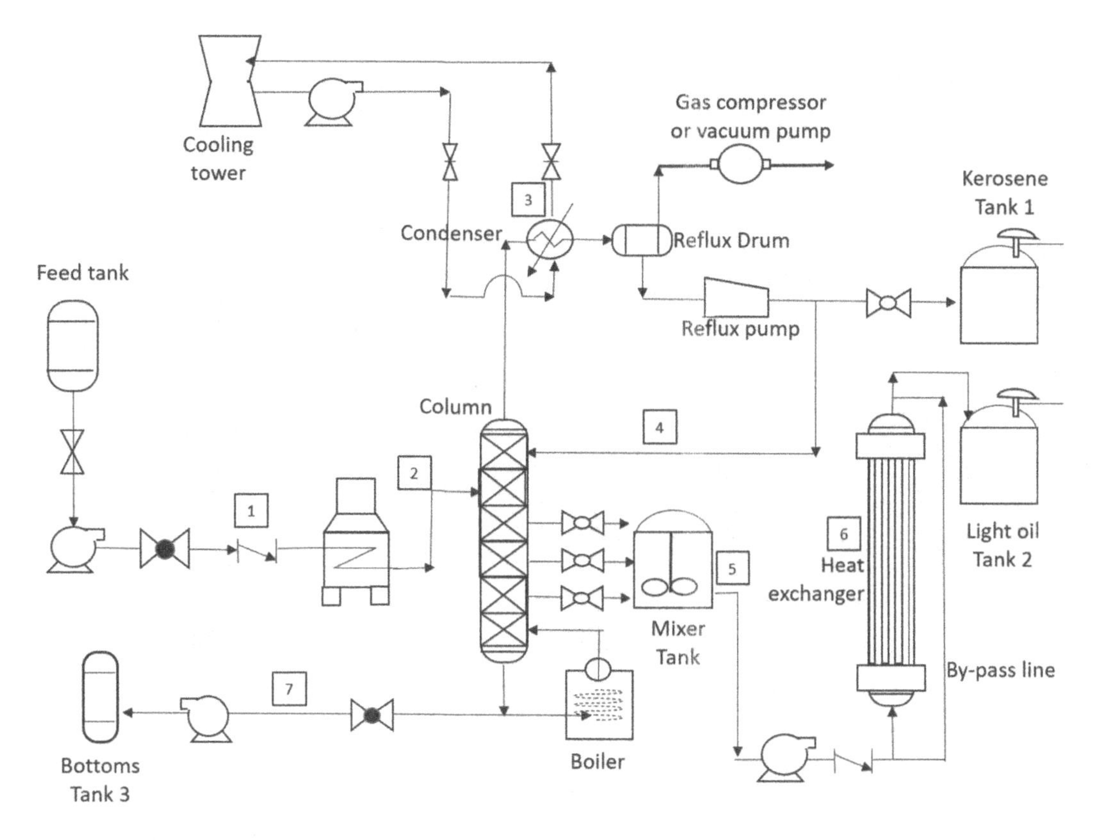

FIGURE 7.8 Process Flow Diagram for a crude oil distillation plant.

This process has now produced a light kerosene which is used for jet fuel, a medium to heavy gas oil and a heavier residue which may be fuel oil. Other products may be produced from the different separation column stages but this is the simplest description that highlights how a crude oil refinery operates, and shows the use of many of the symbols explained in this section.

7.2.3 P&ID Diagrams for Automated Control

The P&ID now repeats this process but inserts the instrumentation as shown in Figure 7.9. It will be noted that where there is a flow control valve, now current and pressure inverter (I/P) sits in the line along with a flow controller (FC) and a flow transmitter (FT) to provide data locally and to the remote control centre. For a simple system such as this, there are now six independent local control systems that have been added.

The valves, tanks, and pumps are now labelled, so that if there is a problem with any pieces of individual equipment during operations, that specific component can be commanded to open or close with a record made of these instructions taken both locally and remotely.

In addition, this diagram will now appear in the main central control centre, where if there is a problem, the instrumentation should pick it up and cause the equipment to flash on and off, to bring the main control centre's operators' attention to the problem [(8)]. This

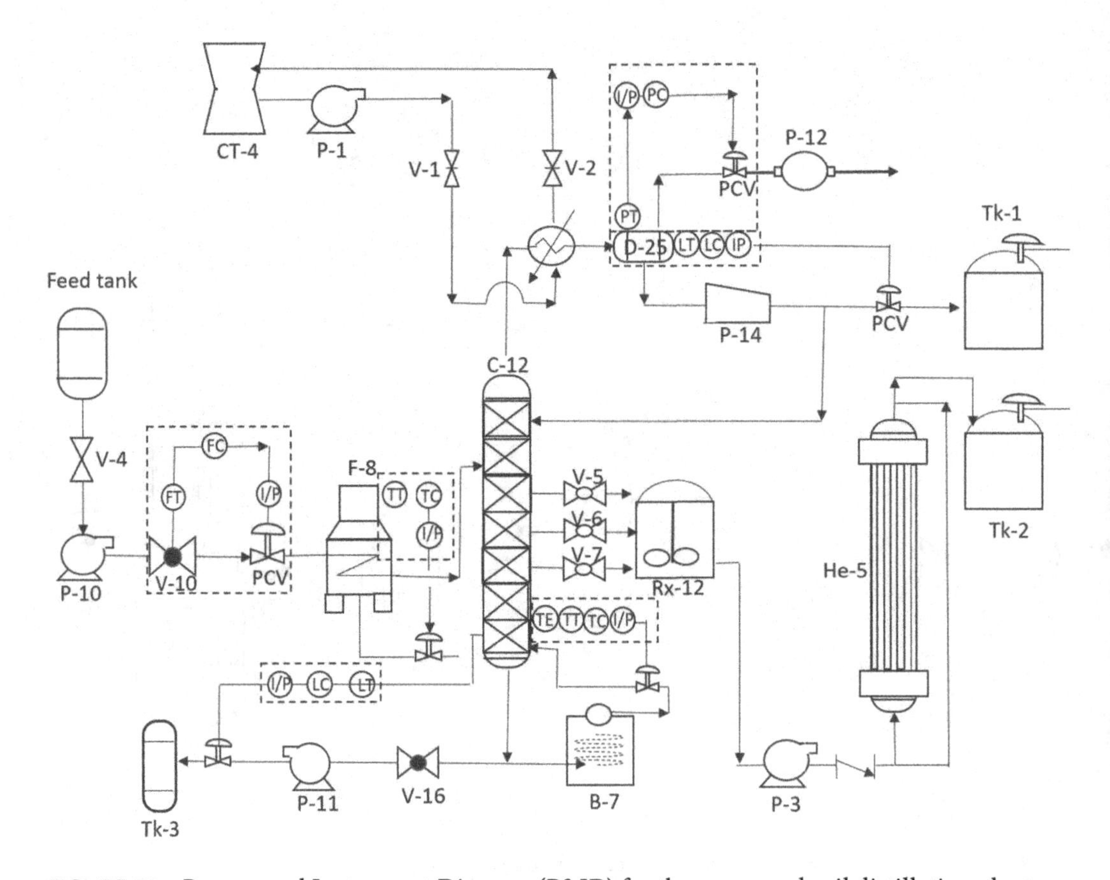

FIGURE 7.9 Process and Instrument Diagram (P&ID) for the same crude oil distillation plant.

will then result in the operator taking remedial action, as well as looking at the history of operation of that component and going through the details of how and why it may have failed. Of course, as in all realistic applications, sometimes it is the monitoring equipment that fails, as opposed to the component being monitored.

When fitting the monitoring instruments, it often becomes necessary to replace a conventional valve with an adjustable regulator valve. This enables remote flow adjustment and indication of issues, as with the regulator taking the normal valve position for the feed to Kerosene Tank 1.

In P&ID diagrams, *level control instruments* shown in Figure 7.10 typically comprise of an *I/P current/pressure inverter*, which is a centrifugal pump driven by an induction motor and pressure transducer that provides pump pressure. Typically pump power is over 1 Kw with 9 amps output for a pressure of 0 to 10 bar (150 psi).

A LC is used for controlling liquid levels in separators and flood evaporators. It controls the valve, increasing or decreasing flow by activating pressure switches, pumps, alarms, and solenoids. When used with motor drivers, pump controllers, and flow controllers can be used for metering flow.

A *Level Transmitter* (LT) is also needed, and there are six different types of level transmitter which are:

I/P current with pressure inverter (courtesy Electroil s.r..l) (I/P)

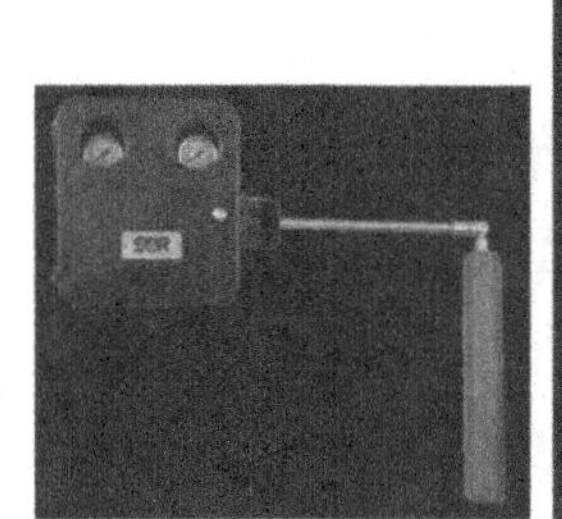

Level Controller LC (courtesy SOR Measurement & Control) (LC)

Level Transmitter LT (LT) (courtesy Burkert Fluid Control Systems)

Pressure Transmitter (Hart) (PT)

Pressure Controller PC (courtesy Parker Hannifin) (PC)

FIGURE 7.10 Different level and pressure instruments.

1. *Radar* – transmitted into the tank, it records the time of the reflection and calculates the fill.
2. *Ultrasonic* – which operates using sonic frequencies in the same manner as radar, but usually cost less with less accuracy.
3. *Guided microwave* – which is an EM signal transmitted down a rod and its reflection returns up to the receiver.
4. *Magnetic type* – which uses a buoyant float containing a magnet in an auxiliary column, and the float movement is measured by a different magnet.
5. *Capacitance type* – in which a non-conductive medium between two electrodes stores the charge.
6. *Hydrostatic* – which measures the pressure of liquid at a fixed point for constant density liquids.

Clearly it becomes horse-for-courses, and much depends on the accuracy required of the level measurement, and the liquid properties being measured, with a few examples of this equipment being displayed in Figure 7.10.

A simple instrument often used is the *pressure controller*. This is similar to a valve, except it is automatic in operation, by using *poppet valves*, *springs*, and *diaphragms* to control output pressure (typical of poppet valves used to control gas in a car engine). Figure 7.11 shows the basic principle of how this works.

In this case, the inlet pressure is required to be reduced with a final manual adjustment if necessary, to produce an outlet pressure. At the inlet, the gas meets a poppet valve which

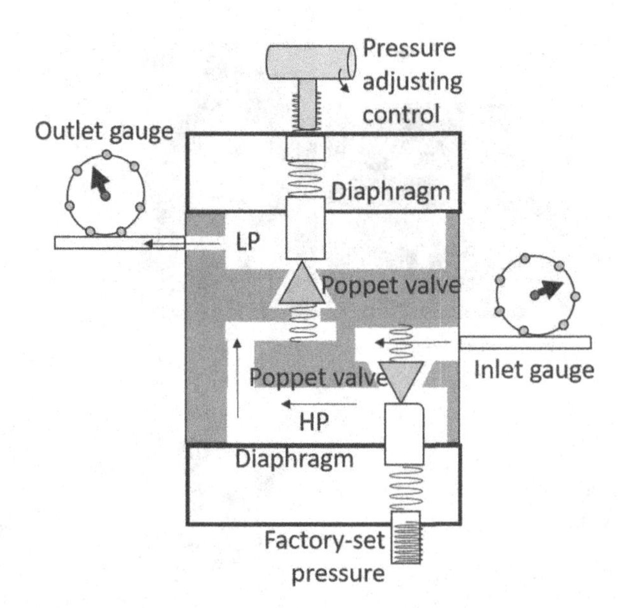

FIGURE 7.11 A pressure controller uses poppet valves, springs, and diaphragm to control pressure.

is a tapered steel rod being seated in a tapered compartment and held in place by a spring at one end and then by a solid support column at the other end which in turn is held on a diaphragm. The position of the support column is set at a fixed point in the factory to allow a predetermined amount of gas pressure past the inlet poppet valve seating.

The gas then passes to an outlet poppet valve which has similar support as the inlet, whereas in this case the pressure is fine-tuned by an external pressure control handle. By adjusting this handle, the outlet pressure can be reduced further.

In its motor vehicle application, a single poppet valve would be seated so that a small amount of petrol vapour is allowed through the valve *seat* – which is the idle setting for the car. Then the pressure adjustment (shown here as a handle) would be the accelerator, which when depressed would open the valve to allow more vapour through the chamber (and the spark plug to have more gas to ignite) allowing the engine speed to accelerate. Flow control valves work in similar manner, using a simple poppet valve construction in which the valve has a handle or is motor-driven into a position across the fluid pathway, to reduce or stop the flow of a gas or liquid.

Figure 7.12 provides examples of some of the commercial flow control valves and flow/temperature transmitters, each with their P&ID symbol alongside them. Consider each of these components being used in Figure 7.9 and the basic concept of local control systems soon becomes a complicated task.

The *Flow Control Valve* is designed to control specific liquids or gas at particular flow rates. The Aalborg *liquid flow controller* will have different operating conditions from the Vogtlin *gas flow controller.*

A *Flow Transmitter* can be *ultrasonic*, *paddle wheel*, *EM*, or *turbine* types and provide data to the recorders. Liquid or gas turbine meters use a turbine wheel to convert flow rates

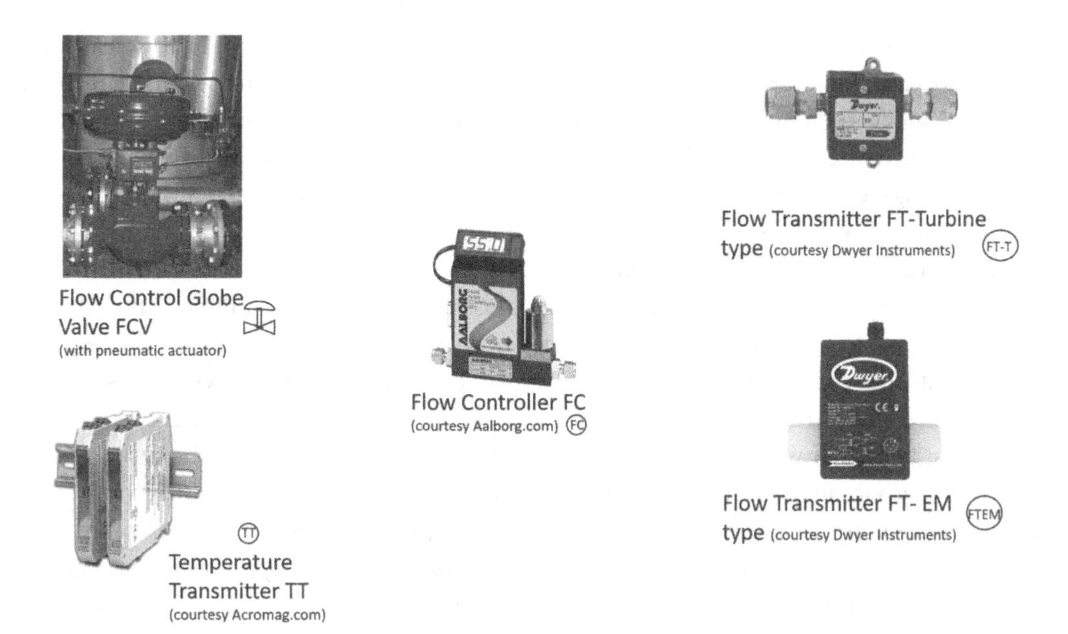

FIGURE 7.12 Typical commercial flow control and transmitters.

to linear DC voltages. The EM sensor type uses conductivity to measure flow rate so that if the fluid has some level of conductivity, its flow will be measured.

7.3 AUTOMATING A CONTINUOUS PROCESS

When automating a continuous process, it is the engineer's job to figure out what technology should be used. There are many technologies available and the reality is that the automation process is all about improving efficiency for increased profit [9]. In some cases, where location of a product is needed, it may be that a simple *infrared* or *laser* beam across a production line conveyor belt may be adequate and cheaper than an *ultrasonic transmitter/receiver* (which provides speed and distance to the product). Then the engineer needs to consider how signals will transmit on the local *Bluetooth/Blu-Ray* or *wifi*, and which of these may require more maintenance – all of these points are a contributor to cost in one form or another.

7.3.1 Issues in Continuous Processing Operations

We have so far considered that automation mainly affects the production line of goods where some form of modification is being made to goods. Examples of this form of continuous production may be in a post office sorting centre, a food manufacturing facility, a pharmaceutical lab, a newspaper mill, a car manufacturing plant, and an oil refinery or a gas processing plant, as we discussed in previous Sections 7.2.2 and 7.2.3.

But continuous processing can be more diverse, such as at open or underground mine sites. In this case, the advent of *new technology* due to the increasing speed of networks has meant the take-up of technologies to increase the production of ore or coal that is mined.

That is, the open-cut mining process consisted of drilling a site, blasting it to release the surface cover, digging out the broken ground and soil, and loading this on a dump truck (aka *haul pack*). The truck then carried the ore and soil around the mine site tracks to a general point where the ore could be washed for loading onto trains or ships.

A major issue was the wear and tear causing constant mechanical failure of trucks which could carry as much as 400 tonnes of ore over short, variably inclined high-dust distances. There was also an issue that the truck drivers were highly paid, and would have to stop work due to any minor truck mechanical issue (including the failure of their truck stereo-music system). Around 2010 (when networks were becoming faster), the open-cut mining industry considered how it may automate its mine site starting with these high-maintenance dump trucks which were the greatest contribution to lost productivity.

The first approach was to look at which part failed most. Often it was the cab air conditioning that vibrated apart (and with no air conditioning the drivers would stop work), and so vibration was the first issue to be solved. After applying various vibration sensors to the trucks, it was recognised that unfortunately vibration would always be a constant problem especially when dumping 400 tonnes on the tray of a truck – but the driver could be replaced instead. So automation in the mining industry began replacing dump trucks with driverless trucks. The issue then became more manageable, in terms of having forward, reverse, and side cameras built into trucks and eventually removing the cab.

In doing so, it meant there were soon fewer drivers, with less health and safety incidents. Trucks no longer required air conditioning but instead required robust electronic control so that they could automatically steer a pre-planned course along twisting mine site roadways. They would operate constantly without interruption except when fuel and cooling waters were required to be replenished. In addition, the ability to monitor truck performance and use the data to predict issues allowed automated predictive maintenance to commence, thereby markedly reducing the downtime of trucks, and of course removing the need for a driver workforce. Instead, an observer in the central control room could take action when a truck was predicted to fail, and the number of failures reduced. Many small mines still operate trucks with drivers, but most of the major, large mines use automated driverless vehicles and generally their profitability is much higher than in previous zero-automation times. So the major issue is – how do I automate a process which is completely or partly manual?

7.3.2 Approach to Automating the Control of a Process – Heath Robinson Approach

The Victorian era and the industrial revolution of the 1800s and early 1900s saw a remarkable increase in factory machinery to help men (and women) perform menial tasks, in order to increase productivity. By the 1920s, there was a growing middle class of people in Europe who worked in factory management. Most of the machinery was mechanical or steam-driven with the introduction of electronics yet to be developed during World War II.

One result was that this new middle class of management in the 1930s often found black humour in sarcasm towards their factory machine practices as well as their politicians. This led to cartoon magazines like *Punch*, which satirised political and lifestyle themes that

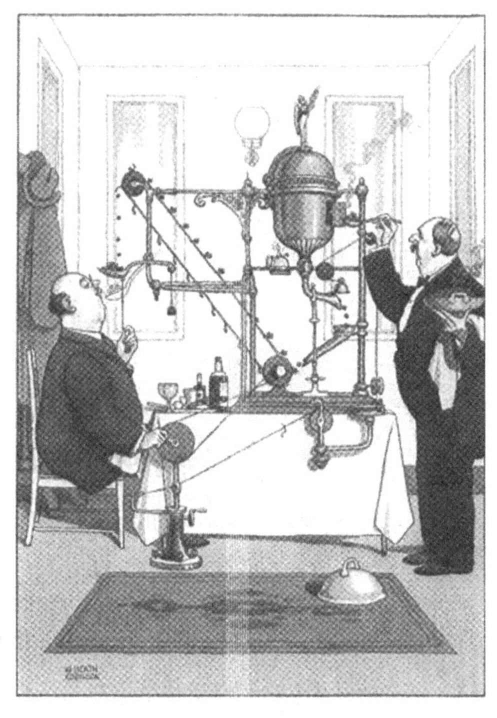

FIGURE 7.13 Photo of Heath Robinson and one of his 'automation' ideas for eating peas. (Original published in *The Sunday Graphic* 1/9/1929 as 'Pea Eating Extraordinary. An interesting and elegant apparatus designed to overcome once and for all the difficulties of conveying green peas to the mouth' - courtesy of G. Beare.)

made people laugh. *Heath Robinson* in the 1920s–1930s, was a newspaper cartoonist who saw fun in automating the most basic realities of life. He developed a series of cartoons that showed simple methods of automation using string, winders, G-clamps, and all manner of everyday articles that could be found in the home.

The main satirical issue of his work of course, was that when he automated anything, it required more labour than the original manual task - and so the take away line was that automation was often more labour-intensive than the practical task being performed. Anyway Figure 7.13 is a photo of the great man at work, with one of his most famous 'early automation' cartoons. In this cartoon, he suggests a method to build an automated system to feed peas from a hot pea tureen into a rather portly gentleman's mouth. Of course, gentlemen of the 1930s had servants and a butler. So he analysed the continuous pea-eating process and drew its required automation with the following important aspects:

1. The pea supply must be fed into a hot hopper so peas are at the right temperature.
2. The pea size must be limited to the maximum number that can be fit on a spoon, and so there must be some form of pea-filter.
3. The peas must be fed into the diner's mouth with a spoon controlled by the diner.

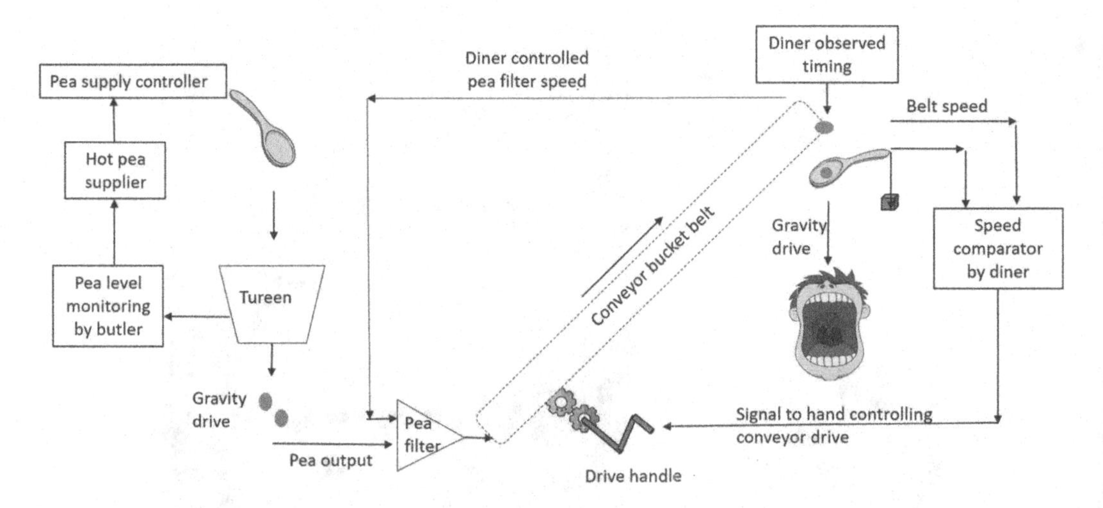

FIGURE 7.14 PFD of automatically eating hot peas – what is missing in the cartoon?

4. There was a limit on peas filling a spoon, since gravity was to be used for tilting the spoon down allowing peas to enter the gentleman diner's mouth.
5. The pea-supply speed must be synchronised with the speed of the conveyor belt buckets.

Viewing the cartoon, we can draw the PFD for it. In Figure 7.14, the PFD shows how the butler arrives with a platter of hot peas, and the butler serves the peas with a spoon into the silver tureen. The amount of peas in the tureen is monitored by the butler so he alters the number of peas depending on the quantity in the tureen, which creates the first feedback loop.

The peas fill the tureen and pass one-by-one down an open pea slide arriving at a conveyor belt hand-driven by the gentleman diner. As peas arrive at the base of the slide ready to transfer to the conveyor belt bucket, the gentleman diner must monitor and adjust the conveyor belt speed to synchronise with the time of arrival of each pea dropping into each bucket. This is the second feedback loop.

That same conveyor belt speed then provides individual peas to a spoon balanced by a weight, in which case the collection of peas must be heavier than the weight for the spoon to drop the pea into the gentleman diner's mouth. The diner therefore monitors the spoon movement in order to catch each pea in his mouth and presumably, chew and swallow it – the third feedback loop.

While this is silly in practice (because in reality the butler might as well feed peas directly into the gentleman diner's mouth), at its publication time in 1930 it was a political comment that the use of a butler to assist gentlemen go about their day was unnecessary and an overrated practice. It would likely fail anyway because it would be very hard to have the gentleman diner synchronise the supply of peas to the conveyor belt of buckets, while monitoring the spoon dropping peas into his mouth. And what if the process was too fast for him to consume all of the peas in time for the next spoon's supply?

7.3.3 Basic Hardware and Software Control Requirements

For most process control operations, there are specialised monitors/controllers and feedback pieces of equipment that have been developed for this purpose, with many being referred to as *condition monitoring instruments*. Such equipment uses all of the technologies mentioned earlier [10].

They must be able to monitor the flow or level determined by their *set-point value*, and then provide a digital signal output to a local recorder if needed, in a format that can also be transmitted to a central control. This signal must be able to be fed into a local comparator so that the level or flow controller can change the process conditions.

The central control must be able to observe all local controllers, and be able to make automated changes as a function of what the optimised process is deemed to be. Central control must also be able to send data to remote recording sites and be able to make predictions of potential system failures. These should be linked with the predictive maintenance program and include planned shut downs for maintenance. Ideally, there should be a plan ready for when major events occur such as major loss of power due to fire or storm.

7.4 THE DIGITAL TWIN (AKA GHOSTING/SHADOWING) AND SIMULATIONS

For many years since the development of computing in the 1970s, industry has sought to simulate a process to allow better understanding of the variables involved. The classic example has been the use of a flight simulator. First known to develop a flight simulator in 1929, US pilot *Edwin Link* built a static aeroplane trainer out of a metal box that rested on inflatable bellows to give the impression of tilt and roll of a plane. It had real instruments and was useful for night flying training when a canopy was placed over the windows.

Eventually it was found useful for night flying in poor weather conditions, and the *US Army Air Force* initially bought six to train pilots in night flying, and 10,000 were developed during World War II training some 500,000 pilots. When commercial pilots began attending lectures in flying a plane with many paying passengers on-board, the airline companies in the 1950s added TV sets for visuals, sound, and improved the movement to produce the flight simulators of today.

Since then with the development of faster computers and networks, it has become possible to simulate many aspects of life, including in the gaming industry which commenced of course by simulating race car driving as a game. But simulations have extended into the simulation of all industrial processes [11].

"it has become possible to simulate many aspects of life, including in the gaming industry which commenced of course by simulating race-car driving as a game. But simulations have extended into the simulation of all industrial processes."

7.4.1 Development of Simulators

During the earlier explanation of how control systems work in this book, we compared the output of some process with the desired (aka *reference or set-point*) output of a device or component. If the output is not correct we use a *feedback loop* and adjust the *transfer function* so that we can control a device or process. This is the fundamental component of local control systems.

We also showed earlier in this chapter that any process can be described using a PFD, and that by adding instrumentation to this, a P&ID can be developed which has all of the instruments needed. But to get to the point of understanding which control system or instruments are needed in preparation for purchasing them, it is necessary to know the working input and output parameters of the instruments so that appropriate transfer functions can be used in them.

For example, consider the case where we have a conveyor belt with jam jars undergoing a process such as putting the label on the outside. In this case, the jam jars may have just had lids put on as they travelled on an earlier conveyor belt, so the belts have to be synchronised to make sure jars passing from one belt to the next do so at the correct speed.

If we know that too many jam jars passing the label fixing point indicates the process is too quick, we can slow the belt speed. We can do this by putting an infrared beam across the belt to count the number of jam jars passing over a defined period in time. We can then change the belt speed so that the jam jars are moving at the correct rate.

Using a PFD, we can have inputs to a comparator of:

1. The previous lid-placing jam jars' belt speed and
2. A second input of the number of jars passing a point on the label conveyor belt, with the difference being related to the conveyor belt speed mismatch.

This comparator mismatch in speed is not fed directly to change the speed of the jar label conveyor belt, because the speeds of the two belts may be quite different (due to the different operations performed on the jars). There is also the fact that a belt cannot go too fast (otherwise the jars may fall off a corner if the belt goes around a bend) and so a maximum speed must be set.

The P&ID will therefore have an infrared beam across the belt which when broken gives a count output to the comparator which relates to speed for comparison with the previous jar lid belt speed (see Figure 7.15). The *comparator* has a *transfer function* that takes the difference between the two belt speeds and multiplies this by a speed factor to adjust the belt speed. The output from this *comparator* then goes to an *amplifier* which increases the belt drive voltage to a level the motors need, and also determines if the speed is greater than the maximum speed allowed. If so, the amplifier reduces it to the maximum set speed value. The output then goes to the jar label conveyor belt to adjust its speed.

We can simulate this process with a computer program that has the belt speed values differenced, followed by a multiplying factor to adjust the speed. This difference is compared with a maximum belt speed and the final difference is used to either leave the speed

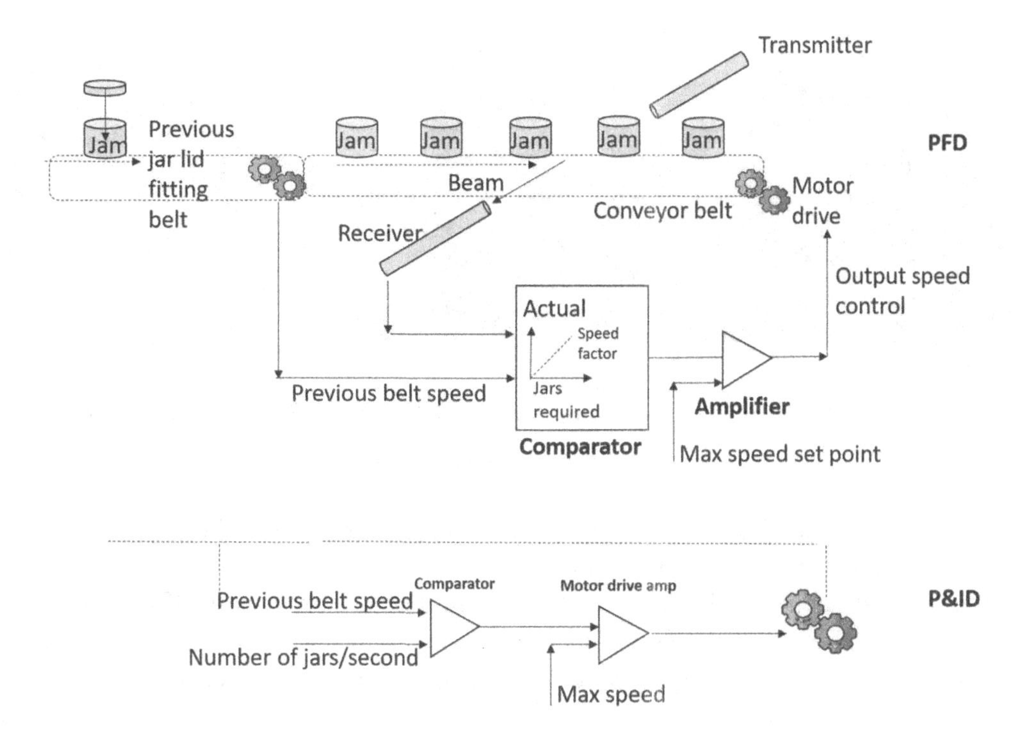

FIGURE 7.15 PFD and P&ID for conveyor belt control.

where it is or reduce it to the maximum speed value. The advantage in doing this computer simulation is that we can feed into the *comparator* any values we wish while maintaining the maximum speed value, thereby simulating the process. We can put the process tools (comparator and amplifier) on the screen as an emoji representing the component. This becomes very useful in hazard detection, and is a major tool used in HAZOP operations (see Section 7.1.4).

7.4.2 Numerical Simulation

Having the P&ID control diagram on a screen allows the observer to be a devil's advocate and input values that may be beyond the range expected, to observe how these change the functionality of the process control system. Various alarms may be written into the computer program so that components that are being driven beyond their design criteria may flash with different colours to show their status, or they may change size in order to bring the observer's attention to them.

The point of the simulation is to act in the same manner as the control system equipment, so that HAZOP trials can be done, and people can be trained in what to do in unexpected situations, just had originally been intended by Link's *Trainer* aeroplane. A typical simulation training session is shown in Figure 7.16 where a photo shows an operator class being trained on how to handle fires and gas leaks in an offshore gas processing plant. An interesting point here is that the instructor is seen on the screens at the far end of the room (and they are in Aberdeen, UK), while the students are in Perth, Australia some 12,000 miles away (truly remote operations). The figure on the right is a screenshot of the

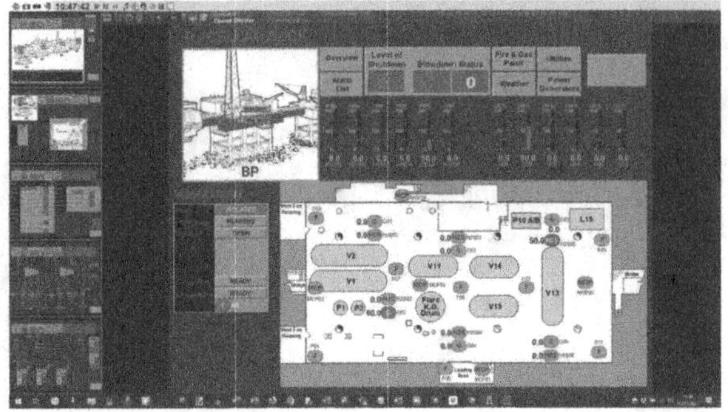

FIGURE 7.16 Photo of simulation training session and screen offshore platform simulation. (Courtesy: PiSys Co.)

simulation in which a fire is set by the instructor (in Aberdeen) and the students have to respond in the correct manner (in Perth).

This HAZOP simulation has a fire starting on one deck of a multi-deck offshore platform (controlled by the instructor). As a number of alarms sound, some of the tanks or valves overheat and flash to show they are on fire and it is spreading. The trainee operators have to ignore the alarms and work in a controlled manner to not only put the fires out, but also to call management to determine whether some parts of the rig should be evacuated. Once the instructor is satisfied that the trainee operators can handle the pressure with a steady and logical temperament while making the correct decisions, they are allowed to go to the rig.

Simulations like this continue to be run internationally in all areas of industry and when it is considered how many permanently positioned offshore platforms there are in the world (around 800), it is clear that process simulation for HAZOP control is a major industry.

7.4.3 Running a Simulation in Parallel with a Process System – The Digital Twin

It has been explained how running a simulation of a process operation during a HAZOP can help simulate the process with a failure involved. Instead, if we now input values of the real-time processes into the HAZOP simulator, we could then run both the simulator and the real-life process operation at the same time. This is known as the *Digital Twin* because it is running the process in parallel with the real-life process, using all of the inputs and giving outputs the same as the real-life system. It would of course be an expectation that any failures in the system would then occur at the same time as its *digital twin* simulator, but the simulator does not include failures of hardware. Consequently, we need to build in failures into the HAZOP simulator.

If we now compare each individual output data-set of the real-life system with each individual output of the simulator, we can use the simulator output as a reference compared with the actual output data, to constantly check if the system is performing correctly. For example in Figure 7.17, vapour is leaving the top of our oil separation column. It enters a condenser which provides sufficient cooling to allow it to condense any vapour content by heat exchange in a cooling tower, so that the pure vapour is now in a *condensate* form (a

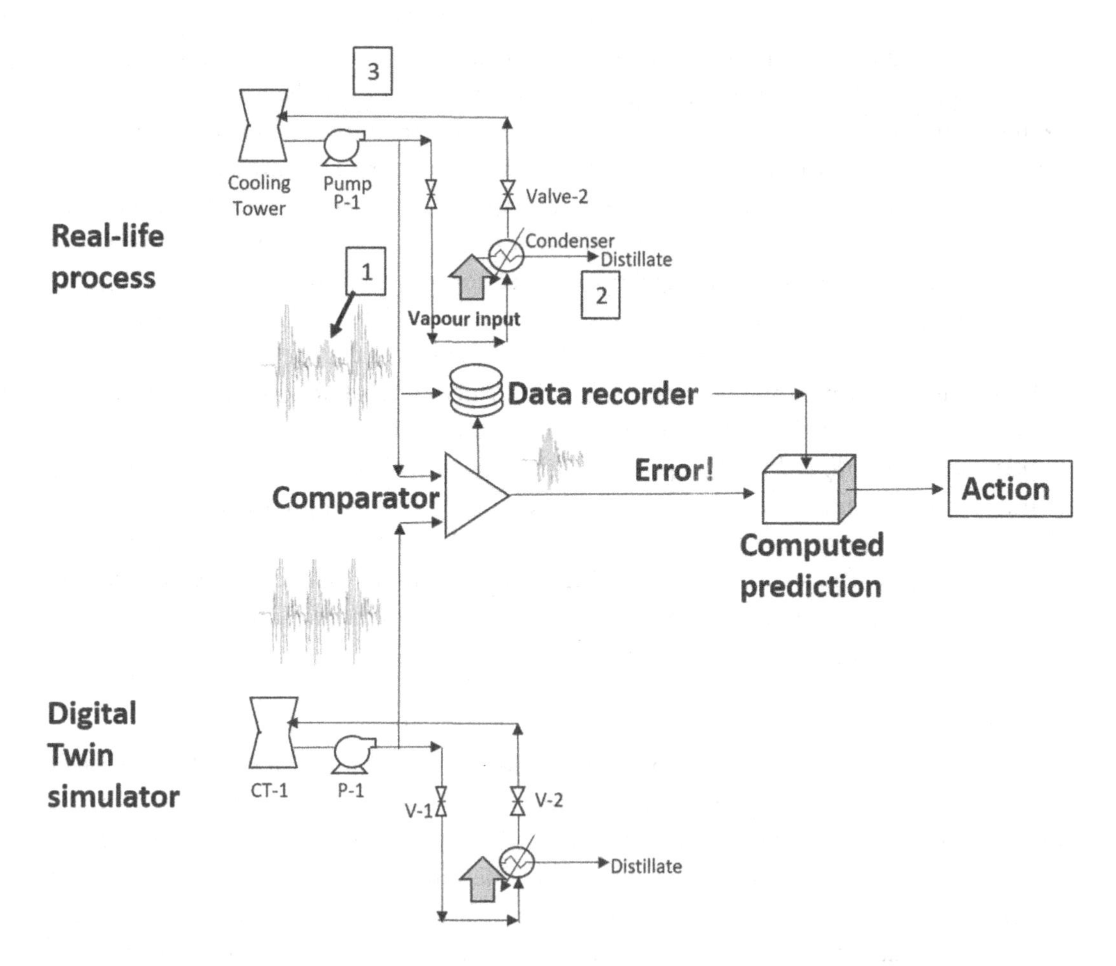

FIGURE 7.17 Real-life process compared with digital twin simulator gives an error output which compared with history of these errors, causes an action.

fine fuel similar to kerosene) product which can be further refined before collection and storage as a product of the process.

If the supply from the cooling tower pump becomes inconsistent (pump P-1 works well for a short time, then stutters reducing output before running well again) as shown in the data set in the figure at point 1, this could cause the condenser to inadequately cool the input vapour so that the distillate output at point 2 would be an erratic mixture of vapour and distillate, which would cause problems for the processes downstream of point 2, as well as inlet to the cooling tower at point 3. Consequently, we can expect that the outputs from the condenser would change also if the condenser is not adjusted by an action from the monitoring control centre.

In order to solve this problem, we feed the erroneous pump P-1 signal into a comparator circuit, and compare this with the reference signal from the digital twin simulator. The comparator observes there is a difference between the two input data sets and flags an error to the operator monitoring this (in practice, this may be observed as the P&ID pump flashing on the screen along with an alarming beeping noise). This error output from the comparator is fed into a computer which is programmed with the action to perform in the event that this particular pump P-1 fails, such as changing the cooling level provided by the condenser.

The pump data is also recorded, so the computer now takes past recordings and cross-correlates the data with previously recorded data (see *pattern recognition* Chapter 3) to determine if this latest failure has been systematic over the last week, month or year. Consequently, the computer makes a prediction of how random or consistent this failure has been, and then informs the control centre operator if the pump needs to be replaced.

Some smart computer programs will even start to search the local company warehouse to see if there is a replacement pump, or the internet to find the cost of a replacement pump and their delivery times. The operator is then informed of what actions to take from this point, and how serious this failure could be for the system overall.

This now is full *digital twinning* and automation, in which a component is gradually failing and the system has taken action as a result of a data-based prediction. This *predictive maintenance* has the beneficial effect of reducing downtime and costs. Many continuous processes have this capability, but in reality there are more that do not, which rely too much on operator decision making. If the operator does not have all of the facts, then a correct decision may be rare and in fact, not possible to make.

7.4.4 Blockchain and Its Use in Administering Developments

Blockchain is a trendy word which most people don't understand. It derives from the fact that there are lots of small zones or blocks which when used together in a process, can be considered to be a *chain of blocks. Blockchain* is software linked to the internet, but can be considered as a simple financial spreadsheet in which the final total profit/loss statement is a result of all of the different cells in a spreadsheet adding to get the result. Each of these cells is located at a different position in the spreadsheet and does not influence another. Access to the spreadsheet is over the internet, but access can only add new information and therefore everything is recorded on the spreadsheet without correction.

Consequently, this notion of the inability of any one cell to influence another becomes useful in the financial world, in which having separate blocks each with a history unable to be changed by any other outside interference, leads to improved transparency of a transaction's history, and therefore presumably reduced corruption in the spreadsheet (i.e. business corruption). It has become a possible mechanism which allows transparent transactions for financial systems in banks, casinos and share trading of *cryptocurrency* (computer-based digital money, which normally has no record of money movement/ownership and is sometimes used in illegal operations and money laundering).

In our continuous process of Figure 7.9, we have lots of small confined areas where we gather data and when compared with our *digital twin* simulator we take an action if there is an issue. The data in any confined area such as that of Figure 7.17 can be considered as a single *block*, with the continuous flow of data having the output of one data set leading on to another (as in our simulator). This data flow then becomes the *chain of blocks* and we could consider the process flow of data as a *blockchain* process [12].

Industry has not yet taken full advantage of what this means for the continuous process system, but in reality, when ordering replacement components, *blockchain* concepts can kick in when verifying whether replacement components are of equal quality to those that are failing, when the replacements were manufactured, from what materials, and so on.

The dumping in Europe and the United States of poor-quality cheap steel manufactured in China during the late 2000s, caused many issues with industrial replacement components failing (such as valves that failed by jamming when undergoing temperature changes). The use of blockchain registers or warehouse accounting of replacement component history may in many cases, have avoided this issue.

7.4.5 Cyber Issues

A brief comment here about cyber issues, in which foreign software from the internet can maliciously replace our normal operating software, which then causes our control hardware to do different things from what it would normally do. This form of malicious software is known as *malware*. It is reality that we normally record and store all of our data and its controlling software locally on computers or in the *cloud* (i.e. somewhere else). If someone hacks into our computer system and replaces the controlling software with their own *malware*, it is possible that the operation of any autonomous system may change which can then cause catastrophic results.

Alternatively, *malware* may be used to monitor the performance of some service and at an appropriate time, to change its operation. In extreme cases, such *malware* has found its way into US nuclear facilities and power systems. Russia, China, and Iran are often accused of having teams of programmers who do this constantly for their own advantage, trying to break into other country's systems where disruption can cause local havoc. One wonders how China's *Huawei 5G* networking company managed to develop with little home-grown electronics development industry (or patents in the area) yet they became a world-scale 5G manufacturing and distribution company.

There is a new industry in cyber security that is intent on stopping the entry of *malware*, which employs programmers to code up defensive software (which recognises the malware and deletes it immediately), but sometimes it seems they can't keep up – *malware* finds many forms. So while we sit and watch TV, be aware that somewhere in the world, *malware* is trying to shut some power supply down or cause false alarms which are not being openly reported. It is all cloak-and-dagger stuff.

7.5 EXERCISE

Exercise in control systems.

In the crude distillation plant P&ID of Figure 7.18, valve V-1 is an isolation valve controlling the flow rate of the distillate from the cooling tower CT-4, which is arriving from the exchanger at a steady flow and pressure.

However, it starts arriving in short, sudden bursts at peak pressures twice as high as average. The system can handle this occasionally, but we need to determine why this is happening, and put a control system in place to warn us in future as well as fix the failing pump or valve.

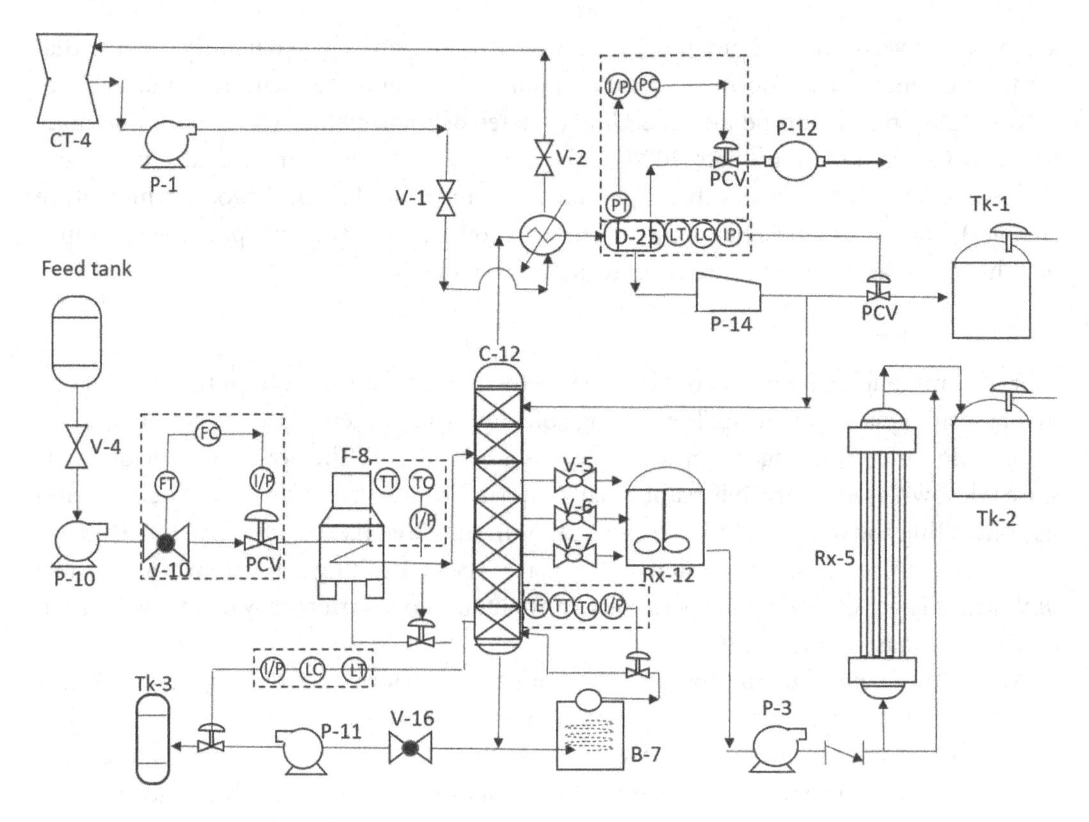

FIGURE 7.18 Exercise – P&ID of crude distillation plant.

What control system do we need to put in place, in order to monitor and warn of impending issues, such as the failure of pump P-1 or valve V-1?

FURTHER READING

1. Feedback systems – what are they?: https://www.electronics-tutorials.ws/systems/feedback-systems.html
2. What is automation?: https://www.ibm.com/topics/automation
3. Recursive filters: https://www.analog.com/media/en/technical-documentation/dsp-book/dsp_book_Ch19.pdf
4. ADC conversion: https://splice.com/blog/analog-to-digital-conversion/
5. What is the cloud?: https://www.cloudflare.com/en-gb/learning/cloud/what-is-the-cloud/
6. Maintenance: https://en.wikipedia.org/wiki/Maintenance_(technical)#Preventive_maintenance
7. Predictive maintenance: https://www.onupkeep.com/learning/maintenance-types/predictive-maintenance
8. Control rooms: https://www.consoleconcepts.com.au/index.html
9. Automation economics: https://www.economicshelp.org/blog/25163/economics/automation/
10. Condition monitoring instruments: https://www.erbessd-instruments.com/articles/condition-monitoring/
11. Simulation of industrial processes: https://wwtopia.com.au/simulation-of-industrial-processes-for-control-engineers-philip-j-thomas/book/9780750641616.htmlw.book
12. Blockchain applications: https://blockgeeks.com/guides/blockchain-applications/

CHAPTER 8

Technology of Household Appliances

This chapter discusses all of the technologies found in household appliances, with the exception of the *light emitting diode* (LED) which can be found in the Introduction Chapter 1, since it has now become the most basic component of modern-day living. When we consider the kitchen, there is now a plethora of devices most of which are controlled by electronics and micro-computing devices, and most of which were developed during the affluent technological era after World War II.

8.1 THE MICROWAVE OVEN

Developed initially by Mr Spencer of *Raytheon* in 1946 for heating food in restaurants and on planes but far too expensive for everyday consumers, it was commercialised by *Sharp* in 1970. The microwave oven is ubiquitous and many households could not live without it in the kitchen. But few people realise how it operates and the level of technology it takes to produce microwaves. First to the basics – microwaves are transmitted into open space, and bounce through it vibrating electrons that are spaced apart particularly in liquid, and this heats the food. See Figure 8.1.

Now, let's see how this is done!

- A *step-up transformer* takes mains voltage and steps up to 4 KV input to a magnetic tube called a 'magnetron'.
- The *Magnetron* filament is heated by voltage and outputs electrons whirled around by two ring magnets, causing microwaves to be liberated at around 2.45 GHz. The magnetron consists of a cathode (which is like a rod running down the centre) with the anode around it. Putting a magnet underneath causes a magnetic field to be generated in the space between the anode and cathode. Applying high voltage to the cathode causes electrons to try to jump the gap to the anode, but the magnetic field deflects them so the electrons rotate in the space. Now make slots in the circular anode of

DOI: 10.1201/9781003108443-8

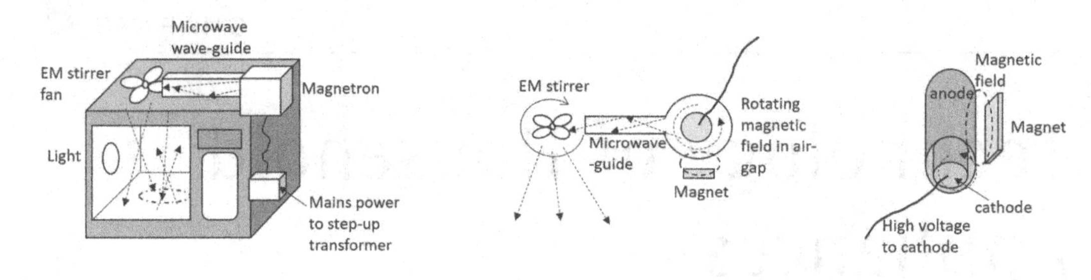

FIGURE 8.1 Microwaves are produced by a magnetron and radiated into food compartment.

particular width and they vibrate with EM microwaves (this is similar to someone blowing into a flute, vibrating the flute's internal tubing to produce the sound when the tube acts as a sound waveguide). The slots are connected to an antenna which is an *electromagnetic (EM) waveguide* mounted over a box.

- The microwaves pass along this waveguide antenna built into box top and meet a rotating fan that deflects them around the box like a stirrer (1).
- A *hot spot* can occur because reflected microwaves may add together (aka *superposition*) but others subtract causing *cool spots*. Because we want our food heated evenly, we put the food on a rotating turntable so it gets an even amount of microwave energy all around.
- The wire mesh across the door has small holes to allow internal light to shine out so that the object being cooked can be seen but no microwaves can get out because the mesh gaps are too small.
- If we put a metal cup in the microwave, the arriving electrons will try to set up a circular *Eddy current* in the metal walls or metallic strips and so the contents of the cup will not receive the electrons, and so will not heat. If we put a coating of metal down the cup's handle it may spark from the top of it just like lightning strikes trees (known as *arcing*) that are higher than ground level. So the strong suggestion is made that there is no point in putting metallic cups or even designer plates that have metallic surfaces or strips in the microwave – because the electrons won't heat any liquid or food. It may be possible to use metallic vessels coated with some form of insulation over the metal, but over time that may wear out and expose the metal again to *arcing*.

8.2 THE REFRIGERATOR AND FREEZER

Equal with the microwave, the kitchen refrigerator and freezers are ubiquitous, with some households having one of each while others have them combined (which limits how much food may be stored for future consumption). See Figure 8.2.

Here is how their background and operation:

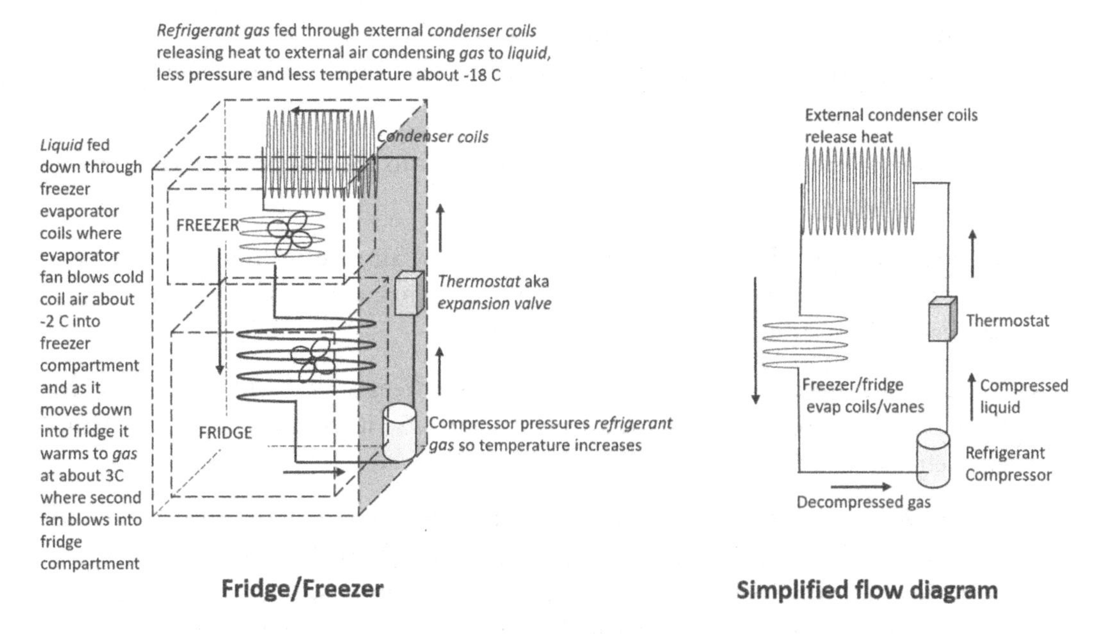

FIGURE 8.2 The fridge/freezer works by compressing/expanding refrigerant gas to liquid and back.

- *Frigus* in Latin means *cold*. Re-frigus means to re-cool, so a *refrigerator* keeps objects (food) cool. Developed in the early 1900s, the *refrigerator* word was shortened by speech to *fridge* with a '*d*' so that the word *fridge* would sound the same in common English when people learnt to write (otherwise it would be pronounced as *frig*).
- The refrigerator is used typically to store fresh food (and beer) between 0°C and 3°C. Freezers are set at -18°C while chillers are close to 0°C. The milk and soft drink compartment may be up to 8°C. Soft drinks like colas like to be at 3°C so that they can be held at their bubble point to allow them to fizz when opened. Any colder or hotter and there may not be maximum fizz.
- When refrigerant gas is compressed by a compressor to high pressure and it rises in temperature (like the hand pump when pumping up a flat bicycle tyre). This pressurises the refrigerant gas into a liquid which is fed through an expansion valve (aka *thermostat).* The *thermostat* controls how much refrigeration liquid is now released – if the temperature needs to be lower, it releases slightly more liquid.
- As the liquid passes through the condenser coils on the outside of the fridge unit, the heat is radiated to its surrounds (liquid cooling down), after which an expansion valve is used to reduce the pressure and allow the liquid to expand by absorbing the heat inside the freezer. The piping, sometimes called *evaporation coils,* may include a fan blowing the cold air around the freezer compartment to keep the freezer chilled.
- This expanding liquid is then piped into the fridge where it is allowed to expand further in a different series of pipes or thin panel vanes, but in doing so it is removing

heat from the fridge air. A soft blower may be directed over it to keep cooling the fridge compartment but in most cases, the thin panel vanes are adequate (minimising power needs).

- Eventually the refrigeration liquid is directed back to the compressor to restart its cycle.
- Suction at an opened door is caused naturally by hot air entering during times when the door is opened, and as it rapidly cools, it reduces in volume and the reduced internal pressure causes suction at the door as it closes.
- An automatic *freezer defroster* may be used where freezer ice-build is excessive, but these may allow frozen food to unexpectedly defrost so it is often preferred to manually defrost instead.
- *Cycling defrost* in most freezer compartments of refrigerators minimises the build-up of ice so reducing the requirement for defrosting. This is achieved by the temperature fluctuating between a lower and higher amount so that any ice build-up is prevented. A clock circuit can simply change the thermostat by about 3°C every 6 hours for 30 minutes during the normal thermostat on/off process. Cyclic defrost freezers therefore tend to operate at a slightly higher temperature than a freezer without this feature.

8.3 THE COOKTOP, INFRARED CERAMIC VERSUS INDUCTION

Apart from household gas, solid fuel (wood) burners and camping (methylated) spirit-stoves, the most common other technology used for heating kitchen pots is the electrically powered cooktop using electric heat rings. However, these are now being replaced by new technology ceramic glass cooktops (a glass top is much easier to keep clean than the thick electric heat rings) and are either directly heated through infrared radiation or are induction-driven.

- *Infrared Ceramic cooktops* consist of a large plate of tempered ceramic glass which has four or more circular radiant halogen lamps (containing a halogen gas with a thin electric coil) beneath it, as opposed to the standard robust thick electric heating coils. If you consider the halogen lamps are often used as bright motor vehicle headlamps, instead these halogen lamps are made to give out heat rather than light. The ceramic glass, which is a poor conductor of heat, has been especially heat-formed to be able to be heated to allow expansion without breaking. Consequently, where the halogen lamps are positioned, infrared heat is radiated, passing directly through the glass to the other side. You can see the infrared lamp heat as a red glow against the black background.
- The *infrared* heat radiates through the ceramic glass surface to heat the pot or pan base, as well as its food contents in similar manner to microwave stove heat discussed earlier. This means that the glass gets hot to touch dominantly because the pot or pan

base is in contact with it rather than from the direct infrared waves passing through the glass. There may be minimal low-level ceramic heating by the infrared radiation (glass being a poor heat conductor so very small heating effect).

- The standard stove electric rings heat a pan base directly, and use a steady current flow to give a constant temperature, which is controlled by a thermostat. By comparison, the infrared lamps radiate their heat more efficiently and so have a thermostat that quickly switches the lamps on and off constantly to maintain the temperature.
- Because the black glass is hot but it isn't obvious, manufacturers often put a red dot indicator at the front controls of the glass-top to show which ring area is still hot (Figure 8.3). Ceramic cooktops have touch-sensitive controls or can have old-fashioned knobs which some prefer because the touch-controls sometimes don't readily respond [(2)].
- I*nduction cooktops* use a similar ceramic glass top but this time the base of the pot must be of a strong ferrous nature – high in iron content (wrought iron and some stainless steels – if in doubt test it with a magnet). A copper coil is placed beneath the cooktop and powered, which produces *an electromagnetic (EM) field* around it. The *EM field* passes through the glass top and will continue on through any iron material that can conduct an *EM* current. In turn, this *EM field* passing through the iron induces *Eddy currents.* This was discussed earlier in Chapter 6 (Figure 6.20) explaining that the *EM field* effect was used in magnetic payment cards, security walk through metal detectors (Figure 6.22), ground-probing radar (Figure 6.23), and household electricity meters (Figure 6.26).
- T*he Eddy currents* rotate within the pot base causing an *EM force.* Because the force can't move the pot (as it does with the rotating electricity meter disc), it causes electrons to collide and generate heat, thereby heating up the base and by conduction, the contents in the pot. An advantage of this is that the glass cooktop does not heat up as much as the glass ceramic cooktop and only heats by conduction in contact with the iron base of the pot. This has the safety advantage that once turned off, the

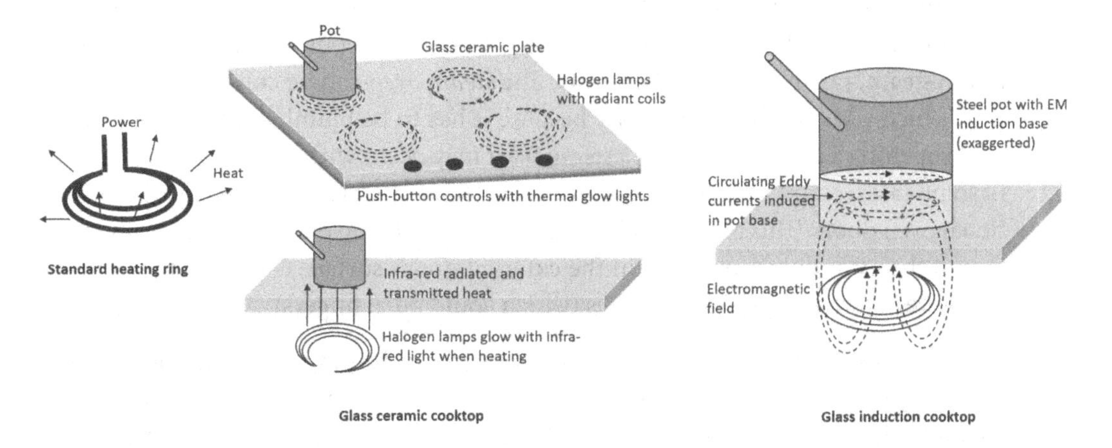

FIGURE 8.3 Standard heating ring versus glass ceramic and glass induction cooktops.

induction cooktop immediately cools when the pot is lifted off whereas the *ceramic* cooktop retains some heat. It is also argued that induction provides speedier heating than ceramic, but hard to prove since they use different cookware and it may be more dependent upon how fast the cookware conducts the heat.

- The downside of the *induction* cooktop is that it doesn't work with normal pots and pans, and instead a whole new set of pots and pans have to be purchased [3]. The infrared cooktop does tend to generate more heat within the glass compared with little heat from the *induction* cooktop, so in principle, there would be no need to keep a kitchen exhaust fan going when cooking has finished with the *induction* version. So some would say *induction* is more economic to use – however this does not take into account that new cookware is required to be used with *induction*. The induction cooktop can cause digital thermometers to fail but old fashioned analogue liquid thermometers still work well.

8.4 THE STEAM OVEN AND BAR-B-QUE

Ovens are generally split into conventional gas and electric convection types. They may be *self-cleaning* aka *pyrolytic* (*pyro* being Greek for 'fire') in which oven temperatures exceed 500°C to incinerate grease and spilt food on oven walls. Alternatively, they may be *catalytic* in which the oven walls are chemically treated so that incineration of food on surfaces will occur by oxidisation to ash at temperatures over 200°C. The downside of catalytic ovens is that if cooking does not exceed 200°C, they are not cleaned!

However, new technology has introduced the *steam oven* to us. These were introduced around 2015, to solve the problem that convection ovens have – that of producing very dry heat, which can cause excessive baking and drying of food, particularly seafood, steamed rice, and vegetables. With a wet heat, meats can be cooked in their own fat without adding further fat.

- The *steam oven* (Figure 8.4) can have a water canister inside the oven cavity which heats the water to steam ready for injection, and can be fan-forced to circulate heat as a normal oven. Water may be injected into the boiler by a pump and oven heat turns it into steam, raising the oven humidity. The *steam oven* can be connected to a water line, which adds further complexity because it has to be plumbed in. Consequently, when opening a steam oven door a steam blast may occur with steam rushing out and in some cases, could replace the hot blast sometimes observed when opening conventional oven doors. This blast can also be a result of air being pumped over the internal face of the oven door- to maintain the external door's surface cool thereby avoiding excessive external door temperatures (which could burn fingers).

- The downside of *a steam oven* is that it needs more cleaning than a *pyrolytic* oven, since they don't work in terms of self-cleaning. The *pyrolytic oven* self-cleans with the application of high-level heat to burn off residual fats etc. Combination *steam* and *pyrolytic* can be purchased and all can be fan-forced (i.e. a fan circulates heat and/or

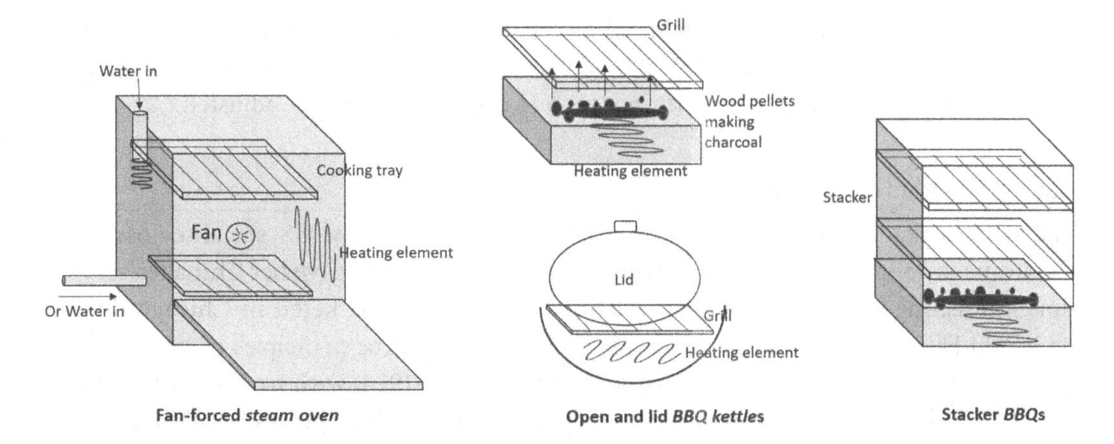

FIGURE 8.4 Steam oven, open and lid BBQ kettles and Stacker BBQ.

steam). Steam can be steadily injected or there are options to have regular 'puffs' of steam injected [(4)].

- Many new ovens can be connected to the *wifi*, so that the ovens can be remotely programmed to cook at times and to whatever temperature is needed, without the need to be in the kitchen. Of course, this assumes the food has already been loaded into the oven.
- *The Bar-B-Que* (BBQ) used to be a simple open campfire in which coals (charcoal) were used as the heat-base beneath food, giving it a smoky flavour. Then open *gas or electric BBQs* with gas-fired coals beneath a grill became popular because they were more controlled and offered heat that was predictable and cleaner to use. When plates were used alone the gas heated the plate rather than the coals, and the smoky flavour was lost.
- Then electric element BBQs with lids came into vogue, with wooden pellets available which could be fed over the electric element, thereby giving the burnt wooden charcoal smell while cooking the food (of course, you can buy chemically-based liquid additive to give this charcoal smell and taste). The benefit of these electric BBQs is that the BBQ lid which contains a built-in thermometer can be put over the food making it into an oven. This then allows temperature to be controlled locally or remotely using a *wifi* connected digital thermometer smartphone application.
- When the BBQ base is reasonably deep (enough to house a whole chicken), it is sometimes referred to as a *BBQ kettle* (the deeper the kettle, the larger the amount of meat it can accommodate) [(5)].
- Often a *stacker* is used with *BBQ kettles*, which is simply an additional vertical spacer to allow an additional shelf or rotisserie to be fixed on top of the normal BBQ grill (if a *rotisserie stacker* is installed, the excess fats from the rotating meat above can then drop on the meat beneath it, acting as natural basting).
- The benefit of using *wifi* connected digital thermometers with a smartphone when cooking is that food may have one or more thermometers inserted into it, so that the

wifi transmits this temperature (and/or humidity) data to the smartphone. The phone then provides the cooking data in real time to allow remote temperature monitoring and changes in BBQ programming (such as cooking time) to be adjusted.

8.5 THE DISHWASHER

The first workable hand-powered dishwasher was invented by US socialite Josephine Cochrane in 1886. She didn't like the mistreatment her servants gave her China tea service during washing so invented the *Lavaplatos.* This was first marketed in Chicago during its World Fair of 1893. The first invention which had all of the principles of today's dishwasher was developed by UK physicist William Livens in 1924, with front door loading, a wire rack holding the plates and a rotating sprayer. The first European invention using an electric motor was in 1924 by German company *Miele*, but dishwashers were commercially unsuccessful until fully publicised from 1950 onwards when they included a filter to catch most of the large-sized detritus before it was washed away (Figure 8.5).

- The washing cycle starts with water entering the machine's water heater that runs it up to about 50°C to 60°C at a pressure of about 50 psi (machines often have hot water piped in and the heat and pressure used is dependent on the cycle selected from between gentle wash to hard wash – heating up to 75°C and 70 psi if needed). The machine therefore contains a heater and water pump which operate at the correct levels depending on the chosen cycle.
- The hot pressured water then passes into the upper and lower rotating sprayers, where it is sprayed up, down, and across the dish space. The water collects in a sink at the base and passes through a filter, to then be reheated and re-sprayed again. This continues to occur with water recirculating through the filter collecting the largest detrital

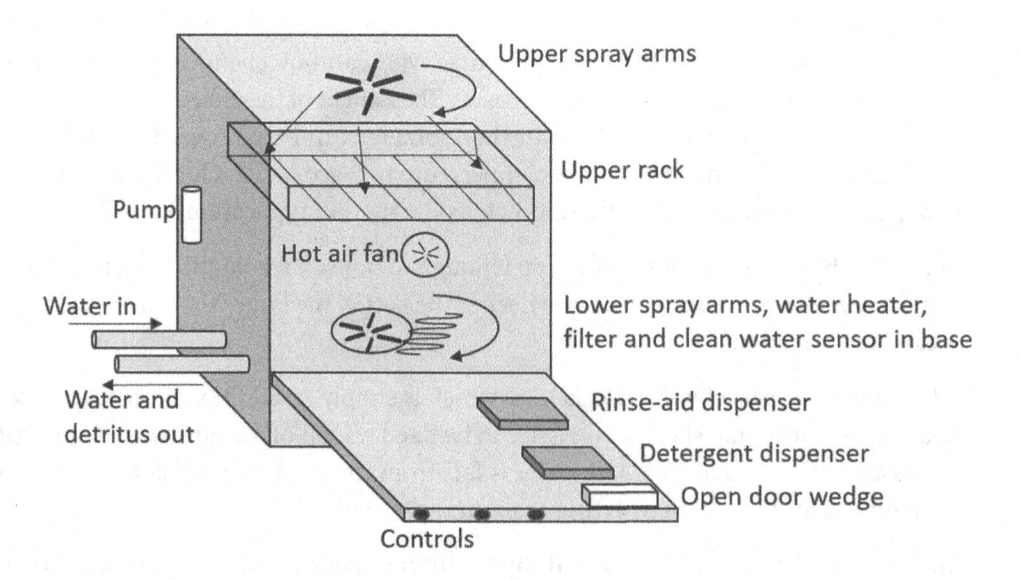

FIGURE 8.5 Dishwasher.

particles. The filter may also contain a sensor with a light or infrared cell built-in, and is used to determine how clean the water is. If light shines through the water as it leaves the filter, it means that it is clean, otherwise it still contains heavy detritus.

- The door will house the push-button controls as well as a *detergent dispenser* which may take detergent powder in a compartment with its own automatically opening door, but sometimes the detergent does not fully dissolve especially if the water for some reason is cold, and consequently today's dispensers tend to house detergent either in the form of a dissolvable pod or liquid. The dispenser door will open automatically dependent upon the program selected.
- Once the hot pressured water has been sprayed on the plates and the filter sensor has determined the plates are clean, a *rinse-aid dispenser* which is housed in another compartment in the door will automatically open at a set time. This rinse-aid liquid on contact with water, causes plates to become wetter – that is, by reducing their surface tension (making water droplets flatten on their surface rather than allowing them to stay as rounded water droplet beads) [(6)]. This helps the drying process and reduces the final water streaks on glass and dishes.
- A hot air fan may be included in the machine wall, which ducts hot air through the washing space causing the water to drain off more readily and the dishes to dry faster. The rinse-aid compartment is filled only occasionally, since only a small amount of rinse-aid liquid is used each time. Some plastic containers being washed may not dry completely, because they are *hydrophilic* meaning they don't respond as well as other materials to rinse-aid and may retain water droplets (as opposed to *hydrophobic* which means they repel water – and respond to rinse-aid). Small quantities of vinegar have the same effect as rinse-aid except they don't contain the polishing agent that rinse-aid products have in them.
- A recent innovation has been the addition of the end-of-drying-cycle open door-wedge. At the end of a normal washing cycle, some excess water droplets may still remain on dishes and cups. Although they are hot enough to evaporate remaining drips, the level of humidity within the washing space may not allow normal evaporation. Consequently, we can simply open the door ajar for long enough (about 15 minutes) at the end of the final washing/heating cycle to provide dry air into the washing space and allow residual steam to escape.

8.6 THE AUTOMATIC WASHING MACHINE AND THE CLOTHES DRYER

The washing machine has been around since slaves washed Roman soldiers clothing, but the first rotating drum patent is attributed to Henry Sidgier in 1782 England, with the first commercial *washing mill* being sold in England from 1790. However, these manually operated machines did not become popular until motors were fitted to them and marketed in 1907 by *Hurley* Company of Chicago. The basics of automatic filling with water, agitating and spin-drying did not happen until 1937 when *Bendix Corporation* commercialised it, and by the late 1950s, they became accepted household appliances. As has been often quoted, there now is more computer power in a standard washing machine than there was onboard the *Apollo 11* module which landed on the moon. Typical schematic drawings of how the *top loader* and *front loader* washing machines work are shown in Figure 8.6.

This is how they work:

- The washing tub is a perforated shell and is independent of the washing machine frame so that the tub can be rotated around, back and forth, as well as spin providing centrifugal force at much higher speeds. The tub can be vertical (*top loader*) or horizontal (*front loader*). *Front loaders* tend to be preferred because their horizontal spin tub containing a heavy wet laundry load is more easily controlled and economical to operate than the vertical tub. Space is needed above a *top loader* for opening the lid, then loading and unloading laundry, whereas it is needed in front of the *front loader*, which again is sometimes more readily available and convenient.
- Water is input from a cold-water tap and heated within the machine up to as much as 95°C. This heat level could be controlled by a thermostat – see Chapter 6 (Figure 6.4). A *front loader* may take only half the amount of a *top loader* to be filled, because there is limited access to the top of a *front loader*, therefore requiring less water.
- An impeller (which is mounted on the wall or base of the tub) or *agitator* (which is a post containing fins passing up through the centre of the *top loader* tub) has a rotating motion either with or against the tub rotation direction, to provide the mechanical movement needed for washing. Some machines have a vibrating plate or wash-plate at the base instead of an impeller, in which the plate vibrates the clothes during a wash. However, these have a tendency to excessive wear and tear of laundry.
- Contra-rotating tubs were tried (by *Dyson* UK) in 2004 which were considered to give a cleaner wash in a reduced time, but were too expensive to sustain in the tight washing machine market. As an alternative where space is limited (such as in European laundries), top loaders can have horizontal rotating drums which are held by bearings on each side of the static walls, and reduce the amount of space needed being narrower than the conventional *front loader*.

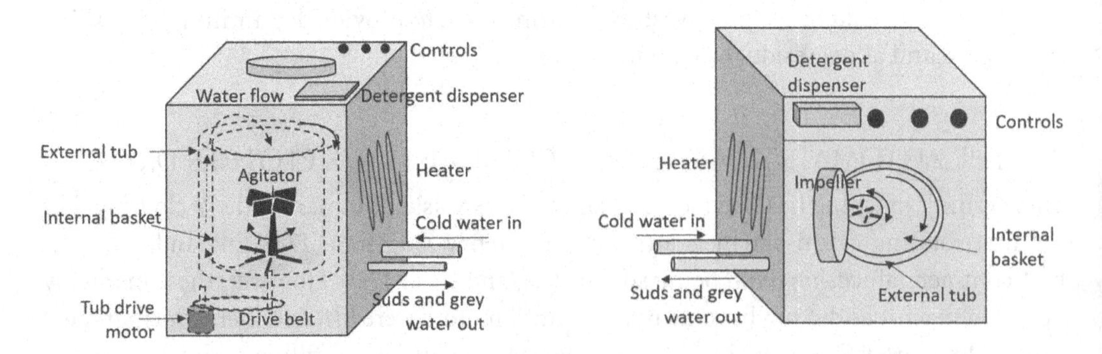

FIGURE 8.6 Top and front loader washing machines.

- A perforated basket is filled with laundry, and positioned inside the water-retaining tub which may be driven by a pulley belt attached to a motor, or can be driven directly by the motor.

- In *top loaders*, a finned agitator is at the base whereas the fins are built within the tub on *front loaders*. In *top loaders*, the agitator pushes water outwards through the laundry and up the sides of the tub and down from the top towards the agitator again as the laundry spins horizontally. The agitator occasionally reverses direction to reverse the direction of water flow. *Hoover top loaders* have used an impeller mounted on the side of the drum to spin laundry through the water, but had the disadvantage of not being able to take as large a load as those with bottom agitators.

  In most machines, the basket rotates in one direction with the agitator rotating in the opposite direction, or stands still. There is a fast spin cycle to spin the clothing dry (using centrifugal force to move water outwards) but this is inadequate to completely dry the laundry. An attempt was made to resolve this issue by one manufacturer who developed a machine that had two *top loading* tubs side-by-side (so that wet washing could be removed from the washing machine and put into the spin-dryer next to it) but because side-by-side meant twin tubs, this took up a lot of space thereby reducing the volume of washing that could be done at any one time (and in reality, some preferred to hang laundry out on washing lines anyway). This type of *top loader* washer/spinner was eventually discontinued in favour of separate forms of dryer because the washing still needed to be dried after the spinner had done its work.

- In *front loaders*, the drum spins using gravity to drop laundry through the water, to make the water pass naturally down and through during the washing process. This process requires less detergent. Before the basket starts to rotate, some laundry will absorb water upwards actually starting the washing process naturally. Consequently, some rotation is expected at the start of the wash allowing the entry of water to wet the clothing. The rotation increases in speed when the spin cycle is trying to spin-dry the laundry to a wring-dry level, but as with the top loaders, washing still requires additional treatment to fully dry.

- Detergent and softener dispensers can take liquid, powder, or capsules (pods). Some machines allow water and air to mix with the detergent before it is sprayed into the washing. Lint filters are generally fitted which filter out lint during the water cycle.

- Clothing should not be packed tightly into the *top loader* since this can limit the water circulation, and cause incomplete rinsing. Overloading can imbalance the rotating tub causing the potential for burnt-out drive motors. A benefit of top loaders is that there is no major need to replace lid seals since the lids take little water pressure. By comparison, the *front loader* must have a well-sealed door, but has the advantage that a thick window allows the viewer to see the washing as it is rotating. A rubber or plastic bellows also seals the *front loader* to the rotating drum, which is not needed in the *top loader*. The bellows can collect dirt or mould over time, so some machines have a special cleaning cycle when laundry has been removed, to allow bleach to be run through the machine at periodic times.

- When it comes to troubleshooting a problem, some machines have phone apps that can be used to listen to the sound of the machine, and then it provides advice to the consumer on what is to be done to solve the problem [7]. Others require a photograph to be taken and messaged to them, which then provide feedback on what to do.
- A benefit of the *front loader* is that washing can only be packed into the accessible space, because, once the door is closed, there is space between the door and rotating laundry, and also between the laundry and the upper drum which is difficult to reach when loading the machine from the front. However, a major issue therefore with the *front loader* is the drive drum bearing. Because the *front loader* rotates heavy wet laundry in a horizontal cantilever action, the drive bearings take all of the weight. Over time these fail, and the amount of effort required to replace them renders it easier to purchase a new machine rather than repair the old machine. This is obviously built-in obsolescence.
- Washing machines can also perform drying operations, but in reality in order to dry laundry, hot air must be passed through the wet linen reducing moisture content, whereas the high humidity of many of these washing machines does not allow this to occur. Therefore most often we have a need to have a separate dryer. Some machines do come combined with two parts – a washing machine at the base and a dryer above it.
- There are also combination washing machines and driers which have been around for 50 years. The disadvantage of these is that only one load of washing can be done at a time, and they are generally more expensive.

The clothes dryer (aka *tumble dryer* and *heat rack dryer*) comes in a number of simple forms:

- The most common form is the *tumble dryer* (Figure 8.7) in which a barrel containing clothes to be dried is rolled while heat passes through the barrel, heating and removing humidity at the same time – the heat being controlled by a *thermostat* (see Section 6.1.2). The barrel is often perforated so that hot air can pass through, and the clothes are allowed to roll and tumble through the air with gravity allowing heat to pass through them.
- Air is fed across the heating element and down through the dryer as items to be dried are rolled around. The hot air is sucked out through a lint filter by a suction fan, and the barrel is controlled usually by a motor-driven pulley belt. The heat is controlled using a *thermostat*, so that the heater is switched on and off as a function of the ambient temperature in the barrel.

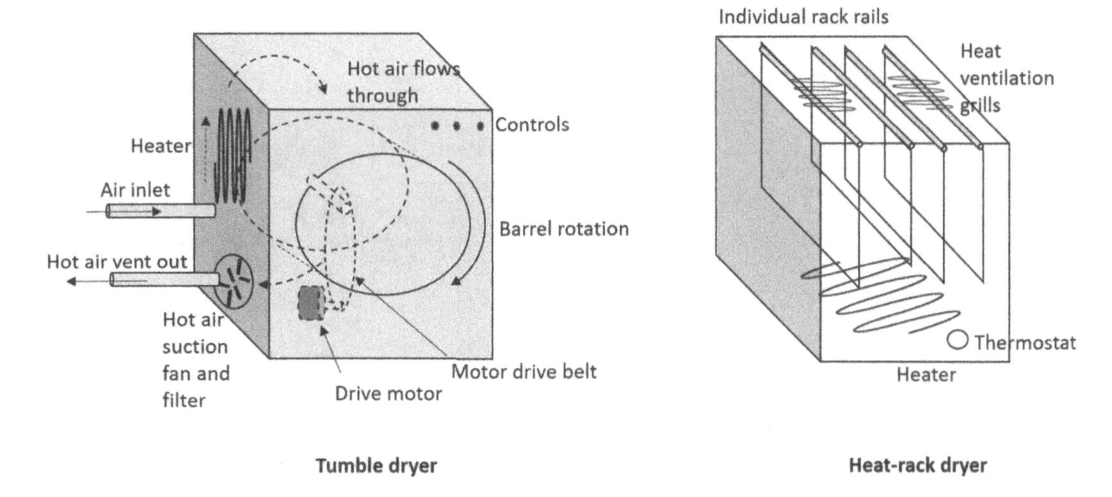

FIGURE 8.7 Tumble dryer operation versus simple heat rack dryer.

By comparison, the *heat rack dryer* is a very simple and a relatively cheap steel box with a heater at the base, allowing hot air to rise up through a rack of individual rails from which washing is hanging down imitating an outdoor washing line. In this case, the high-humidity hot air passes up through *ventilation grills* in the box top and lid. There are no moving parts to the *heat rack dryer's* operation – and only the *thermostat* controls the heat. The *heat rack* is often used in countries where it is generally cold and wet.

8.7 THE ELECTRIC/OIL/WATER HEATER VERSUS REVERSE-CYCLE AIR CONDITIONER

A standard *electric heater* has a simple wire coil or element (Figure 8.8) that heats as a function of the amount of current passing through it, and may have a blower fan that blows the air out through a grill. The level of heat is controlled by a *thermostat*, just as for the *heat rack dryer.*

- An alternative approach is to have the heating element immersed in oil in a container, which then heats up the container and either warms the surrounding air through conduction heating, or is fan-blown away from the vessel using hot air convection. The hot oil generates an odour that must be ducted away up a chimney in some manner to remove it from the internal house area.
- The housing vessel can be a simple box with hot air being blown out of it (but because the oil is hot, it slowly evaporates and a smell of boiled oil or kerosene can result). Alternatively, water or steam is used in the form of convection heating of the air around a house using a steel structure containing heat-exchange fins (which ironically, are called heat *radiators* but they actually work by convection rather than radiation).

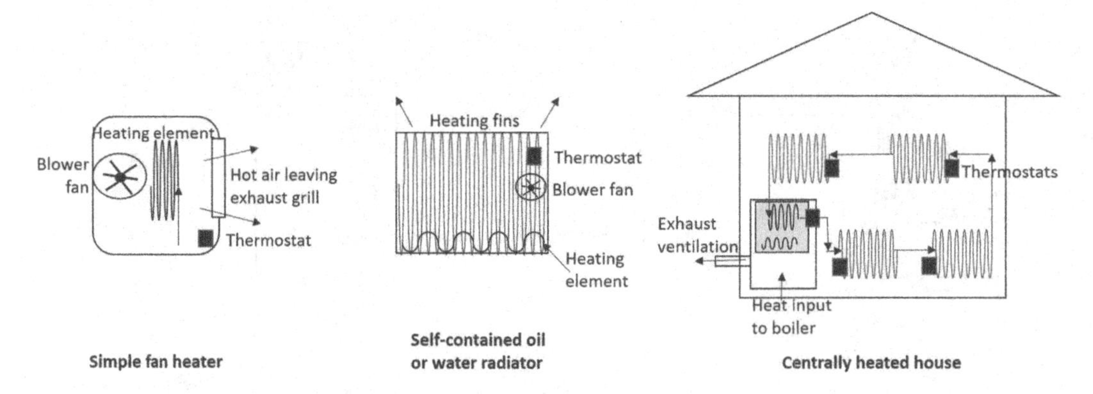

FIGURE 8.8 Different forms of heaters commonly used.

- The water or steam is transferred from a heated boiler (by either an electric element or external heat) through pipes to each radiator. There is a *thermostat* at the boiler as well as at each individual radiator (alternatively, each radiator may be controlled by inlet and outlet valves limiting how much new water may be taken into each of them) see Figure 8.8.
- When a house is fully heated by such radiators, it is known as the house *central heating* system. If the radiators are somewhere vertically at a lower level than the boiler, they may need some form of pump to push the water or steam into the first radiator to start the flow of hot fluid. Alternatively, the hot water may be able to naturally start the flowing process as it heats and expands into the piping system (responsible for the phrase *piping hot*) [(8)].

The *reverse-cycle air conditioner* is a totally different concept.

- The air conditioner is based on the concept that when we compress a fluid, it heats up (see Section 8.2) In this case, instead of the external condenser coils releasing heat to the atmosphere, the coils are inside a steel container and a fan blows the hot air out through the front vents (Figure 8.9).
- Similar to the refrigerator, the cooling of air on the outside of these evaporation coils causes cooling of the air around the coils or vanes, and sometimes evaporated water forms which has to be ducted out of the air conditioner unit.
- The air conditioner is said to be *reverse-cycle* because it operates on a normal refrigeration cycle as well as a heat cycle (so the process is reversible). This time a single thermostat controls both processes, and it should be noted that the refrigerant piping system is closed, so when there is *loss of gas* (aka *degassed*) or poor performance by the equipment, it is usually due to a poor quality pipe weld becoming cracked. Once the weld is fixed, the refrigerant system can be *re-gassed.*

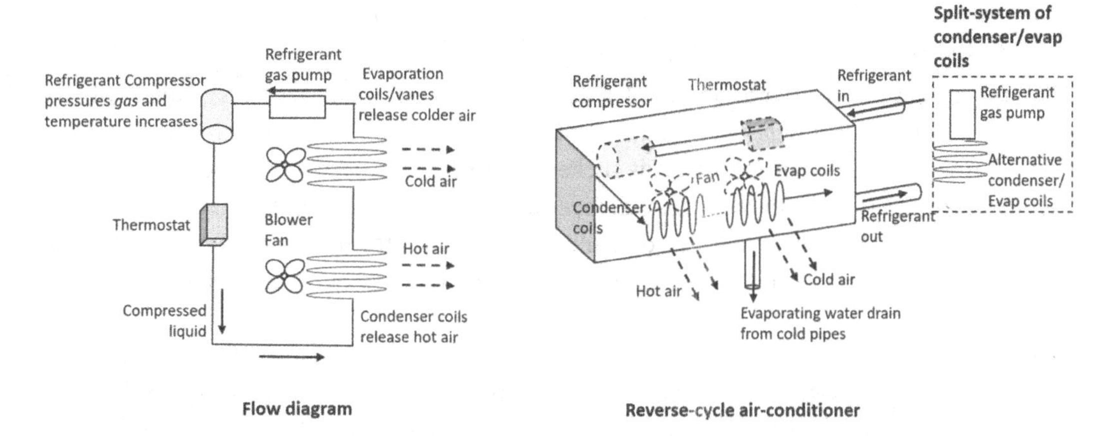

FIGURE 8.9 Reverse-cycle air conditioner (aka *split-system*) uses the same principle as the refrigerator.

- The evaporation coils/vanes and condenser unit (known as the *heat exchanger*) may be outside of the room it is cooling/heating built together as a single *heat-exchanger* unit, to allow cool air to be pumped into the room while the hot air is pumped outside of the building, and vice-versa when operating in the heating mode [9]. This is known as a *split-system* since the cold/hot air outlet unit is inside the room whereas the *heat-exchanger* is mounted outside the building. Of course, the original air conditioner halfway point was a 'window hung' system which had the outlet fans inside the room window with the evaporator and condenser coils outside the window in the open air (these are still very common in older buildings around the world).

> "The evaporation coils/vanes and condenser unit (known as the *heat exchanger*) may be outside of the room it is cooling/heating built together as a single heat-exchanger unit, to allow cool air to be pumped into the room while the hot air is pumped outside of the building, and vice–versa when operating in the heating mode. This is known as a *split-system*."

- Of all types of heater/coolers, the *split-system* air conditioner is considered to be the most efficient and cost-effective of all heater/coolers with a benefit being that one *heat-exchange* unit sitting external to a house can be used for all cooling and heating room outlet units in the house. They are simply connected together by pipes so the refrigerant continues from outside to inside the house in a closed-loop system. Office blocks have large *heat-exchanger* units mounted on their roofs (with large exhaust fans pointing skywards) rather than hanging out of windows like the early window hung air conditioner design.

8.8 THE VACUUM CLEANER – SUCTION/VIBRATION VERSUS BARREL CYCLONE VERSUS ROBOT

The earliest vacuum cleaner to be invented appeared in 1860 by Daniel Hess of Iowa, which was called the *carpet sweeper* and had a hand-pumped set of bellows which caused a suction. Connected to a rotating brush, it sucked dust up into a steel container. Since then, vacuum cleaners (aka *Hoover* because *Hoover* company was the main commercial developer after World War II) have evolved, using an electric motor instead of hand pump and added removable bags rather than the steel container, and then become bagless.

- The basic design of the *Hoover*-style vacuum cleaner was to have a motor just above ground level providing suction via an impeller to a small oblong area that housed a spinning brush roller. These were at the floor level, with their controls at hand-pushing height, and so this version became known as the *upright vacuum cleaner.* The motor used a rubber belt to power the impeller and the brush roller (in some models they have their own motor) so that as the brush roller ran over a carpet it pushed dust in the direction of the suction forces so that it could then be sucked into a bag. The bag was removable but it was heavy, and it was always very dusty when emptying (Figure 8.10).
- It was noted that if the suction power increased, the suction could lift the carpet off the ground and if the brushes were designed correctly on their roller (in a helical manner) they could actually lift the carpet and drop it, thereby vibrating the carpet and improving the cleaning efficiency – *Hoover* had a song jingle 'Hoover beats, as it sweeps, as it cleans'. However, improving the cleaning efficiency increased the weight of the bag.
- To simplify removing the heavy bag and the dust which was inevitable when emptying the bag contents into a bin, disposable paper bags were then added to the bag

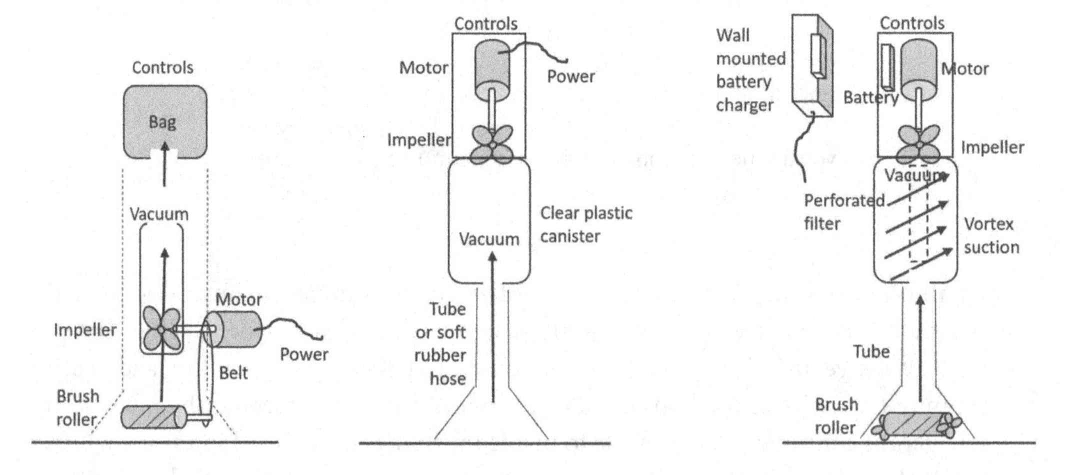

FIGURE 8.10 Vacuum cleaners have developed from simple uprights to complex *handheld cleaners.*

chamber. The bag was easily removed by zipping the bag chamber down and unplugging the bag.

- Some companies added carpet cleaning mechanisms replacing the basic vacuum or brush roller with a water and detergent tank, and while these were popular – particularly those built by manufacturer *Kirby* which were very strong – they were more expensive to purchase and had the expectation that the user was prepared to interchange parts – which often the user was not – so the market for these types of cleaner is small.
- When disposable vacuum bags are full the user can be unaware and the vacuum stops working efficiently. Also, it was accepted that the rubber motor drive belt could fail due to wear and tear, which became a maintenance issue. To solve these issues, the next step was to replace the bags with clear plastic chambers or *barrels* which enabled the user to see typical household dust/dirt and hair being sucked into it. This allowed a quick visual check of the contents of the barrel. Importantly, a check could be made that no valuables had been sucked up from the carpet and the plastic *barrel* could be emptied of dirt when observed to be full.
- This development removed the motor drive belt to the rollers, which meant that vacuuming now relied solely on vacuum pressure, since there was inadequate vacuum to drive the brush rollers. Also, because impeller fans can get dirty and clogged up with dust passing through them, they were then placed at the end of the clear plastic *barrel*, with a filter positioned directly before the impeller. Happily, the user could now see the dirt passing into the clear *barrel*.
- The idea of having a plastic expandable tube or hose was also considered as useful, because instead of the motor/barrel/brush roller being on the end of a stick as with the upright vacuum, the end of the hose could be moved around separate from the motor/barrel which would make vacuuming easier. Now the motor/barrel could be put on rollers on the floor or carried on a person's back (as is commonly used in commercial cleaning). This allowed more versatility, with the mobile barrel/drum in one location (e.g. at the bottom of the stairs) while the vacuum wand could be pushed across the stair steps at all levels – a feat the *upright* could not manage. Having the vacuum at the end of a long plastic wand with expandable hose caused loss of vacuum, and the loss of brush rollers due to there being inadequate vacuum to drive them also.
- The next step was to attempt to make the clear barrel more efficient to improve the suction strength of the vacuum. By modifying the walls of the barrel design, the air suction could be made to act in a *vortex* manner and suck at an angle (rather than straight) into the circular barrel, which was an action well known in industrial buildings and sawmills where cyclonic vacuum systems were used. In this case, the air would rapidly spiral up the inside barrel wall with a centrifugal force sucking the lighter dust towards a centrally located perforated barrel filter, and the heavier particles staying on the inside. It is argued that this form of vacuum allows better packing of hard particles, leaving more power available for increasing the suction as well as more outer storage-barrel space to take dust.

- Initially tested out on *upright vacuum cleaners* in 1979 by James Dyson of the United Kingdom, the *vortex* was sufficiently strong to drive a brush roller, with the *vortex* turning the roller around. Now adding a hard tube wand (rather than the flexible rubber tubing) between the motor/barrel and the brush roller at its end, the clear barrel to be used with a hard plastic wand and brush-roller could be built into one single unit light enough to be operated by hand and with the development of rechargeable battery power – could now be independent of the power cord.

Remove the long wand and you have a short stubby handheld vacuum cleaner that can be taken anywhere assuming the battery is fully charged. The handheld version needed a charging station that could be wall-mounted so that the handle fits snugly into a slot in the charger on the wall and the cleaner could be stored, hung conveniently against the wall as it charges up.

With the advent of computer programming and battery power, the development of the *robot* vacuum cleaner became inevitable. Initially developed for mass production around the turn of the millennium (2000), the first mass-produced *robotic vacuum* was produced by *Electrolux* with numerous brands now available (Figure 8.11).

- These vacuum cleaners are programmed to run around the floor frequently bumping into furniture, and both suck and horizontally brush dirt into an on-board container which must be emptied at some stage. They work well on wooden floors but have issues where floor heights change such as when there is carpet on the floor. Consequently, they are mainly used in households that have wooden or tiled floors with no carpets, and are most useful for picking up animal hair from the floor plus occasional crumbs.

- The on-board dustbin fills up and needs more power due to its increased weight as it picks up more dust. Hence, it's useful more for dust and hair removal rather than anything heavier, and it always has to retain some power so that it can find its way back to the power dock.

- *Robots* are relatively low powered since their on-board battery power is limited. However, they do make up for this by being remotely controlled by the user. When their battery power is low their computer uses *infrared* scanning (discussed in Chapter 6) to detect their docking station to which they automatically return to recharge. Small wheels help them run around since they have direct contact with the ground.

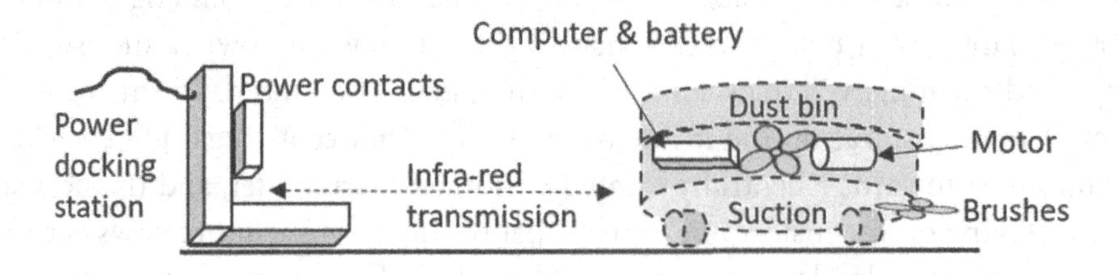

FIGURE 8.11 Robot vacuum cleaners steer back to their power docking stations using infrared.

- Innovations have included the ability to automatically dock to empty their dustbin and then return to vacuuming, while some robots can *laser-scan* (see Chapter 6) and map the floor, to allow more efficient cleaning next time it goes vacuuming. In some cases, their on-board computer system also detects the removal of objects which were previously mapped (such as a dog getting up and moving to a different spot), which would then allow it to vacuum the area where the dog had been. While their dustbin is very small, the important point is that their use is limited by the inability to negotiate different floor levels and the presence of carpets in a house can dismiss their use. However, they are ideal for any house that has a dog and no carpets.

8.9 THE HOT WATER SYSTEM – ELECTRIC IMMERSIVE STORAGE VERSUS INSTANTANEOUS HEATING

The hot water system found in the family house generally relies on heating large tanks of water ready for use (aka *storage tanks*), which are sometimes referred to as a *boiler.* These water *storage tanks* can accept heat from outside such as when heating the boiler's external surface with gas, or heating the stored water inside the tank with an electrical element immersed in the water. The technology of heating a *storage tank* using an internal element is compared here with the technology of using individual elements without the *storage tank* (Figure 8.12) [(10)].

Immersive storage tanks use an electrical element immersed in the water like an element in an electrical kettle:

- The *electrical element* is often positioned at the base of the tank, and is insulated from the water but able to pass heat into the water. Hot water rises from it and cold water is normally fed into the base of the tank.
- The hot water rising causes minor heat turbulence in the tank, with cooler water sinking down the sides of the tank to be reheated.

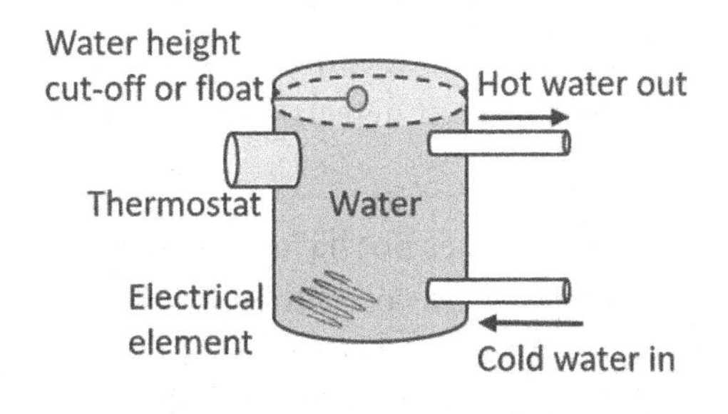

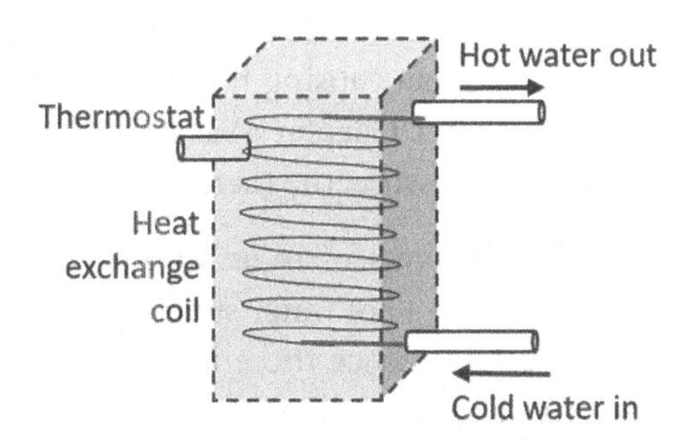

FIGURE 8.12 Immersive tank storage heater and the instantaneous water heater.

- A float or similar water cut-off instrument is positioned at the top of the tank which closes a valve and stops further water from entering once the tank is full of water.
- A *thermostat* (explained earlier) is positioned near the water exhaust pipe, so that once the water reaches a set temperature of between 50°C and 70°C, it switches off the power supply circuit to the heating element.
- A failure of this type of water heating boiler is that the tank is being permanently heated, which can be expensive to maintain. An advantage of this boiler is that there is a ready supply of hot water. However, the tank may be large in volume and clad with heat-insulating material, therefore needing adequate space in a house to operate. Also, it is possible that the heating element may corrode or lose efficiency. In cases where the water has a high calcium carbonate content, it may precipitate out of solution and settle around the element resulting in calcite build-up on the element and eventual failure. To avoid this a sacrificial anode is sometimes placed in the water, adjacent to the heating element.

Instantaneous hot water heaters, by comparison, work on a similar principle but do not have the water storage tank (Figure 8.12):

- Cold water is ducted in at the base of a long heat exchange coil within a wall-mounted cabinet, with the element electrically isolated from the water, but allowing heat to pass to the water. The element consists of a wire inside a covering that has poor electrical conductivity but high heat transmission.
- The water, under pressure, is heated as it passes up through the inside of the element.
- It passes through a thermostat positioned in the line, and out of the top of its cabinet. The thermostat is often set no higher than 50°C, since instantaneous heaters are generally used for showers and hand-washing (temperatures above 50°C may scald the skin).
- Water flows as a function of pressure, so that if a hot tap faucet is opened and the water pressure within the heater element reduces, cold water starts moving through the element causing the thermostat to operate. The thermostat can have a pressure sensor (see Chapter 6) as part of its operation, so that when it detects a reduction in water pressure, the thermostat functions.
- An advantage of an *instantaneous water heater* is that it does not have to constantly maintain a hot water supply in a tank, so maintenance is less costly than an *immersive tank*. Since there is no tank and this heater is small in size, it needs far less space and can fit neatly inside a wall cavity or underneath bathroom cabinets – and allows a single heater per shower or sink if needed. Another advantage is that if the *immersion tank heater* fails, the hot water supply to a house stops, whereas there can be multiple *instantaneous heaters* that stops the dependency on a single heating element keeping stored water hot. The *instantaneous water heater* is ideal for

apartment living where there is limited space, in which individual heaters can be placed underneath each sink.

- The problem of calcite build-up applies equally to the *instantaneous heater* coil (which potentially could block up) but it is much easier to replace than an *immersion tank*. The other issue with instantaneous heaters is that they may readily turn off when the hot water demand is low (so water pressure is higher than when faucets are fully open), which can cause the instantaneous heater to switch off during times of minor adjustment (causing water to run cold). The solution may be to reduce the thermostat turn-on pressure, but this may be set by the manufacturer and not adjustable.

8.10 THE HOT WATER KETTLE – IMMERSIVE ELEMENT VERSUS INDUCTION

Hot water kettles which are operated using electricity come in two standard forms, that of *immersive element* type and the *induction* type:

- The *immersive element* type has been the standard version of electrically heated kettle for decades. Similar to the *immersive hot water heater* explained in the previous Section 8.9, an electrical element (which typically has an internal heating coil with a surface coating made from a material that is a poor electrical conductor but a good heat transmitter) is submerged in the water and electrical conductivity to the coil heats the coil and hence heats the water surrounding it until a thermostat senses the steam rising above the boiling water (100°C), and opens the heater's electrical circuit (Figure 8.13). The *immersive kettle* can have a separate cradle-base attached to the power cord, which allows the kettle upper canister to be removed from the cradle and taken across the kitchen if needed.
- By comparison, the *induction kettle* operates on the basis of induction heating (discussed earlier in Chapter 6). The *induction kettle* comes in two parts, with an upper

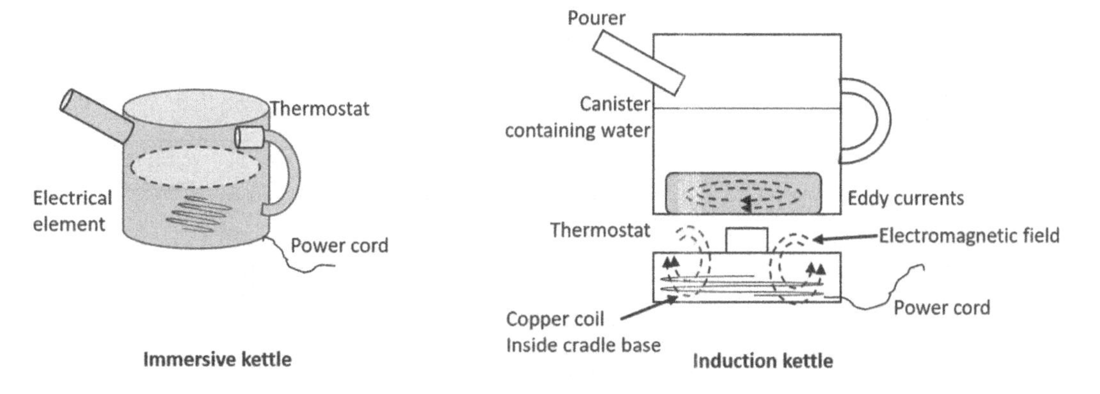

FIGURE 8.13 Immersive kettle and the induction kettle.

part being the conventional water holding canister that has a heavy base made from plastic or metal, but containing some form of ferrite that can be magnetised, similar to some types of steel. It has a separate base containing a copper coil. When power is applied to the coil in the fixed base, it induces an *electromagnetic* (EM) circuit which passes across the air gap to the kettle canister metallic base – causing *Eddy currents* to flow in a circular manner and heat the base (just as it does with the cooktop of Figure 8.3). A *thermostat* is built into the base to cut off the electrical current to the copper coil when the water starts to boil. Removing the canister from its base cuts off the *Eddy currents*, stopping the canister base from heating further. The *Eddy current* effect can be seen if the canister is made of glass and the water is boiling – if the canister is rotated on its base (while the water boils), the water will be seen to continue to boil from the same locations where the *Eddy current* heat is strongest.

8.11 DRIP COFFEE VERSUS PERCOLATOR COFFEE MACHINE AND THE ESPRESSO COFFEE MACHINE WITH PODS

The normal drip coffee machine is based on passing hot water or steam through ground coffee beans (often in a paper bag or basket for filtering). The exiting hot water now has the taste and aroma of ground coffee, with the single dripper shown in Figure 8.14. It was noted over many centuries since the 1500s that the more pressure and steam/hot water that is passed through the coffee, the more the different flavours that are produced and that steam under various pressures can produce a surface froth. So the operation of the drip coffee machine therefore became more complex as a range of greater pressures (more than just gravity with drip coffee) was applied developing the operation of the *percolator coffee pot* and the *espresso coffee machine.*

The *percolator coffee maker* recirculates boiling water through the coffee grounds, and collects in a reservoir after which it is recirculated again. *Percolation* is the word given to the process of a fluid passing through a permeable layer, changing chemically by the process as shown in Figure 8.14.

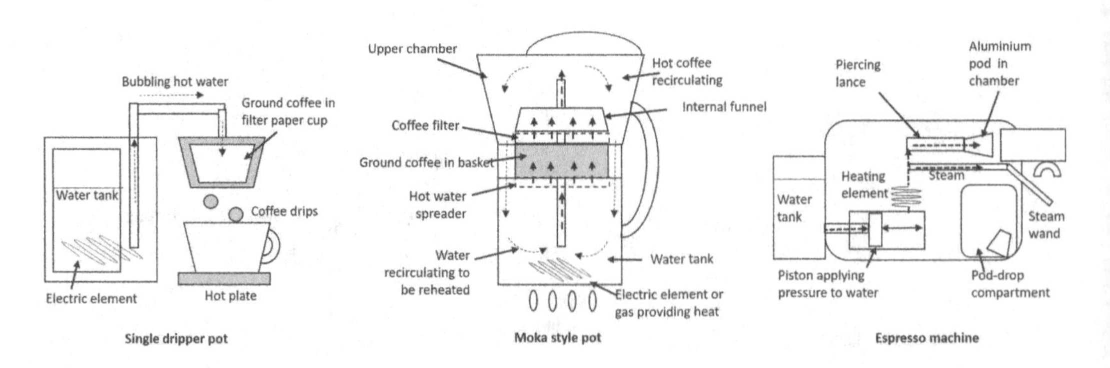

FIGURE 8.14 A single dropper pot, the Moka machine, and the espresso machine.

Here, the *Moka style of pot* (the name comes from the Italian city *Mocha* which popularised it in the 1600s) is presented – this is a similar action to the normal commercial coffee maker:

- When water boils (getting its heat from either an electric element inside or heat from outside a water tank), it releases bubbles of steam which expand, increasing pressure. The water then rises up a centrally located narrow stem and passes through a perforated spreader causing the bubbling steam to pass up and through the ground coffee held in a basket, saturating it evenly under pressure.
- As it exits the basket in the upper chamber, the temperature now has reduced back to boiling water which passes through a perforated filter, over the basket edge and drops by gravity from the upper chamber through the container back to the lower chamber, to start the boiling cycle all over again. Filters can be put anywhere along the track to remove larger coffee particles. The coffee gets to the point where it has expanded so much that it starts filling a cup with coffee in the upper chamber.
- The number of times water is recirculated is limited to how strong the coffee should be, and to what extent a surface froth is desired. In most cases, a single pass or recirculated pass is adequate, since an excessive number of passes made the coffee bitter. Each pass reduced the amount of water available.

The *espresso coffee machine* used in coffee shops allows different pressures of hot water to be pumped through the coffee basket, dependent on the strength of the coffee to be served [(11)]. It also has a separate thin steam wand, which when put into milk (for the production of *cappuccino* or *latte*), provides extra steam to oxygenate the milky coffee which results in more froth. To cater for all types and flavours of coffee is a simple process, but it can become complicated by the need to have different amounts of coffee strength and the need for a milky coffee (cappuccino or latte). This led to the addition of the *steam wand.*

By contrast in 2000, Swiss company *Nestle* introduced and popularised the *Nespresso machine*, a much quicker and simplified version for making all forms of good quality coffee. It works on the basis of providing aluminium *capsules* or *pods* (containing ground coffee which are available containing many different blends). The pods are simply placed into a chamber on their side (Figure 8.14) after which hot water may be passed through at precise pressures as follows:

- The basic machine has a water tank, pressure pump, electric heating element, and piercing lance or handheld coffee cup (aka *portafilter* because it acts as a mobile cup and filter) – a steam wand may be added.

  When the machine is turned on (by pressing a button on the control panel) with a pod in its chamber or ground coffee in the *portafilter*, water in the tank flows into a pump, which builds the pressure up to around 130 psi. Note that the *portafilter* requires the ground coffee to be of a certain density, otherwise either the pressured hot water will not spread through the portafilter (and hence produce less output but very strong), or if it is too light, the hot water will find a rapid path through the coffee grounds and produce a weak coffee. This is why the ground coffee in the *portafilter* is manually tamped with an aluminium piston and vibrated before the *portafilter* is inserted into its cradle.

- The pressurised water is then released through a heating element (not dissimilar from the instantaneous water heater shown in Figure 8.12) where it is heated to around 93°C.
- At this point, if it is of the *pod type* a piercing lance having three piercing heads (like a trident but in triangular form) at its end, is pushed into the pod breaking into the ground coffee, and the pressured hot water is then injected through the lance into the pod – the pod is pushed against a grating containing sharp pins which puncture the thin flat aluminium face of the pod and the pressured hot water then flows through and out of the pourer. Sometimes this does not necessarily work well, and if the pod is not quite seated correctly, some of the coffee grounds will escape in the water producing a grainy (sandy) coffee which is not nice. Also it is possible that over time, small coffee grains can collect in the coffee chamber and be flushed through later by the hot water. If it is of the *portafilter* type, the pressured hot water passes through the *portafilter* and down to the spouts. Only tiny particles of coffee are passed through with the hot water.
- This type of action produces a coffee with a small froth head, but if more froth is required such as when making cappuccino or latte, it is possible to use a separate *steam wand*. However, most simple espresso machines do not have this attachment whereas most commercial coffee shop machines with a *portafilter* and *spouts* do have a *steam wand* since they have to cater for all forms of coffee bean types.

8.12 THE HOME PRINTER AND SCANNER – 3D PRINTING

There are two types of *home printer*, which are the *inkjet printer* and the *laser printer*, with the *inkjet printer* being the most common technology in the home (it is cheaper to buy but produces less prints) while the *laser printer* is more for large quantity production (and is more expensive) [(12)]. Note that in the following explanation, it does not include the rapid mechanical operational movements which are taking place to make the process happen:

The *inkjet printer* originally invented by UK physicist Lord Kelvin in 1867, was properly developed into two types of printer in the 1950s: the *continuous inkjet* and the *Drop-on-Demand* (DOD) inkjet.

- The *continuous inkjet printer* is often used in marking packages because it provides a continuous stream of ink in which a high pressure pump squirts liquid ink from a tank through a tiny nozzle.
- An *ultrasonic piezoelectric* crystal (Figure 8.15) can provide very short pulses of pressure down the nozzle to break the stream into tiny (under 0.1 mm) droplets. The

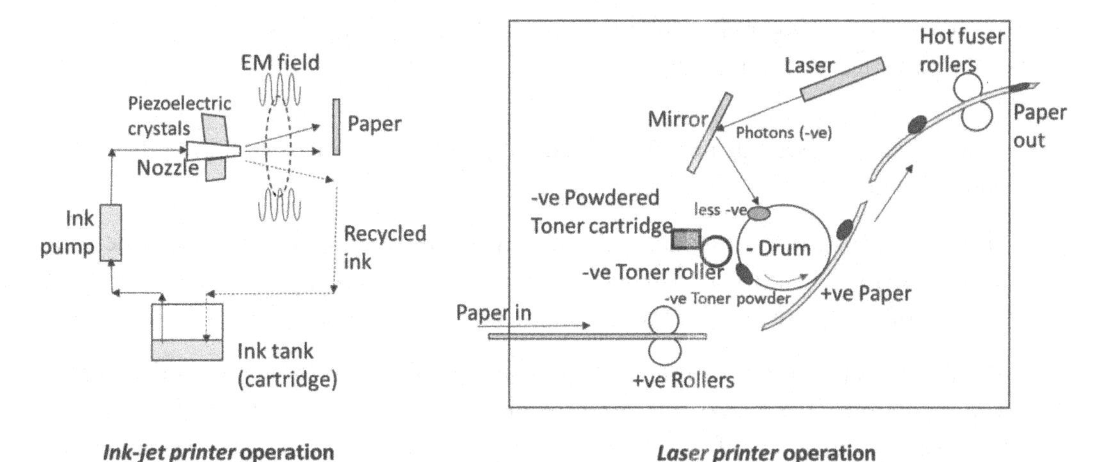

FIGURE 8.15 Inkjet and laser printer operation.

nozzle is electrically charged so that the exiting droplets are charged and as they exit, they pass across a magnetic field that deflects them (like electrons flying through the old cathode ray tube of Figure 1.12) directly onto the object being marked which moves slowly. Just as with conventional rastering of data (Figure 1.13), the droplets are rastered onto paper to form images of words and pictures, with droplets not being used sent on to a receiver where the ink is recycled.

- By comparison, the *DOD printer* can still use the concept of a piezoelectric crystal vibrating a drop out of a nozzle, or it can instead use a heating element to expand and pressure droplets through a hole in the nozzle (aka *thermal* or *bubble printing*).
- Either way, there is no pressure pump, electrically charged nozzle or magnetic field. In this case, there can be one or multiple nozzles dropping ink while the object being printed upon is rastered with respect to a stationary nozzle.
- Consequently the *DOD printer* is simpler in operation but limited to printing operations where the paper moves with respect to the nozzles and their droplets, causing this process to be slower than that for *continuous inkjet printers*. However, the *DOD printer* provided the concept of squirting melted adhesive droplets onto a surface, which was then the forerunner to the concept of *3D printing* which will be discussed later in this section.

The *laser printer* uses the principle that rollers, paper, and powders can all be electrically charged at different levels, which can then be used like magnets – to attract powder to adhere to paper. It works on the basis that a laser beam can change the electrical charge of points (pixels) on a drum (that can be electrically charged). When paper passes over the drum, the paper is charged with particles which can then attract electrically charged powders to be deposited on the paper. The powder is melted onto the paper by a heating (*fusing*) *roller* (if the *fusing roller* isn't hot enough, the paper feels slightly damp but if it is too hot the paper feels hot but dry):

- A laser beam pulses out negatively charged *photons* at a mirror which reflects the photons onto a *drum*.
- The drum's surface has the ability to be negatively electrically charged. When the negative drum rolls, the spots where the laser's photons are arriving changes charge level becoming more positive (i.e. less negative).
- The drum's surface then passes over a *toner roller* (which is a powder depositor). The toner roller has picked up charged toner powder from a toner cartridge, and because these toners are less negatively charged as they stick on the *toner roller*, the difference in charge means they are electrically attracted to deposit on the drum, having left the *toner roller*.
- While this is happening, paper is fed over strongly positively charged rollers and on to the other side of the drum.
- As the positively charged paper comes into contact with the *roller* with a more negatively charged set of powder particles, the powder is then attracted to the paper's surface, after which the paper then rolls through a pair of *fuser rollers* which apply heat to melt the powder into the paper.
- Any residual powder on the drum is removed by a scraper bar which is even more negatively charged, and the process continues.

The *3D printer* was developed from the *DOD printer* as explained earlier when the DOD tested the use of different types of inks including plastics and metals. As explained either a piezoelectric crystal using pressure or a heating element is used in the *DOD printer* to direct a liquid through a tiny nozzle, and the resulting drop is directed at a particular location on a medium – typically paper. Over the years, different liquids were tested in which the heating element (aka *thermal printer*) could be used with liquid which was more often thicker than ink.

The development of *computer-aided design* (CAD) software for designing and controlling automated machines brought about the ability to move the printer nozzle in different horizontal directions, while the surface being printed could be moved in any other direction.

So it became possible to take a 3D CAD model and section it (in the computer) horizontally so that the contours of a 3D model could steer the injector nozzle around the contours. It also became possible to move the surface to be printed in other directions, typically vertically. Having the nozzle move around a contour map squirting droplets out over a surface, with the surface moving vertically, meant that a contour slice of the 3D model could be laid of plastic or metal, and then the surface could be moved vertically down, and the next layer of plastic or metal could be laid over it, thereby building it into a 3D shape.

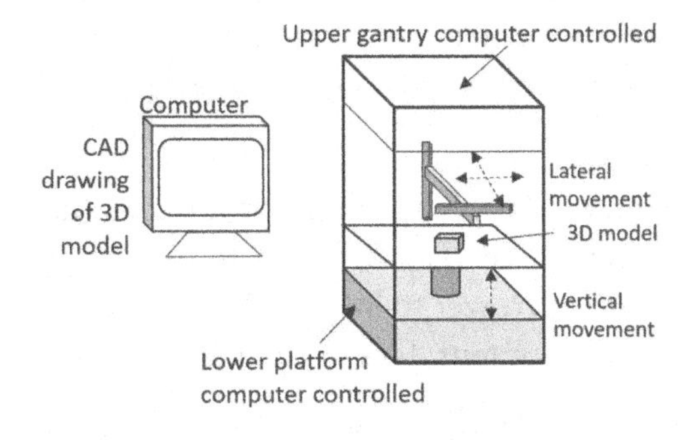

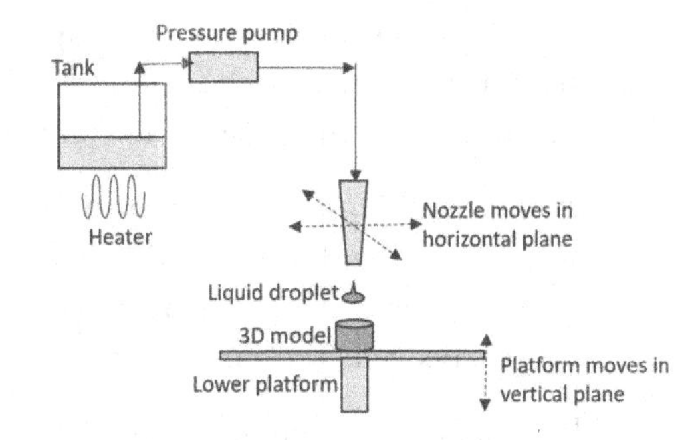

FIGURE 8.16 3D printing (aka *Additive Manufacturing*).

This meant that 3D shapes could be printed vertically and today it is now possible to print metal tools, rubber vehicle shock absorbers, and simulated heart valves using all types of plastic and metallic additives to build up the 3D model. This is why it is now referred to as the industry of *Additive Manufacturing.* But for the home, it is now possible for a few thousand dollars, to purchase a home *3D printer* to build simple 3D models of different objects that are limited by the type and cost of additive materials that are needed [(13)] as shown in Figure 8.16.

- The *3D printer* comprises of a computer and separate 3D printer. The computer has CAD software, which allows a 3D mathematical model of an object to be input – the software can thereafter mathematically slice the model horizontally to provide the *3D printer* with a set of horizontal maps that contour the model from top to bottom.
- The *3D printer* has an upper level which houses the electronics to drive a gantry containing horizontal arms, with the ability to move laterally in x and y directions, computer controlled by the *3D CAD computer* software. It also has a lower platform that can be raised or lowered vertically upon which the material is to be deposited. This lower platform is also CAD controlled.
- Referring to the flow diagram of Figure 8.16, the material is placed in a tank and heated to a specific temperature and liquid state. It is then pumped into a vertical nozzle at pressure, and the timing is such that at specific x and y locations, liquid droplets are squirted out of the nozzle to be deposited in particular positions on the lower platform plate, so that after each contour map of liquid has been dropped, the lower platform moves down a specified distance. During the time from liquid placement to the next droplet release, the preceding liquid will have dried and set hard.
- The cycle is repeated but this time, the droplet is now dropping onto the previously hardened droplet, and so the droplets combine in the vertical direction so that a model is built vertically.

- Liquid droplets can be dropped anywhere on the lower platform plate, but in order to make a complete model, at some stage individual components will be expected to be linked with the main model to complete the 3D model.
- In industrial *3D printers*, this process can go on for days depending on the size of the model being built, and eventually the completed model may still need to mature or solidify by being placed in an additional furnace or freezer to have the final material features needed. The simplest model can take three or four days to complete.
- The industrial benefit of *3D printing* is that, provided CAD measurements are available along with appropriate materials for the printer, any single object can be 3D printed in any location, which is useful when specific unavailable parts are needed in remote locations of the world. For example, if an offshore platform valve ruptures, a replacement valve may be manufactured on-site rather than wait for one to be built onshore and then delivered to the remote location days later. In fact, the old part or component that has failed may be used for measurement purposes to *3D print* a replacement component. There is a limit to the size of any components that may be made (depending on the size of the printer), and large components bigger than a cubic metre are hard to produce – typically something about 0.3 m^3 (1 cubic foot) is the limit.
- A major issue is the type of ink to be used for the particular application. Metal ink comes in all forms of alloy mixes. Rubbers have different strengths, elasticity, and manufacturing temperatures. Plastics are also being used, of sufficient elasticity that they can be used to manufacture heart valves and replacement veins. However, the more complex a 3D object is, there more complex the 3D printing of it and the longer it takes to manufacture correctly. A single-layered model may take a week to manufactured, but a multi-layered model (i.e. a model which has one layer printed, and then a second layer is put over the first layer, etc.) can take months. A heart pacemaker may be made containing plastic to form soft tissue with an embedded electronic circuit as the pacemaker made of different plastics and metals. There is a lot of complex chemistry involved in these inks, which is an industry of its own.

8.13 THE HOT AIR FRYER AND THE THERMOMIX

Because some foods (like chicken, French fries/chips) taste good if the outside has a crusty layer but the inside retains its moisture, we use fat or oils which have a high boiling point to rapidly heat the outside of food (up to about 180°C) and create a crust without cooking the inside – a process known as *frying*. But fat contains glycerides which can add fatty acids to our body, unhealthy fats we don't need. Instead, it would be good to cook food using something which does not involve fats but heats the food up rapidly to give a fried crusty outer layer without cooking the inside [14]. This can be done in a fast-acting fan-forced hot air oven or a *hot air fryer*. Basically, a *hot air fryer* as shown in Figure 8.17 has the following features:

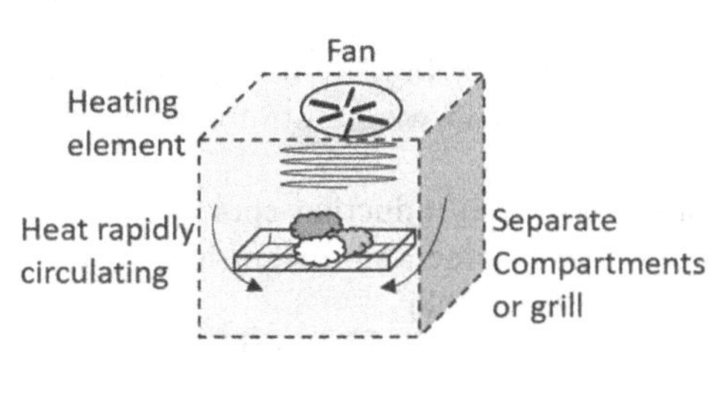

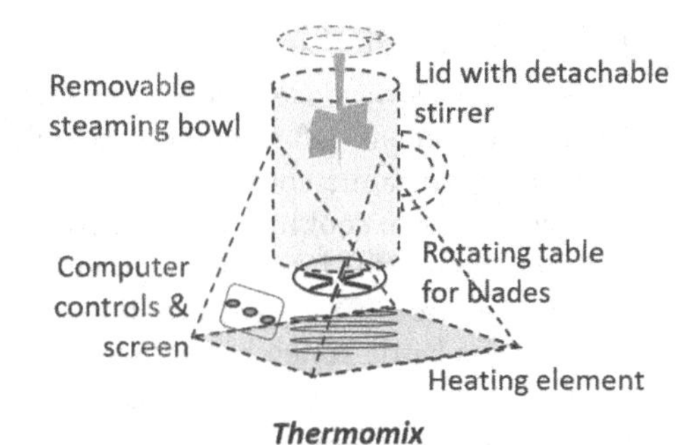

FIGURE 8.17 The Hot Air Fryer and the Thermomix.

- Fast operating electric heating element on the top with a fan directing the heat downwards.
- The food sits in a compartment or grill raised in a manner that the food sits off the fryer floor and away from the walls, with compartments having separate sections for cooking different foods.
- The heat gets up to high temperatures quickly, and is circulated around the food, heating it on all sides equally, with the natural fats and oils of food being cooked internally, or in some cases being uncooked if the heat is insufficient.

The Thermomix was patented by German company *Vorwerk* in 1961, to mix, chop, grate, stir, grind, caramelise, steam, mill, and juice food. It has a heating element in its body, cutters, and rotating table so is effectively a minicomputer controlled *mixer* or *blender* which can also be used to weigh food as a weighing scale that works from a touch-sensitive screen. The removable lid has a paddle or stirrer attachment.

The minicomputer is loaded with different menus to allow an easy choice of food production provided the food ingredients are loaded at specific times, and owners can pay for online recipes since it has *wifi*. It comes in modular form with some interchangeable parts. The *Thermomix* does not replace *the hot air fryer* since current models do not exceed 120°C; however, *Thermomix* has the ability to steam which is important in most foods [15].

> "The *Thermomix* does not replace the *hot air fryer* since current models do not exceed 1200C; however, *Thermomix* has the ability to steam which is important in most foods."

It has a removable steaming bowl (which contains four blades at the base) which, when locked into the housing, has its blades driven by a rotating table within the *Thermomix* body. The four blades have four heights for different chopping levels and rotate at between 40 and 10,000 rpm. A *Turbo* button allows the mixer to go immediately up to top speed for short periods.

FURTHER READING

1. How do microwave ovens work?: https://www.britannica.com/story/how-do-microwaves-work
2. Pros and cons of ceramic and glass cooktops: https://www.homestratosphere.com/pros-cons-ceramic-cooktops/
3. How induction cooktops work: https://home.howstuffworks.com/induction-cooktops.htm
4. Steam ovens 101: https://www.goodfood.com.au/good-living/home-and-design/steam-ovens-101-what-are-steam-ovens-and-how-do-they-work-20170824-gy3h8j
5. What is a kettle kone?: https://www.bbqspitrotisseries.com.au/What-is-a-Kettle-Kone-How-Does-it-Work-and-Whats-Special-About-It
6. How does rinse-aid work?: https://stovedoc.com.au/blogs/dishwasher/how-does-dishwasher-rinse-aid-work
7. Machine phone customer support apps: https://zapier.com/learn/customer-support/best-customer-support-apps/
8. Origin of 'piping hot': https://wordhistories.net/2016/08/10/piping-hot/
9. What is an evaporator coil?: https://www.aceplumbing.com/plumbing-faq/what-is-an-evaporator-coil-and-how-does-it-work/
10. Electric storage and instantaneous water heaters: https://www.energyrating.gov.au/products/electric-storage-and-instantaneous-water-heaters
11. The nine best coffee and espresso machines 2021: https://www.thespruceeats.com/top-coffee-espresso-combo-machines-765533
12. Inkjet v laser printers: https://www.brother-usa.com/inkjet-vs-laser-printers
13. What is 3D printing?: https://3dprinting.com/what-is-3d-printing/
14. Hot air fryers: https://www.goodhousekeeping.com/appliances/a28436830/what-is-an-air-fryer/
15. Thermomix: https://thermomix.com.au/pages/get-to-know-thermomix

CHAPTER 9

The Future of Analytics and Automation

In this book, we have discussed the recently developed *smart* technologies which rely on changes in different material's physical properties, and are increasingly used due to the increase in computer and network speeds. Current computers have now almost reached their maximum speeds and we eagerly await the introduction of the *quantum computer*, which will be a totally new innovation providing incredibly fast computer speed which will start the development process off again [1].

> "we eagerly await the introduction of the *quantum computer*, which will be a totally new innovation providing incredibly fast computer speed which will start the development process off again."

We have considered how the fundamental physical attributes of a material can be used to develop a *smart* device, and in particular applications using the simple vibrations of both sound and electromagnetic waves (both used in *smartphones*, TVs, and robotics). We have also discussed how simple computer *pattern recognition* works using *cross-correlation* methods. This can equally apply across all manner of applications, not just credit card or face recognition, but also for automated parcel delivery and other equivalent automated devices.

An early take-up of technologies in sport has resulted in team selection being biased towards those who have better performance *analytics* and consequently, sport has led the field in the take-up of technology with real-time analysis and performance predictions.

It's hard to judge the potential for *infrared sensing*, *avatar* usage, *electromagnetic levitation*, high-frequency *sonic vibration*, *microwave* applications, and *pattern recognition* [2]. However, we can make some reasonable assumptions about the physics we know and the potential technologies that may use them.

DOI: 10.1201/9781003108443-9

9.1 SMART APPLICATIONS TO 2D AND 3D IN SPORT AND SOCIETY BENEFITS

When we refer to *smart*, we mean clever or intelligent. Generally, that suggests a technology that appears to think for itself, to immediately assist the user of that technology. Because electronic equipment (phones/computers/iPads) will be almost infinitely fast when the *Quantum (Q-)* computer (Chapter 6) arrives, computing and refreshing screens which are already faster than we can blink, will appear to be done immediately. This makes the presumption that the faster electronic speeds of *Q-computers* will be used to increase network speeds. The first industry to be involved in rapid computing is likely to be the computer gaming industry again, followed by sport linked to the gambling industry (which is because of the huge sums of money that these industries attract). Consequently, it is easy to speculate about which areas will make great leaps in user technology.

9.1.1 Some Applications of 2D Analytics to Sport and Their Consequences

We discussed how sports, and particularly baseball, was first to develop predictive analytics. Using statistics of baseball players and comparing player performance, it was possible in the early years of the 20th century to determine which players were best in terms of batting and pitching averages (Figure 5.12) and who should be substituted when not playing well. If these statistics were relevant during such a match today, they would be displayed on-screen during the game, and as more strikes occurred in the game, so the statistics would immediately change. In most games, the quality of a player's performance is assessed during play, and for tactical reasons, players may be replaced by other players depending on their current and historical performance. We can expect to start to experience this technology in about 2040:

- *Pattern recognition* – The ability to use real-time *pattern recognition* will allow a player's action in real time to be displayed alongside that of a well-known player in a similar running position [(3)]. This is being done presently in single athlete games such as a tennis player serving a tennis ball in which the serve of one tennis player can be compared with a previously recorded historical serve by a well-known world-class tennis player. However, the ability to perform *quantum-computing instantaneous* (QCI) pattern recognition will allow the viewer to be shown the identical action immediately from existing data, alongside both sets of analytical figures (speed of serve, arm action, height above head etc.). The viewer will have the option to be able to view a repeat of the action from any angle immediately [(4)]. Technical details of the serve (arm hinge-point, wrist metrics etc.) would be displayed at the same time. The viewer can have control over which statistics are to be displayed since some stats can get in the way of watching a good game.
- *Simulated 3D from 2D cameras* – Having multiple cameras (including the *Spidercam*) in a sports arena already allows a simulated 3D view, but a *QCI* viewer will have the ability to be able to constantly view the simulated 3D view on a screen by using a joystick or by moving the viewer's head to create a simulated 3D view. For the sports

viewer, this would provide a view of the flight of a ball's trajectory including along its direction to see if it spins and curls in one particular direction or another - which may provide an indication of the skill of the player. Since football sports boots now allow a ball to be kicked and spun in a favoured direction, it would be possible to observe the flight of the ball from the boot, over or under the cross-bar, and provide the various statistics that go with the ball's trajectory. The viewer would have the ability to play the action back and request particular stats about the ball spin, etc. This is presently done to a very limited degree in some sports, but with *QCI* control it will be incredible to see.

- *Real-time pressure quantum analytics* - The clothing worn by sports people would be sensitive to pressure, so that not only would the number of steps and direction that a sports-person would take be recorded (and transmitted), but also the forces during the game on that player could be calculated as well as the force given to any action by that player [(5)]. These forces would link to player energy and consequently exercise training and work out routine as well as diet. A knowledge of the statistics of impact of these forces when one player impacts another (e.g. a tread on an opposing player's foot) could then be used to determine whether the impact was intentional, and may lead to an automatic yellow/red card or other player disciplinary action, without any ruling from the referee [(6)]. This would then take borderline cases of player contact out of the referee's qualitative judgement and into an immediate *QCI* quantitative decision.

- *Wafer-thin quantum monitor TVs* - We will be watching TV monitors in the future which will be credit card thick and equally light in weight [(7)]. This will allow hanging the monitor on a wall using suction pads, and the monitor magnetic frame would connect with the data transmission network. These would also provide power to the TV using a local house EM field (similar to how electric cars can be charged by driving over a hidden magnetic grid, and how a *smartphone's* battery charger works) or local built-in solar panels.

- *Data spectral analysis* - The 2D TV frequencies could be managed like we conventionally select channels on a remote controller, so that all world stations would be received and also all spectral content could be displayed. For example, the *infrared* spectrum could be viewed during a game by the coach, to observe if any player is becoming hot or has issues with a high impact bump to the body - the *QCI control* would be programmed to inform the coach of this issue. The computer monitor could then display a wind-back action box to show the incident, while looking at the player's personal body data to see if there were blood pressure changes or other personal body crisis developing at the time of the impact.

  But it would not be on the playing field alone that such spectral analyses take place. If an observer is monitoring a landscape view of a bush fire, the different frequency spectra could be chosen so that the viewer would be able to check where the fire's hottest and coldest points would be. The viewer could zoom in and out of areas of interest shown on the monitor. Touch controls, as are presently displayed on some sports monitors to replay the action, are used on the common monitors of today, whereas *QCI* monitors will allow the user to replay any action at any time by a touch of their remote control, or by mind control if a specialised *VR headset* were used.

- *Real-time reality predictions* – 2D predictions will be possible on a 2D screen (such as a *smartwatch*, notepad, or laptop), to monitor the direction a wildfire was moving and animations could be played out, after automatically checking wind speed, direction, and predicted weather pattern. Land map spectral data which provides soil moisture content would have been automatically checked and included in the predictions. In wildfire situations, such data would become invaluable for determining optimum fire areas to be suppressed by water bombers and predictions would be automatically provided to all occupants in the area wearing *smartwatches*, so that they could be advised of the optimum departure direction from where they are if needed.
- *Seeing more with smart quantum eyeglasses* – Simple light-weight eyeglasses would have built-in transmitters and receivers, to allow the viewer to observe data on the screen [(8)]. By pushing a button on the side or by voice control, data would be displayed on a virtual screen (as *heads-up* data is presently displayed on jet fighter pilot screens). The data or images will appear to be at a short distance in front of the viewer, and by moving the head the viewer could change perspective from the equivalent of reading a page on a *virtual screen*, to looking at a distant object (this will be done by the eyeglasses monitoring eye movement). While voice control would be the most likely viewing command control, a keyboard or *virtual* keyboard could be viewed (as in viewing a computer keyboard) that allowed the observer to type or push potential buttons in the air to bring up or enhance the appearance of a desirable object (but then it is very likely that instead of a keyboard we in future will just use voice commands).

  For example, if an internet page is required, an internet logo would appear on the *virtual screen* and after the viewer has pushed the logo or spoken the name of the logo, that page site would appear. If a video-conference is to be held, this could occur without removal of the eyeglasses, and any number of images of people could be viewed on the *virtual screen* because they could be individually expanded (zoomed in). These glasses would respond to individual eye ball changes so that when the eye lens becomes older, the glasses automatically compensate to better focus the lens. A further benefit is that if the eyeglass wearer had a speech impediment, the glasses could be trained to read the eye movements, which could then be translated to speech (as with *Hawkins computer-speech* but with the difference that this speech would be similar sounding to the voice of the wearer). If the viewer was deaf, the eyeglasses would be programmed to read lips and any form of sign language, and translate this to vibrations representing normal stereo hearing, with the same sound and tone, by vibrating the ear bones.

- *Construction without training* – Training courses and industrial product guides would make use of *augmented reality* by providing *YouTube*-type movies in the eyeglasses of how pieces of equipment can be dismantled/rebuilt, and the user would be able to take apart or rebuild equipment with minimal formal training and the maximum of external assistance [9]. As a result, the practical knowledge of the world's population would improve dramatically.
- *Home-based working* – The ability to attend instructional courses and meet people without being in their presence, as has been demonstrated during COVID-19, indicates that university and college buildings will be a thing of the past and unnecessary except when using laboratory spaces for research and development. The workplace will transform from major office blocks to work from home 'cottage-industries'. Only manufacturing, maintenance, and public utilities plant will require large housing in factories.

9.1.2 Some Applications of Quantum 3D Analytics to Society

The ability with QCI technology to rapidly compute statistics in 2D will also apply in 3D sports, with its own social consequences. The difference between 2D technology and 3D technology is that 2D technology provides the user with a view of action, whereas the 3D technology giving image depth makes the user part of the physical action. In 3D sports, the ability to run a 3D simulation is just an add-on of depth, making the 3D result seem very real even if it is in fact, false. The difference in this case is that the viewer has the chance of personal involvement. For example:

- *Feeling the occasion* – The difference in 3D is that images of any scene are now shown in 3D, in which the QCI glasses provide the 3D image when used with the 3D virtual screen. In principle, we would be able to walk down any street in the world, turning the head, and feeling part of the action without actually being there. Imagine an event in which a prize is being awarded on the world stage – for example, the Oscars or a Nobel Prize. Wearing the 3D glasses, it would become possible to stand next to the person being awarded the prize, and if associated soundtracks were fed through the eyeglass vibrations, it would be possible to have a sense of being a part of the occasion. This would work for historical circumstances in which 3D game movies would be made of events such as the lunar module of Apollo 11 landing, the arrival of the Pilgrim Fathers onshore US, epic battles fought in Europe during the Roman Empire expansion and how daily life was for an Egyptian Pharaoh 4000 years ago. With the glasses on, it would be possible to stand on the moon and be there when Neil Armstrong steps down the ladder onto the moon. All of the view appears on the eyeglass *virtual screen*, with a vibration suit being worn as part of the action if body vibrations are needed. You are on the moon so smell is not an issue, but now particular frequencies from the glasses to the nose will provide a chemical interaction in the nostrils, producing the experience of smell – which doesn't happen on the moon. But the smell of a nice, tasty pizza is possible – without the real food being available.

- *Cooking 3D food* – using QCI technology it is possible to use *augmented reality* with special food gloves, to make cakes and smell their aroma. Learning to cook becomes something you can do without a kitchen provided your glasses are programmed correctly, and you wear the right haptic cooking gloves which give the sense of handling real food without the mess (or soiled dishes, pots and pans – hey no washing up!) [10].
- *New chemical formulae* – it is presently possible to observe computer simulations of what happens when some chemical compounds are mixed with other chemical compounds, which steers the development of new chemical products. We have all seen 3D animated COVID-19 virus images wandering around in 3D space, and how different formulae of vaccines may affect the virus leading to an immune response. However, wearing the *haptic gloves* and the 3D QCI eyeglasses, the *virtual screen* will provide a 3D perspective of the virus which can be touched by the gloves giving the researcher a sense of their strength and elasticity. The viewer could take a vaccine image (with all physical statistics having been input to the computer) and attach it to a virus to see how well it adheres.

  The concept here is that there are formulae available which describe all natural reactions of both vaccine and virus, and all we are doing is putting them together to observe the reactions. Of course, the *quantum computer* would be sufficiently fast that it would give the result before we had time to put them together in the virtual world anyway, but this example just demonstrates the ability to develop new medical and equivalent industrial tools, through the use of 3D *augmented reality* using *haptic gloves* and *QCI virtual* eyeglasses. It follows that there are numerous disciplines in which such virtual work can be done in the future, with the advent of *quantum computing* electronics.

9.2 GAMING AND SIMULATIONS CHANGING SPORT

Gaming and their simulations, the first to take up conventional computing to provide games to the public, will be first responders to developing *quantum* game studios where the game player can enter a room specifically chosen for the coming experience. Imagine a seat in which a racing car driver sits, located in a small zone where a racing driver *haptic suit* is worn. Insider the suit, the driver sees through the virtual goggles and smells the odours such as fuel and hot tyres. This is a true *virtual haptic experience* which is now a *QCI experience.*

9.2.1 Simulations in Real Time Allowing New Games to be Developed

The new *quantum computing* world interfaced with *haptic experiences* ushers in a new era of sensing *augmented reality* [11]. For example, this type of gaming allows people to take part in a real race without stepping into a real racing car. During a real event such as a Formula One race, the view, sound, and smell of these racing cars on the starting grid will be broadcast live to the gamer, who is you sitting in the driver's seat of what appears to be an F1 racing car (in the *QCI Experience* room). The smells and sight around the driver add to a sense of realism, and the fact that well-known racing drivers are seen smiling at you from the cockpits of their vehicles, adds credence to it all.

> "The new *quantum computing* world interfaced with *haptic* experiences ushers in a new era of sensing *augmented reality*."

As the race is about to start and after putting the *haptic suit* and *QCI glasses* on, the gamer steps into something that seems to imitate a racing car, grabs its steering wheel and revs the car engine in a *QCI-real* event, without ever leaving the room or having the danger of personal injury.

The race commences, the corners are taken, and a roll over occurs – but while the sky appears to roll around this car as it tumbles, the driver senses the rapid compressive forces on the *haptic suit*, smells the tyres burning and feels the suit becoming hot. The vision through the googles is one of the earth and sky rolling around. Eventually the car comes to a standstill (the actual car is sited on a gimbal so that there is a sense of gravitational change which allied with the *haptic* senses from the suit pressure, make it all very realistic) the right way up.

These games can be extended into flying planes, engaging in World War I airborne dogfights with the Red Barron, or just driving a double-decker bus through central London with the selection of any form of light or heavy traffic available (in an accident, you can select a conversation or argument with the driver of another vehicle or bus, which could be *Michael Schumacher* with a typical London Cockney accent). At last, those who want to be a train driver in the *Flying Scotsman* train travelling from Edinburgh to London can sit in the train-drivers cab, feel the train vibrations, smell, and sound while watching the stoker shovelling coal into the boiler if needed.

9.2.2 Reality 3D Sports Using Haptic Sensor Suits and Automated Refereeing

During a real game, the players would wear thin suits containing the *haptic sensors*, much like those discussed so far. However, this time the suits are calibrated so that any particular pressure on the suit is transmitted to the team physio who is monitoring the player data. Boots and gloves are worn, so that any impact on the boots or gloves can be computed. If any values exceed a specific value deemed to be damaging to the body, appropriate alarms are sounded so that the coach and physio/doctors know one of the players may be singled out for damaging tackles.

Because the suits are transmitting their positional and other data, the player's position is constantly automatically monitored – but this position monitoring is far more precise than ever before, since millimetre accuracy of the receiver's locations is being transmitted.

In football (soccer), when a player kicks the ball the location of the boot is known and a determination can be immediately made of whether a player is offside or outside of the line of play without the need to check with a referee or linesman (aka *assistant ref.*). This approach would be an extension of the existing VAR conditions, making the game faster and far more precise in decision making.

As a player runs across each line, a brief pressured sensation could pass through the haptic suit so that the player would know his or her precise position with respect to the lines on the field. This would improve player tactics since they would automatically have a sense of where they were on the field with respect to areas of the field. Equally, a player who is offside would feel a sensation in the suit which would be an indication of the part of the body that made the player offside. This would naturally cause an adjustment to the way the player would position him/herself.

In rugby, baseball, American football, lines are no longer needed for refereeing on any pitch because the precise location of every part of every player is broadcast live to the watching audience, which can sit watching the game live through the *QCI eyeglasses* and can request a play back of any action just on their eyeglasses. Of course, lines will be needed just for player judgement of the ball's position with respect to the line.

Because the pressures are being monitored continually on all player bodies, when one player comes into contact with another, the amount of impact and the form in which the impact is made, is automatically determined as acceptable, or a foul, and a yellow or red card given. Instead of a referee blowing a whistle, an audible alarm would sound and the haptic bodysuit could stiffen causing specific players to slow down, hesitate, or stop.

There would be a set of automated penalties for a player making a foul with a specific impact type and level, with the punishment provided from the database. The punishment would be shown on a screen – if the player seeing this was sent off (having a red card) but refused to go, the bodysuit would contract – squeezing the player and for every second the player hesitated, it could award a sanction. This could be a ban for a number of games. Of course, these bans could be reviewed by a panel of experts a few days later if needed. Whichever way it was done, the computer would replace the referee.

This approach to refereeing any game would take the human element and subsequent controversy out of any professional game and increase its speed. In the event of on-field controversy, the captains would be empowered to work out what to do with the aid of the computer [(12)]. Recommendations of what to do would be provided on their *QCI eyeglasses*. In a worst-case scenario, if the players did not stop kicking the ball, the ball could be signalled to deflate, only to be manually replaced when required.

9.2.3 The Avatar, Soccer with Pele, Cricket with Bradman, Baseball with Babe Ruth

Games could be watched in the stadium at the ground, or in the *virtual haptic experience* game centre, where the audience could take part in the game as mentioned earlier. However, this time the team could be filled with *avatars* of great players of their sporting era, such as playing alongside *Pele* at his prime, bowling to the world's best batsman *Don Bradman* to try to get him out, and pitching at *Babe Ruth* in a scene reminiscent of the 1920s *New York Yankees* baseball ground.

Because the *haptic bodysuit* could be worn by anyone anywhere, and the 3D *QCI eyeglasses* provide a 3D sense, sound, and smell of being anywhere, it then could be possible to have any form of *avatar* in a person's company, unseen by anyone else [(13)]. Of course, the *avatar* could engage in two-way conversation and it would be possible for a second person to join the conversation with the computer providing the same *avatar* to both people. Taking this to the next step, it would be possible for a live audience to meet with a chosen *avatar* to discuss issues of the day, in which the *avatar* could be programmed by another person with specific answers to specific questions. This would make live entertainment quite different and be the first step towards *avatar* robots operating alone with their own programmed reactions.

We could for the first time, simulate being at an event which we could not have attended. For example, we could dance at the *Cavern Club* in the early 1960s while the *Beatles* played live, or be on the stage at *Woodstock Rock Festival* in New York in August 1969 dancing alongside *Jimi Hendrix* playing the legendary *US National Anthem*. We could stand next to *Winston Churchill* in his bunker during the *German Blitz*, or stand in the *Capitol* next to *John F. Kennedy* during his January 1961 inauguration (and then shaking *haptic* hands!). The potential for using *avatars* in a *virtual haptic 3D* space would be remarkable, very instructive, and an incredible learning experience of the hardships and joys mankind has suffered/delighted in over the centuries.

9.3 SMART TECHNOLOGY CHANGING INDUSTRY

9.3.1 Communications

Most people have a mobile (cell) phone. Most people in the future will be wearing the *3D QCI eyeglasses* as a piece of normal clothing, and these will be used as the mobile phone, hearing device, and vision aid, all in one. They will be controlled by voice or mind and will link with computers around us using *wifi* and/or shorter-range *Bluetooth* broadcasting. In the *QCI* era, communications will seem to be immediate with no latency. There will be no network drop out issues with there being multiple towers with multiple telco providers all operating with *QCI* electronics.

Rapid communications will allow us to work at home, in the office, or restaurant. Today we are used to occasionally seeing busy people walking around with an earpiece apparently talking to no one but in the future, this will be commonplace and we won't consider it is an unusual or a 'busy' sight, because we do it ourselves during times of rest and relaxation, and business. This means that fast communications allow us to work constantly and it will be hard to get away from it. Business meetings with colleagues and friends will be using the *QCI eyeglasses*, which will allow us to have the image of those attending the meetings displayed on the virtual screen. The database will have an image of the meeting attendees and their broadcast message or discussion will then provide a 3D animated version of the person talking. Alternatively, they can still sit in front of a normal 3D TV monitor as may be done today, if live viewing is necessary.

Our *3D QCI eyeglasses* allow us to have nights out, write a document, draw a picture, or whatever we wish to do using the virtual screen. Whatever we do can be recorded, and our body data can tell us how fit we are for any action. Government has brought in controls so that the sales and advertising media cannot break into our daily life unless previously approved by us. Equally, we can contact vendors of specific goods and services, because we can call-out an item, and these will show up on the *virtual screen*.

9.3.2 Traffic and Transport

Due to increased greenhouse gases in the atmosphere, mankind has been moving more towards sources of power that do not produce such gases. These include solar, wind, hydro, hydrogen, and of course nuclear. Given that an electricity grid will be powered mainly by these sources in 2040, we can expect the major user of power – to be transportation of people – using electricity. It follows that we can expect battery power storage abilities to improve, but not to be the prime mover in public transport. Instead, the potential to use highly efficient *electromagnetic* (EM) energy fields would be excellent.

For example, consider the use of today's high-speed rail systems, as used by the Japanese *Shinkansen* railway (Figure 6.27). The rail has an *EM field* which the train uses to lift off the track (maintained in place by a central steel guide), and drive forward with little frictional force apart from travelling through air. This principle can be applied to public road transportation in which the top surfaces of all roads have a metallic grading or strip which is powered, and produces a strong EM field above the roadway. The electric cars would have an *EM field* that is powered by their battery (Figure 9.1), which in turn is charged when the vehicle is parked in a public car parking lot or simply standing at traffic lights. In these driverless cars, the car moves along a route provided by the computer, following a map that has been input to the computer as the end destination. The car turns corners or overtakes other cars, by using the *EM vectorial force* direction provided by its own *EM field.* Consequently, no central guide rail is needed, as required by the *Shinkansen* [14].

Under these circumstances, the car would follow all traffic signs to the letter, keep within speed tolerances and distances, and traffic in principle would flow efficiently without accidents since all cars would be communicating with each other within their own zone of impact. There would be a reduced need for traffic police, except in cases where pedestrians had caused issues by walking out in front of vehicles – however, such cases would be minimised since the personal phones would vibrate or ring prior to there being a hazardous incident. These phones would be connected to the car computers and announce a pedestrian's presence, preparing the vehicle's EM controller to take evasive action. Possessing such phones would be mandatory, which became accepted practice after the COVID-19 pandemic.

9.3.3 Everyday Living with Robots and Technology

We have discussed how robot vacuum cleaners can be useful in automated cleaning of the house, but in reality they are limited to areas where they can run unimpeded along the floor, and when they meet a carpet edge they invariably stall. They can't climb stairs nor

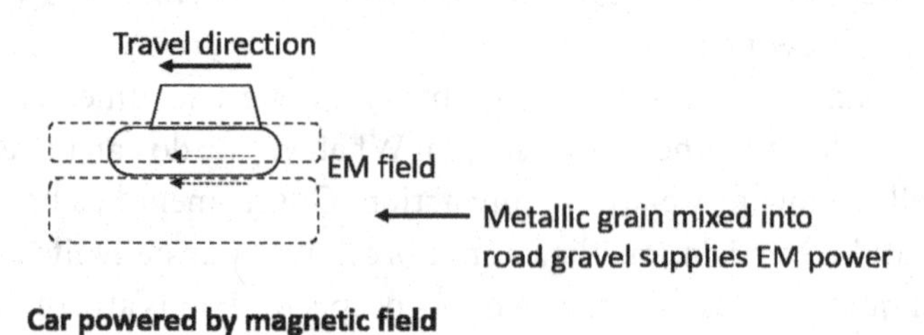

FIGURE 9.1 Electric cars with battery recharged from an EM strip in the road.

can they automatically empty themselves of dust. This provides us with a clear indication of the reality of robotic applications – they are built for limited delivery of services and it will be many decades – possibly the end of this century – before humanity will see a really useful *R2D2* robot that thinks and acts fully on humanity's behalf.

For many thousands of years, mankind has used slaves for the menial tasks of living, which is where robots and automation fits. Presently, automatic lighting is used in many office blocks where lights turn on and off depending on the detection of movement. *Siri* and equivalent *voice recognition* software allow house lighting to be turned on and off at the sound of a voice, and in some cases, houses are being built which use *voice recognition* to open and close blinds/curtains/doors, and so on.

Kitchen appliances are gradually being hooked up to the internet so that if we wish, we can turn the oven or washing machine on from a remote location, and we can already bake a cake via the internet using the abilities of the *Thermomix*. But the difference is that a home computer will control the conditions of the various appliances (such as the fridge and freezer) including central heating and cooling, based on maintaining optimum temperatures and minimising power usage [15]. The majority of power will be based on battery supplied electricity, which is fed from both rooftop solar panels and hydrogen/solar/wind turbines.

Citizens of the world will be savvy with technology, understanding how to use it rather than how it works. Having been brought up to understand *smartphone* icons, most technologies are available to most people, and if a better understanding is needed of them, they can just call up the *3D QCI screen* on their eyeglasses and it will be explained.

The only time work is not home-based is when manual labour is involved. All work except government and public servant operations will be based on short-term contracts rather than full-time company employee status. Contracts will be based on recognizable output, so that delivery timelines are set and failure to meet them will result in either a salary reduction (as agreed in the contract) or salary adjustment with bonuses when each output target is met. This causes a need to succeed more than ever, and working for the government will be classed as average pay but beneficial since a public servant is difficult to terminate. Only government work provides job security and certainty, which the public accepts.

9.4 AN AUTOMATED WORLD AND A ROBOTIC FUTURE

When *quantum computing* and its applications have become a part of the automated society around 2070, major careers can be found in maintaining automation systems and developing new application versions. Tourist space flight is well accepted, and mining companies have space shuttles taking their workers to the mining camps in far-off planets around 2100. The use of the global magnetic field and miniature nuclear power plant provide more energy than is needed, because the population is no more than 12 billion – recognised as a maximum for the earth.

Countries have grey and green areas, that is, where people live, and where there is open space. All machines and vehicles are electrically driven because power is provided by solar, wind, nuclear, and hydrogen. Greenhouse gas levels are reducing but still mid-level because a few third-world countries who use coal for producing electricity from power stations are slow to pick up technologies. There is a strong international movement against these low-technology pariahs. However, by 2100 the coal industry is moribund. People use nuclear, solar, and hydrogen as the optimum energy providers, which is a necessity because the greenhouse temperatures are such that we have passed the limit of 3°C. The UN deems that the world ecosystem will get back close to normality by (2110) when there will be less storms and weather patterns will be back to how they were in 1960.

Education is home-based entirely, because the use of government-provided *3D QCI eyeglasses* provides the ability for each individual to have lectures that are pre-recorded and fundamentally one-on-one with a very inspiring teacher. Trades are easily taught because the *haptic gloves* and *suits* extend to manipulation of objects (such as screwdrivers, wrenches, and other tools).

People are therefore far better educated than previously, but there are still laggard countries of the world where the population is educationally inferior to other countries, simply because of being left behind due to a corrupt government and politicians. However, few countries are in this category, because the use of *blockchain* accounting has reduced their number.

Government systems are much more dominated by individual strong character men and women elected on the basis of popular simple solutions to everyday problems. They argue governments must have strong leadership in order to maintain technological progress of the population. Without such progress, those countries become technically weak and the technically stronger countries take advantage of the weak countries as is normal. Africa still has a way to go to catch up to be average technologists.

An example of the difference between people of different countries is that tourists from the technically stronger countries can travel at a reasonable cost from London to Sydney in a two-hour public space flight, but the less technically strong countries are economically challenged and cannot afford such high costs [16]. They still use the old jets to travel. Apart from space travel for the experience of space (and its junk) and to set up new planetary outposts, the travel industry is still very much earth-based.

People live on average until they are over 100 years old with old-age being considered as 150 years old, because the medical industry has the knowledge to forecast and limit the impact of medical issues on the body. Wear and tear of the body which had in previous millennia been a problem, is no longer since gene therapy removes many of the ageing factors during the time when the foetus is in the womb.

Everything is predictable because we have been recording data sufficiently by now that even long-term issues are predictable and within recognised error limits. This applies to the effects of climate change, how to live in a remote site with difficult prevailing conditions, how to use *cryogenic* technology for long-range space travel, and what to do when new worlds are being discovered. By this time, we have recognised the potential for planets to support human life but they are in other galaxies, just not in our own galaxy.

Consequently, long-term space flight requires *cryogenic sleep* through a long flight, with the spacecraft crew and new settlers treating all travel as 'the usual 6 hour trip' (just below the speed of light) to the far-off planets. Everyone who travels in space is technically well educated by today's standards, and travel for a reason, not for simple tourism.

9.5 THE SMART JOBS ASSOCIATED WITH FUTURE TECHNOLOGICALLY CONTROLLED PROCESSES

The automated world is very smooth and not physically demanding – more *tech-know-how* demanding. To have a good life, it is a requirement to be able to adapt and fit into society with a knowledge of how to use *smart QCI* technologies appropriate to your profession [17]. By 2070, those in the educated world, having constant access to these technologies from birth, will consider those without as being inferior. Battles will continue between companies and countries for superiority and the thoughts of their people systems, but the battles are more software-based than physical. The leading technology countries have software police, and quiet wars often rage between countries without the people's knowledge (not dissimilar from today).

People eat and sleep using technology. Life is relatively easy if you have technology, with household appliances automated, and work is done six days a week if you want to improve yourself and your business. It is acceptable to go to outdoor camps for long weekends where there is no technology, to experience how to meet and discuss topics normally with people who are not family, and to restore mental well-being. The old British concept of a 'Butlin's camp' or school camp is very much alive and frequented by people who just want a simple period in their excessively busy life.

There is general recognition that we live on 'A' planet that must be maintained and improved if possible – there is no planet 'B' and when we die it is commonly believed that we have served our time with no afterlife. We maintain the planet by using our technologies to predict future issues, by advancing science which is the new God, by developing new tools and software for sale, and maintaining farming, fisheries, and wildlife. Those who live without technology are considered to be living in poverty – it's a relative thing.

The future jobs are in developing new technologies, writing new software, monitoring, and maintaining existing technologies while understanding the scientific background of one's chosen field. The fields of development are in energy storage and application, food technologies, the automation of medical systems and devices, the control and maintenance of automation systems both at home and in the manufacturing industry. So *STEM* (science, technology, engineering, mathematics) topics are the accepted practice with the arts considered secondary.

> "Jobs derive from the ability to develop good *STEM* knowledge. Little is needed to be understood about actual computing, provided one knows about applications and their use. This means that applications programming is a basic tool, which can be used in all industries. A knowledge of maintenance of *automated systems* is desirable, while the ability to perform well in *augmented reality* operations (games or a job) is prerequisite if the gamer is to play and win."

Jobs derive from the ability to develop good *STEM* knowledge [18]. Little is needed to be understood about actual computing, provided one knows about applications and their use. This means that applications programming is a basic tool, which can be used in all industries. A knowledge of maintenance of automated systems is desirable, while the ability to perform well in *augmented reality* operations (games or a job) is prerequisite if the gamer is to play and win.

FURTHER READING

1. Quantum computing 101: http://quantumly.com/quantum-computer-speed.html
2. EM levitation: https://www.instructables.com/DIY-Electro-Magnetic-Levitation/
3. Pattern recognition: https://www.journals.elsevier.com/pattern-recognition/
4. Football technology: https://football-technology.fifa.com/en/innovations/VAR-at-the-World-Cup/
5. Calculation of force: https://www.wikihow.com/Calculate-Force
6. A soccer red card: https://yoursoccerhome.com/red-card-in-soccer-a-complete-guide-to-what-it-means/
7. How thin can TVs get?: https://www.goodgearguide.com.au/slideshow/273613/how-thin-can-tvs-get/
8. 3D printed smart glasses: https://www.3dnatives.com/en/3d-printed-smart-glasses-to-help-dyslexic-children-020320215/
9. How online training works: https://people.howstuffworks.com/how-online-training-works.htm
10. Virtual objects with haptic technology: https://pluto-men.com/insights/creating-virtual-objects-with-haptic-technology/
11. The best augmented reality smart glasses 2021: https://www.aniwaa.com/buyers-guide/vr-ar/best-augmented-reality-smartglasses/
12. A sports referee: https://www.careerexplorer.com/careers/sports-referee/
13. What is a computer avatar?: https://www.easytechjunkie.com/what-is-a-computer-avatar.htm
14. Vector forces: https://www.ropebook.com/information/vector-forces/
15. Minimising power usage: https://www.choice.com.au/home-improvement/energy-saving/reducing-your-carbon-footprint/articles/five-ways-to-reduce-your-households-energy-use
16. How long does it take to get to space?: https://www.mirror.co.uk/science/how-long-take-space-astronauts-12473232
17. Do we need technology or is it damaging?: https://www.everythingtech.co.uk/2019-2-4-do-we-really-need-technology-or-is-it-damaging-us/
18. What is STEM and why is it important?: https://blog.csiro.au/what-is-stem-and-why-is-it-important/

Exercise Answers

CHAPTER 1 ANSWER

The clock rate or computer timing frequency is given at 3 GHz or 3,000,000,000 cycles per second. Since the sample rate is the reciprocal of frequency (1/f), then data would be sampled every $1/3 \times 10^{-9}$ seconds or 0.33 nanoseconds per cycle. In order to ensure aliasing must not occur (when things on-screen become unsynchronised), if we sample that data we must sample at least ¼ of this sample rate which is about every 0.15 nanoseconds.

CHAPTER 2 ANSWER

Camera 1

- Car Position 1 Distance = 100 m
- Car Position 2 Distance = 71 m
- Apparent speed = 29 m/s = 104 Km/hr = 65 mph.

Camera 2

- Car Position 1 Distance = 141.4 m
- Car Position 2 Distance = 71 m
- Apparent speed = 70.4 m/s = 266.4 Km/hr – **this shows the car has exceeded the speed limit of 110 Km/hr = 70 mph.**

Camera 3

- Car Position 1 Distance = 102 m
- Car Position 2 Distance = 86 m
- Apparent speed = 16 m/s = 57 Km/hr = 35 mph.

CHAPTER 3 ANSWER

3.5 Exercise

3, 1, –2.8, 0 with a binary base unit of 0.1, means that each position is double the next starting with the last value.

So $2^0 = 0.1$, $2^1 = 0.2$, $2^2 = 0.4$, $2^3 = 0.8$, $2^4 = 1.6$, $2^5 = 3.2$.

For 3, this would be represented by 1.6 + 0.8 + 0.4 + 0.2 or the binary number 0 1 1 1 0

For 1 = 0.8 + 0.2 or binary number 0 0 1 0 1 0

For –2.8, this is slightly different because we have minus sign (–), so in this number series we must have the sign bit first. If '0' represents minus (–), and '1' represents plus (+), then it would be for (sign) (binary number) or 0 0 1 1 00. We would then have to put the sign bit in front of all of the other numbers.

For 0, this would then with the sign bit be 0 00000

So the final word is **1 0 1 1 1 0, 1 0 0 1 0 1 0, 0 0 1 1 0 0, 0 0 0 0 0 0,**

which represents 3, 1, –2.8, 0

3.6 Exercise

–24 dB is an amplitude (not in Power which is 2× larger). Because it is a negative value (minus), that means that it is a reduced amount (with 24 dB meaning it has increased).

In amplitude 3dB = ×2, so how many 3 dBs are in 24 dB (i.e. 24/3 = 8). So 24 dB is 2^8 = **256.**

We have really reduced the noise of that car by 256 times!

(If we were talking about how much power had been reduced, this would be 24/6 = 4 = 2^4 = 16, but normally we measure amplitude not power.)

3.7 Exercise

PASSPORT PHOTO

The number series for the (left) passport photo at the gate is 0 0 1 1 1 1 1 1 1 1 1 1 1 1 0 0.
The number series for the (right) photo at the gate is 0 0 0 0 1 1 1 1 1 1 1 1 1 1 0 0 because if a square is mainly white it is 0, but if not it is 1.

When we look at these two photos, we can see that the passport photo is acceptable (which is how the Immigration Office would think). But let's do this automatically by correlation with the first five numbers from the passport photo starting at the top cell going down.

```
0 0 1 1 1           0 0 0 1 1 1 1 1 1 1 1 1 1 0 0
        0           0
        0           0 0
        0           0 0 0
        1           0 0 0 1
        2           0 0 0 1 1
        2             0 0 0 1 1
        3               0 0 1 1 1
        3                 0 0 1 1 1
        3                   0 0 1 1 1
Peak    3                     0 0 1 1 1
        3                       0 0 1 1 1
        3                         0 0 1 1 1
        3                           0 0 1 1 1
        2                             0 0 1 1 0
        1                               0 0 1 0 0
        1                                 0 1 0 0
                                            0 0 0
                                              0 0
                                                0
```

The final peak value of the gate number series is 3.

The total number of the passport series of 0 0 0 1 1 1 1 1 1 1 1 1 1 0 0 = 10.

Therefore final peak value is not equal to or greater than the passport photo half value of 5, so this photo will fail automatic approval and the passenger will have to go to the Passport Control Officer. However, if a larger gate photo number series, say one of about 7 or 8, will do the job – the passport control computer people go to the software and increase the value to a number which passes most people's automated tests. I have been to automated gates where most passengers failed to pass – there was great disruption and the computer people then reduced the peak value required of the numbers to pass everyone!

This example was a very crude exercise, but it demonstrates that a simple cross-correlation of gate photo with the passport photo can be done. In practice, all vertical pixels of both pictures may be run against each other which in this exercise, will give 10 as the peak correlated value, and because this is the same number as the passport photo 0 0 0 1 1 1 1 1 1 1 1 1 1 0 0 = 10, then there will be 100% correlation and the gate will open. Note that the chosen reduced number of the gate photo (5 instead of 15) means that the computation is faster, but that has to be weighed against the queue at the immigration gate counter.

CHAPTER 4 ANSWER

1. We arrange a spreadsheet which is similar to Table 4.3, but it contains a column for 10% bias price increase, and a column at the end which shows average prices per apartment in each year (accounting for the bias).

TABLE 4.4 Table of Apartment Prices Showing 10% Bias

Year	2 brm Sold	Av Price ($k)	10% Bias Price	3 brm Sold	Av Price ($k)	Prices ($k)
2012	2	100	110	3	110	$110
2013				5	120	$120
2014	4	130	143			143
2015				2	150	150
2016				2	155	155
2017	1	130	143			143
2018				1	190	190
2019	3	170	187	1	200	193.5
2020	4	180	198			189

2. Using the table in Excel, draw the graph of apartment sales.

FIGURE 4.16 Graph of apartment sales.

Now compute the least squares on a spread sheet. The following Figure 4.17 shows the methodology, and it is interesting to note that the mean point anchors in 2015 at about the same value as at that time, while the intercept C occurs at $62,800. At first glance this does not seem to be correct, because the data shows that recent values are flattening off and maybe a curved data fit would follow the data better.

However, this does show how a series of later low values (from 2016 to 2020) can result in an early (2010) low value and therefor an unexpectedly excessive predicted value of $250k. It may therefore be preferred to just use the last three data points to get an answer of $200k. That would seem more realistic rather than using the full data set, and it can be

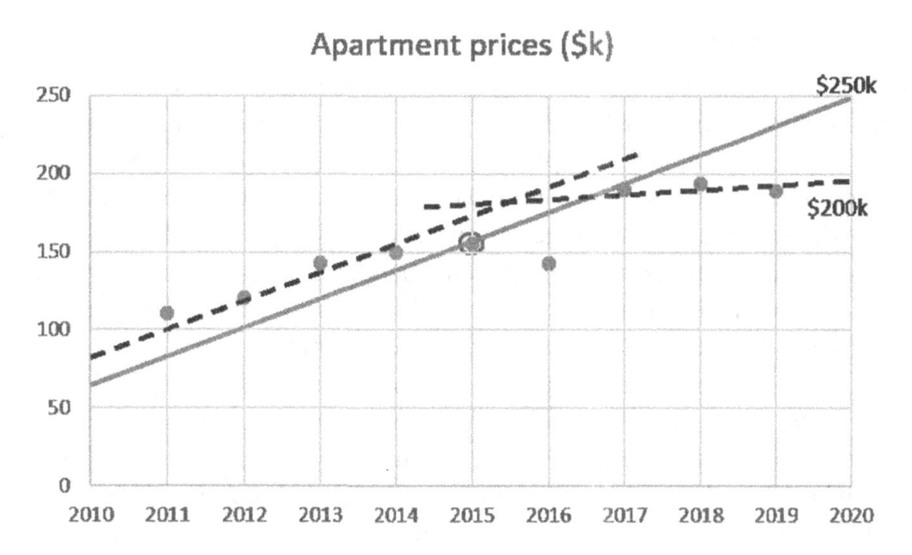

FIGURE 4.17 Graph of possible LS fit shows a broad range of apartment price predictions.

seen, in this example, how we must be careful to decide which data to use and which not to use. More often than not, simple LS approaches are used rather than the more complex approach of applying a curved equation and making adjustments. An LS fit can be done quickly using the last three data points and will give an answer but it likely will be erroneous due to the small input data set of three points.

CHAPTER 5 ANSWER

a. Once produced in New York, the key needs the manufacturer's name, date of manufacture, a serial number, materials used, EM field voltage applied, test details, weight of key, compatible lock, date of departure from factory, shipping agent, approval information from Federal Export Centre, cost of shipping to airport, date of shipment to airport, air-transport carrier details and cost, route the carrier will take, arrivals address in Cairo, customs approvals, customs agent name, date of ETA London.

b. Once aboard the plane travel departure date and carrier, total weight of package, whether package was single or a group package, confirmed arrival date and time in London, sorting date at airport, next carrier and agent if any, warehouse holding package.

c. Cairo arrival date, Cairo agent, customs clearance declaration number and date, destination address, carrier address, date of delivery.

After all this, they found that the key was the wrong model and would have to be reordered (because it was superseded by the manufacturer without anyone else's knowledge)!

CHAPTER 6 ANSWER

Weight – determining the weight of a vehicle often requires some form of weigh-bridge or scales and possibly EM wires under the entry road, but these require digging up the entry-way road. Instead, use a *graphene carbon coating* (Section 6.2.6) which can be applied onto the roadway surface as a thin coating, which requires an electrical connection to a digitizer. When a vehicle stands on the coating, each tyre depresses and thins the coating increasing its resistance (the thinner length has a greater resistance since less electrons carrying the current can pass through the thin portion). The graphene resistance has to be calibrated to the known vehicle weights before it can be used.

Vehicle count – it is popular to count vehicles on public roadways by putting a rubber *air-pressure tube* across the entry road and attaching it to a counter which ticks over the number of times that the tube pressure has been pulsed higher (as tyres squeeze them). The counter can be adjusted to take single or dual axis vehicles. However, while this is cheap, it suffers from constant wear and so is not a permanent vehicle counter. Alternatively, a *laser or LED* beam across the entry will work, but will also count people breaking the beam. Another possibility is a microwave transmitter and receiver (like the *EM* security portal of Figure 6.22), so that as each single vehicle passes the *EM* field is disturbed and instead of it turning an alarm on, it turns a counter on. This requires calibration of the EM field because we don't want people walking through and it counting them – we only want motor vehicles which will reduce the *EM* field by a greater amount than people.

Using a *radar gun* will give the vehicle's speed, while taking a high resolution photograph can put the vehicles license plate number into a database, which thereafter could be used for checking ownership in the case of accidents. The data then sent to the CPU database is the car type, registration license number, speed at entry and exit, weight at entry and exit, and all manner of other statistics that can be assembled at the entry point.

CHAPTER 7 ANSWER

1. Since flow and pressure are the issues, we need to put in place a flow rate monitor circuit, by adding a pneumatic valve flow controller FCV to valve V-1. Equipment must be intrinsically safe, so a valve with a pneumatic actuator along with a flow controller FC that can be set to a specific value, and flow transmitter FT are added as shown below.

2. We need to ensure we have adequate intrinsically safe equipment including the power supply and determine if the transmitter is *wifi* compatible with our remote control operations. Our control room display must cater for the data with inclusion in the control room system alarms. Operators must be aware of this addition and trained in what action is to be taken.

3. Note that just adding a local controller requires a chain of other actions to ensure that data being transmitted is acted upon in control operations. Maintenance must be considered, and how the data may be used for predictive maintenance, rather than preventative maintenance.

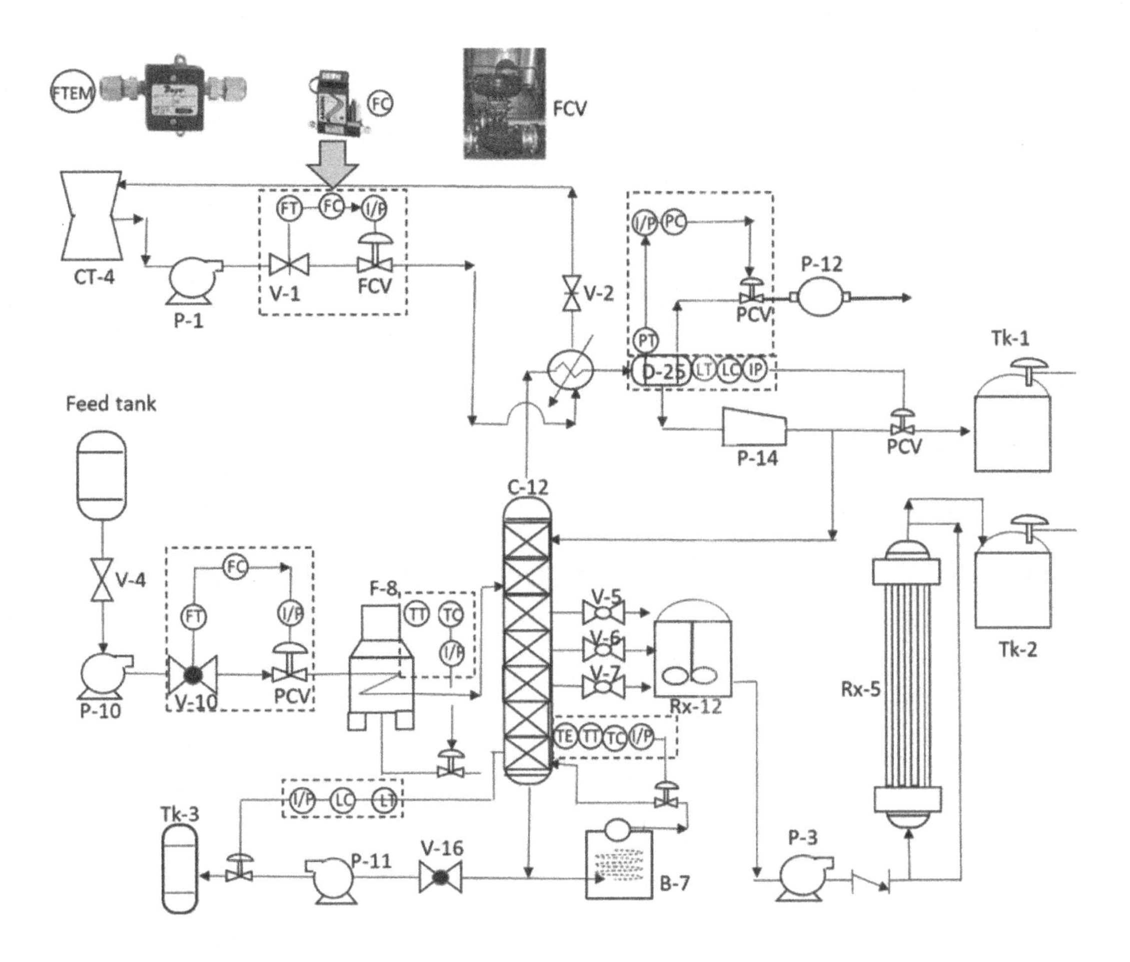

FIGURE 7.19 Updated P&ID of crude installation plant having valve problems.

Glossary of Terms Used

2D: Two dimensional, as in two distances x and y at right angles to each other.
3D: Three dimensional, as in three distances x and y (on a flat screen) and z in depth.
3D Printing: A method of building up a 3D shape, by squirting a liquid to produce a surface-like string, and then building it higher by waiting until it dries and re-squirting the liquid again on top of the first layer, and then re-squirting the liquid again, thereby building it higher again until it forms a wall thereby building a depth of the 3D shape.
1G: The original first-generation mobile phone operating on analogue frequencies.
2G: The second-generation digital version of the analogue 1G but operating using digital frequencies.
3G: The third-generation mobile phone with increased speed.
4G: The fourth-generation mobile phone with increased speed due to improved computation.
5G: The fifth-generation mobile phone with increased speed operating on different frequencies and with transmission towers closer together.
Accelerometer: A tiny device that provides output signals when moved and in particular, provides acceleration data.
Additive manufacturing: The formal name for 3D printing.
A-D converter (ADC): Analogue-to-digital converter is an electronic device that has an analogue input and it produces a digital output which represents the input.
Air-bag: A balloon in a car that automatically inflates (using compressed gas) on impact.
Air-conditioner: A unit usually wall-mounted that provides adjustable cool or hot air.
Airborne Warning And Control System (AWACS): Air force plane that has a large radar fixed to it.
Alarm: A device that either flashes or produces a sound to alert anyone nearby that a set parameter has been exceeded.
Algorithm: An alternative word for 'equation' which some use to sound technical.
Aliasing: The appearance of something which has been incorrectly sampled (e.g. the wagon wheel going backwards).
Alternating current (AC): An electric current that travels with a constant frequency (50 or 60 Hz for household mains power, 400 Hz for aircraft).
Ammeter: An electrical instrument used for measuring current (amps).
Ampere (Amp): The unit for current.
Amplifier: An electrical instrument that increases the values at its input (usually current or voltage).

Amplitude Modulation (AM): Electromagnetic waves that travel in analogue form within a wave envelope, and constrained by the envelope – often used by analogue radio transmissions.
Anaglyph: The low relief on the surface of an object giving it depth, which may be simulated using light.
Analogue: Continuous wave or data flow.
Analytics: The act of analysing data or any other form.
Animation: Movement or the appearance of an object which displays movement.
Anode: The positive element of an electrical system.
Average: The middle of a set of values in which all values are summed and the middle is found by dividing the sum by the number of values.
Apollo: The NASA space mission to the moon.
Apple: The computer and smart device manufacturing company.
Application (App): A small software package activated by an icon on a computer screen.
Association of Tennis Professionals (ATP): World body for men's tennis.
Atoms: The tiniest particles found in all materials.
Augmented Reality (AR): In viewing a 3D representation of an object on a computer screen, AR is an additional object added (augmented) to the 3D object (e.g. a bolt added to a steel beam).
Automation (auto): When a process uses instrumentation to make adjustments to the process without manual intervention.
Automated process control: The controlling instrumentation of a process.
Automated referee: A set of instructions provided to players on a field, which uses data analytics to determine infringements of the rules of the game.
Avatar: An image of a human body form that is not real but computer-generated.
Backwardly compatible: Software that can read earlier versions of software.
Bad angle: The area in which computation of the location of an object has greater error.
Bandwidth: A range of frequencies.
Bar-B-Que (BBQ): A method of heating food, derived from smoking food in the Caribbean (Caribbean word 'barbacoa').
Bar code: A sign placed on an object to allow identification of the object.
Barrel- A tubular object with length greater than circular width.
Baseball: A sports ball-game played in a triangle, which derived from 'rounders' (which was played in a circle).
Battery (-ies): The instrument(s) that store(s) electrical charge.
Bayes Theorem: The maths applied to the theory developed by Reverend Bayes to explain probability of an event to happen, based on prior knowledge.
Beam splitter: A method using mirrors to shift the wavelength of a beam or pulse of light by as much as a half wavelength. Used in interferometers and fibre optics.
Bell curve: A curve that has the shape of a bell, otherwise known as a Gaussian distribution.
Bernoulli's equation: The equation developed by Bernoulli that describes the physics of flow of fluid from one size pipe into a differently sized pipe.

Bimetallic: Two (bi-) metal strips that are welded together to make a single body, which expands differently with heat changes, causing the body to change shape.

Biometric scanning: A scan of features of the finger-tip or eye for comparison with the image retained in a database.

Biosensor: A device that uses living organisms or molecules to detect the presence of chemicals.

Binary integers (BITs): Values based on two integers, '1' or '0' which when formed into a code of many bits represents a value.

Blu-Ray: A small optical disc designed to replace DVDs but with limited success (possibly because they are too small).

BlueTooth: Short-range wireless transmission operating around 2.4 GHz.

Blockchain: A list of records of products containing data on the product, which can be put together like a spreadsheet, in which each block is independent and its origins cannot be changed. Used in financial business to provide transparency of money movement.

Bourdon tube: A tube developed by Eugene Bourdon which is bent, but when changes of pressure occur it unfolds. Often used in instruments to display pressure values.

Breathalyser: An instrument that is used to test for alcohol – contains a chemical that reacts in the presence of alcohol to produce a value of amount of alcohol.

Built Information Modelling (BIM): Computer programs that display 3D images of construction sites, which can then be used to view how construction occurs. Can be interfaced with AR.

Busbar: A power line (or bar) for carrying electricity.

Byte: A group of BITs or digits making a 'word'. A number of bytes (e.g. 10 Kb) is the number that can be recorded and put in a computer memory.

Calculator: An instrument that may be programmed, and used to compute values.

Camera: An instrument that records images of light which can be translated into pictures.

Capacitance: The ability or 'capacity' to store electricity.

Carbon: Chemical element with symbol C and atomic number 6. Non-metallic and usually black in colour.

Carbon Dioxide (CO_2): Colourless gas having a carbon atom joined with two oxygen atoms.

Carbon Monoxide (CO): Colourless flammable toxic gas, joining a carbon atom with an oxide atom.

Cascade: A continuous process in which one individual process is followed by a similar individual process, which has its input being the output of the first process.

Cassette: A plastic container housing two reels, with one reel feeding magnetic tape to the other reel, on which is stored data usually for analogue sound playback.

Catalytic: The action of a catalyst in increasing the rate of a chemical reaction, and when applied to a converter, causes exhaust gases to lose their toxicity.

Cathode: The negative element of an electrical system.

Celsius: The temperature scale invented by Anders Celsius, originally called 'Centigrade' because it graded water temperature over 100 points from freezing at zero to boiling at 100 (see Fahrenheit).

Central heating: The hot water system running in pipes throughout a building.

Central Processing Unit (CPU): The electronics unit controlling a computer process.

Ceramic: A heat-resistive, brittle, corrosion-resistant material, e.g. clay, resulting from shaping and heating ('firing') the material. Found as plastics, pottery and semiconductors.

Chance: The possibility of something happening.

Channel: A route or path made for signals to pass along, used by switches and radios.

Chemiresistor: A material that changes its natural resistance as a function of a chemical reaction.

Chip: An individual integrated circuit or material containing many computer components.

Circuit board: A board containing many electrical components with tracks connecting them.

Clock: A steady pulse that is produced at a constant time to maintain a steady timing of computations.

Cloud: The concept that all data is transmitted to satellites in the sky, which retransmit the data to computers anywhere in the world – resulting in the concept of storing data in the cloud.

Cloud computing: The ability to store and remotely control data anywhere in the world (via satellites).

Cochlea implant: A wire passing from an externally-worn piezoelectric receiver, transmitting electrical pulses directly to the auditory nerve connected to the cochlea, giving an electrical pulse proportional to the sound arriving at the receiver.

Compact Disc (CD): An optical data storage disc developed by Sony and Philips to play digital recordings.

Comparator: An electronics component that compares two or more signals to each other, and is programmed to give an output signal based on that comparison.

Composite: A material or picture that is constructed of many materials or pictures, and displays the overall attributes of the combined properties.

Computer: An electronics device that is programmed to perform calculations and duties using data that has been input to it.

Computer-aided-design (CAD): Software that assists the design and display of a product by performing labour intensive calculations required when a design is changed.

Computer tomography (CT): A device that performs computer-automated calculations of the ray-paths of EM waves through an object, in order to display the interior of the object.

Compressor: A device for compressing a gas or vapour to a higher pressure.

Condition monitoring: The ability to observe (or monitor) changes in the condition of equipment or a process.

Cone: The photoreceptor cells in the retina at the rear of an eye that pass on electrical signals to the brain, as a function of the observed light level and frequency of bright light).

Condensate: The fluid that takes the form of a light liquid from gas being cooled.

Conical: Taken or resulting from the shape of a cone.

Contra-rotating: Rotating in the opposite direction to the normal direction of rotation.
Control: Having the ability to take action as a result of a change in conditions.
Conveyor belt: A metal or rubber belt that transports goods from one point to another.
Cook-top: The top of a cooker or stove upon which cooking utensils are put.
Coriolis: The natural rotation of particles as a function of the vessel through which it passes.
Credit card: A plastic card containing a strip that can be magnetised, used for payment for good and services.
Cricket: A ball game played with a flat-bat on an oval field.
Cross-correlation: The act of mathematically passing one number series across another, in order to obtain a value which is then used for other purposes.
Crude oil: Natural black oil that is unprocessed but has been physically filtered to remove unwanted contents.
Cryptocurrency: A digital currency that can be traded for cash using internet exchanges.
Crystal: A solid material that has a microscopic structure, and can be used to assist some other process (e.g. converting pressure into electrical signal).
Current: A flow of electrons, often passing from a positive electrical point to a negative electrical point (theoretically, the electrons actually travel from negative to positive).
Current transformer (CT): A toroidal coil which when mounted in an electric field, produces an output proportional to the amount of field – useful for detecting the value of current along a line without actually putting an ammeter in the line (instead, using its electric field to feed an output into a meter). Hence, a 'step-down' transformer of mains current flow.
Current/Pressure inverter (I/P): An electrical current or pressure transformer providing an electrical signal out to a controller, which gives a display of the value of current or pressure.
Curve: A graphical change in values which has the shape of an arc.
Cyber: The industry associated with computers, information technology and the internet.
Data: Output information that can be used to determine performance of a system.
Data acquisition: An instrument for recording data, containing the input, analogue to digital (or D-A) converter giving a signal output to its storage unit.
Data analytics: The analysis of data and subsequent translation to meaningful knowledge.
Database: The storage of data that is available to be read at any time.
Data compression: The reduction in space required of a storage medium as a result of reducing the amount of space by repositioning the data in a more compact manner in the area.
Dead reckoning (DR): A method of moving from one point to another by taking the shortest distance route.
Decibel (dB): A scale of values using the logarithmic base of 10, this scale is used when numbers become too large to be manageable. Often used to determine sound noise level, the typical everyday noise found in a house is about 50 dB compared with 90 dB when a jet aeroplane takes off. A doubling of noise amplitude level (×2) is an increase of 3 dB in amplitude and since amplitude squared is power, it is then 6 dB

(i.e. $2 \times \log_{10}2 = 6$ dB) when used for power. So noise levels should be stated whether they are in amplitude or power.

Deep learning: The term used when a process is fully automated (i.e. computer-controlled) so that actions can be performed automatically with no human input. This is the final result after programming for machine learning.

Defrost: To warm an object, raising the temperature gradually to lose the object's internal freezing state and remove surficial frost.

Detergent dispenser: The fluid container in a dishwasher that dispenses fluid that makes the dishes hydrophobic (draining water off plates to dry faster).

Dew point: The temperature point at which a vapour changes to a liquid.

Drive belt: The rubber or steel belt that is driven by a roller or pulley at one end, and is used to drive another roller or pulley at the other end.

Diaphragm: A short piece of elastic material that can change form depending on applied pressure on one side versus the other.

Diastolic: The pressure that occurs when heart muscle relaxes to fill the chamber with blood.

Digital: Data in the form of bits which are digits (or small steps) in current or voltage.

Digital Twin: A computer programmed copy of a process system, which responds in identical manner to the original process system, but does not contain the system hardware.

Digital-to-analogue converter (DAC): The device that has a digital input which it changes to a continuous flow variation of analogue values in current or voltage.

Digital Video Disc (DVD): An optical disc that stores data using magnetic recording methods and is commonly used for video recording and playback.

Digital Signal Processor (DSP): A chip that is built to process digital signals in a specific manner.

Digitizing: The method of taking analogue data as an input and then converting it to pulses of current or voltage representing bits of data.

Diode: An electronic circuit that has a gate which allows and controls the passage of current in one direction only.

Disc: A circular plastic or metallic flat object that, when rotated, can store data either magnetically (by magnetising a track) or optically (using a laser burning a track).

Direct current (DC): An electric current that travels at a constant voltage (i.e. does not have frequency), such as is used in appliance batteries.

DNA: Deoxyribonucleic acid being a double helix that dictates the genetics of a body.

Doppler Effect: The phenomena of a change in frequency as sound (pressure) and light (EM waves) as an object travels towards (compressing frequency higher in value sounding higher in tone) and away (expanding frequencies resulting in lower tones) from an observer, named after its discoverer Christian Doppler.

Doppler shift: The amount a frequency changes as a result of movement of an object towards or away from an observer.

Download: The act of passing data from one location (e.g. the internet) to another location (e.g. a computer).

Drip coffee: The type of coffee machine that passes hot water through ground coffee beans, which produces continuous drips into a cup.

Drop-on-demand (DOD): A controlled method of passing ink continually onto a paper sheet used in printing paper.

Drone: An air-borne vehicle that has propellers batteries controlled from a remote wireless transmitter, allowing the controller to command the drone's actions.

Dryer: A unit that heats and dries wet clothing.

DVD-R: A DVD (see description above) that can be used for recording and storing data.

Eccentric rotating motor (ERM): An electric motor that rotates eccentrically rather than about a single location, due to an offset mass causing a centripetal force to be asymmetric resulting in displacement of the motor.

Eddy current: A rotating current in a material caused by an external magnetic field generated by an external current (used to measure the external current value).

Electric: the passage of electrons causing current to flow and an electric field to be generated.

Electrode: The terminals or probes of an instrument.

Electrochemical: The electric field generated by a chemical reaction (or a chemical reaction resulting from the application of an electric field).

Electromagnetic (EM): A magnetic field generated by the proximity of different poles of a magnet with respect to each other – the field has a force which can be used to move magnetic objects.

Electrolysis: A method in which electrical current (DC) causes a chemical reaction.

Electrons: A subatomic particle that has a negative charge.

Electronics: The physics and engineering technology controlling the flow of electrons through materials to make devices active in some manner.

Element: A part of a component or physical unit.

Ellipse: An oval shape that can be mathematically explained as an oval ring made by an angled or oblique cut through a cone.

Energy: The amount of effort it takes to make an action occur, which may be in units of power, amplitude or joules.

Entangle: Applied to quantum computers, where two particles spin, one spinning up and the other down.

Espresso: An Italian coffee brewing method in which hot water is passed ('percolated') through ground coffee beans.

Estimated Time of Arrival (ETA): The calculated time of a vehicle arriving at a location.

Ethernet: A wired system connecting computers together in a local network so that they can pass data between each other.

Evaporation: The process of a liquid turning into a vapour.

Fahrenheit: The temperature scale invented by Dan Fahrenheit in 1724 in which water freezes at 32 degrees and boils at 212 degrees. His experiments used brine for setting the lower value at 0 degrees, and blood as the upper 100 degree point (and consequently when applied to pure water freezing and boiling the values changed – see Centigrade).

Faraday: The discoverer of electromagnetic induction (demonstrated by building the first electric motor and first electric meter), electrochemistry and electrolysis, who became one of the most influential scientists in history without having any formal education (just Sunday School!).

Fast Fourier Transform (FFT): A mathematical transform developed by Fourier to determine the frequencies of a waveform.

Fibre: Natural or manmade material longer than its width, used in manufacturing other materials such as carbon fibre used in fibre optics.

Federation of International Football Associations (FIFA): The world body governing football (soccer).

Feedback: A loop applied to an amplifier taking the output of the amplifier and feeding it back to the amplifier's input, to either modify the input or compare with the input.

Field Effect Transistor (FET): A transistor that uses an electric field to control the flow of current through it- the input is the 'source', the control point is the 'gate' and output is the 'drain'.

Filter: Either a device with small holes to stop particles passing through it, or an electronic device that stops specific signals passing through it.

Finger print: the surface relief of the pad of a finger – when pressed on a surface may leave an imprint of the relief.

Flag: To provide an alarm or highlight of an action that has occurred.

Flash: An action that is very quick, such as a flash memory which may record and then write that data very quickly.

Flight time: The time taken to travel from one point to another.

Flow meter: A meter that gives an indication of fluid flow through it.

Flow control (FC): An instrument that controls the flow of a fluid.

Flow transmit (FT): An instrument that transmits flow data.

Fluoresce: The emission of light by a substance that has absorbed energy through some other means, such as electromagnetic radiation.

Flux: A flow of energy or particles through, around or across a material.

Football (soccer): A ball game played with the foot, head or body with the intent of making the ball travel into the opposition's goal (net).

Format: An established arrangement or layout standard that allows the writing of data so that others can read it, and vice-versa.

Freezer: The cold air unit that keeps food products constantly frozen.

Frequency (-cies): The rate per second that an electromagnetic or pressure wave passes a fixed point, measured in Hertz (Hz) or cycles per second (cps).

Frequency Modulation (FM): A method to encode an analogue signal to make it have compressed and expanded points along it, which can be received equivalent to a digital transmission.

Front loader: A type of washing machine having a horizontal barrel which can be loaded with washing from the front rather than top.

Fryer: A device that applies heat to food loaded into a wet (oil) or dry container.

Fuel: A material that reacts with others causing a release of energy.

Full High Definition (FHD): A type of monitor screen with high resolution, having pixels that are 1920 wide × 1080 high.

Gain: The increase in rate of current or electron flow through an amplifier.

Galvanometer: An electromechanical instrument used for indicating the value of current and was the first form of electricity measuring meter that was developed.

Game: A competition in which two sides play against each other to win.

Gamma Rays: Waves of electromagnetic radiation emitting photon energy and developed due to radioactive decay occurs in atomic nuclei (e.g. during nuclear explosion).

Gaussian distribution: The change in values of a property which have the appearance of a bell curve, the mathematics for it having been developed by mathematician Gauss in 1809 (as well as the method of *least squares fit*).

Geometry: A branch of mathematics that relates distances and angles of triangles to each other so that a mathematical equation can be used to calculate values.

Geophone: An electromagnetic instrument that responds to vibration by producing an output voltage.

Ghosting: An alternative word for *simulation*.

Glasses: An instrument that is placed over the eyes in order to focus on objects.

Global Positioning System (GPS): A satellite system around the world that provides information on position to a GPS receiver.

Gluons: A term used in quantum computing for the medium or particle that 'glues' two quantum quarks together.

Glycol: An ethylene alcohol having no colour or odour used in antifreeze fluid and in making polyester fibres.

Goal: The end product of an action, such as putting a ball into a net during a game.

Good-angle: The area in which an object's position can be determined with the greatest accuracy.

Goniometer: The instrument used to determine distance travelled by a wheel, and is shown on a dial which has been connected to the wheel through a system of cogs.

Graphene: A very thin layer of carbon consisting of a single layer of atoms in a lattice- it is so thin that it cannot be seen but can respond with properties of a layer covering a surface.

Gravity: The phenomena in which objects are pulled towards the centre of the earth, causing them to have weight (mass).

Graphical User Interface (GUI): Computer software code that puts an overlay on the computer screen so that icons or other graphical items (e.g. menus/windows) can be seen.

Ground Penetrating Radar (GPR): An instrument that transmits and receives radar frequencies at the ground surface, so that any material (e.g. metal) beneath the ground surface can be detected due to it reflecting the frequencies back to the surface.

Guided microwave: A microwave transmitter sends the signal along a wave guide to focus the energy, which may then be either reflected back or left to transmit out.

Gyroscope (Gyro): A device housing a spinning wheel or disc rapidly spinning on an axis. The orientation of the axis is not affected by tilting the device so gyroscopes are used to provide stability and a reference direction in navigation systems (e.g. a compass mounted on the device will maintain its orientation while any axis external to the housing moves).

Hall Effect: The voltage that occurs when a magnetic field is applied perpendicular to the current flow direction.

Halogen: A vapour that can surround an electric filament to make its glow appear much brighter than if it were in air.

Haul pack: A type of dump truck used at mine sites for carrying ore from its mined point to a position where it may be off-loaded.

Haptic: Use of the sensitivity of touch for providing the illusion of substance and force.

Haptic bodysuit: A body covering that contains sensors which act (vibrate) when an electrical signal has been received.

Hard drive: A computer disc drive that is physically mounted in a computer to store all programs, actions and memory.

Harmonics: A voltage, current, or frequency that is a multiple of an operating systems voltage, current or frequency, which is often unwanted since they cause problems.

Hawk-Eye: The computer software that is used to predict the trajectory of a ball, in order to provide vision of a ball and where it would travel if left to continue along its path. Named after the inventor Paul Hawkins and owned by Sony.

Hazard Operability (HAZOP): The methodology used when simulating a hazardous operation, in terms of testing its weaknesses in order to make modifications to overcome them.

Heat exchange: The removal of heat from a fluid by passing cold fluid (or air) over pipes containing the hot fluid, or the reverse act of heating fluid by passing heat over cold pipes containing the fluid.

Heat-rack dryer: A type of clothes dryer that is static and allows clothes to hang on a rack as heat from elements below pass up through the clothes, drying them.

Heath Robinson: Early 1900s cartoonist who made a reputation drawing amusing cartoons of farcical processes which were automated typically using pulleys, string and hand-cranked mechanisms (so in the end were still manually operated).

Hertz (Hz): Named for Hertz, it is the number of cycles per second a full wave has passed.

Heuristically Programmed Algorithm (HAL): The fictional computer that was considered to have high level of artificial general intelligence, appearing in the film *2001: A Space Odyssey.*

High definition (HD): An indicator of high resolution or detail of an image on a monitor screen which with smaller TVs would have 1280 vertical pixels × 720 horizontal pixels, or with larger TVs have 1920 vertical × 1080 horizontal pixels.

Hot air fryer: A device that blows very hot air across food loaded into it performing the function of frying the food without the need for immersion of the food in a liquid (oil).

Hot water system: The system in a building that heats water and then pumps it around the building for the purpose of washing, cleaning or bathing.

Hydraulic: The ability to cause a device to perform an action by applying water under pressure.

Hydrogen (H): The natural element with an atomic weight of just over 1, being the lightest of all elements, and is the most abundant in the universe (approx. 75%).

Hydrophilic: Molecular attraction to water often resulting in being dissolved in it.

Hydrophobic: Molecular repelling of water often resulting in being covered by drips of water having a high contact angle.

Hydrophone: A device that is constructed with a piezoelectric crystal and gives an electrical signal output as a result of being squeezed by a pressure.

Hydrostatic: The state of a fluid at rest and the pressure in it or on it exerted as a function of its position and gravity.

Identification (ID): A means of recognising a person or object and normally given a number or value.

Identikit: An image of a person's face developed by putting different facial features together and normally used by police to identify a person's features according to a witnesses description.

Immersive: The use of 3D computer graphics to give a 3D image that appears to surround the viewer.

Impeller: The rotating part of a centrifugal pump that moves a fluid by rotation (like a water wheel).

Induction: The development of an electric current by a conductor in a magnetic field.

Infra-red (IR): The range of electromagnetic frequencies at the red end of the spectrum that cannot be seen by the eye, but can be used to produce images of heat.

Ink-jet: The process of squirting ink onto paper to produce digital images- the most common form used in the home printer.

Instantaneous heating: The ability to pass water at pressure through pipes that are heated to immediately produce much hotter water without storing it.

Integrated circuit (IC): A chip containing a tiny set of electronic circuits that are embedded in a semiconductor material such as silicon.

Interferometer (-etry): A method to interfere waves with each other (aka wave splitting) so that some waves may be a half wavelength and used in optical transmission.

International Business Machines (IBM): A US company originally founded to build large computers for business, but later found a lucrative market in personal computers (as computers became smaller in size).

Internet: The global system of interconnected computer networks using satellite to transmit and receive data around the world.

Interpolating: Mathematically estimating data points where they are missing, using existing data points as a guide.

Inverter: A device that changes direct current (DC) to alternating current (AC) providing AC power from a battery.

IPad: A tablet computer developed by Apple Inc.

Iris recognition: The transmission and reflection of light to and from the iris at the rear of the eye, in order to correlate the reflected image with that stored in the database. This form of identification is discouraged since the optical transmission may damage the user's retina.

Kettle: Any form of container used for heating/boiling liquids.

Kilo (K): *One thousand* in Greek and adopted worldwide.

Kilohertz (KHz): Frequency of 1000 Hz.

Land-line: The wired telephone cable that has a dedicated cable from the phone to a local phone network connection.

Laplace: A French physicist of the early 1800s whose work was important in maths, statistic and engineering. Known for a mathematical transform.

Latency: The delay or lag in time it takes for a computer instruction to transfer data.

Least squares (LS): A mathematical method to obtain a set of average values.

Level controller (LC): A device that controls the level of liquid in a tank (e.g. cistern float mechanism).

Level transmitter (LT): A device that transmits liquid tank level data.

Level control valve (LCV): A device that controls the movement in and out of a tank to maintain a fixed liquid level (e.g. cistern float valve).

Light: The portion of electromagnetic radiation frequencies we can see with the eye.

Light Amplitude-Stimulated Emission Radiation (Laser): A device that emits light in which photons are amplified to produce a strong thin stream of light.

Light-Emitting Diode (LED): A semiconducting material that emits light (photons) when current flows through it.

Light-Decreasing Resistor (LDR): A resistor that reduces in resistance when light is shone on it (aka *photo-conductive cell*).

Line-of-sight: The ability to see from one point to another, along which radio waves may travel (if a building is in the path, light will not travel through it so there is no line-of-sight).

Linear resonant actuator (LRA): A motor causing vibration in a device (phone) by oscillating a mass in a single direction (a 'voice' coil against a mass on a spring).

Linux: An operating system used by Unix machines.

Liquid crystal display (LCD): A low powered (and cheap) data display using liquid crystals with a back-light (originally used in calculators and digital watches).

Local area network (LAN): A computer connecting network within a building.

Machine learning: The computer programming level to develop automatic responses and depending on an established database (after which is *deep learning* – an expanded program).

Maglev: Magnetic levitation, the name for the first experimental train to use electromagnets to lift the train off its track and then move forward.

Magnet: An object made of ferromagnetic material that produces a magnetic field force.

Magnetic: A phenomena established as a field that is produced around a magnet, especially when electric currents are passed around the magnet through wires.

Magnetic card: A plastic card containing a magnetised strip providing a special magnetic shape or code, which can be sensed by a magnetic reader.

Magnetic Ink Character Recognition (MICR): A machine made to respond to magnetic ink used mainly in the banking industry for processing and clearing cheques and other documents.

Magnetron: A vacuumed tube that generates microwaves by streaming electrons through a magnetic field past cavities which resonate at microwave frequencies, thereby producing microwaves.

Malware: Software that is written to invade useful software and act on what it finds- this may cause corruption of the useful software or it may search databases in order to modify them in some manner.

Manometer: An instrument used to measure pressure acting on a liquid within a U-shaped tube.

Mass flow: The mass of a flow of liquid passing a point over a set time period (with units of Kg/sec).

Matrix: A rectangular array of numbers or elements in rows and columns.

Mean: The statistical average and is the central value of a discrete number set (sum of values divided by the number of values).

Memory: A place or unit in a computer that stores data or computer instructions.

Mercury (Hg): The chemical element (atomic number 80) that is a metallic liquid at room temperature, and has been used for measuring temperature since it expands and contracts linearly when heat is applied to it. However, it is toxic.

Microelectromechanical/electromagnetic sensor/system (MEMS): A tiny micro-device that can take the form of a mechanical device with moving parts like cogs, but when coupled with integrated circuits can be used for many applications (such as acting as a gyro).

Microsoft Disk Operating System (MS-DOS): A computer operating system developed by Microsoft.

Micro-processor: A tiny integrated circuit that contains memory and computer programs, and may control various computer functions.

Microwave: Electromagnetic waves in the frequency bandwidth of 300 MHz to 300 GHz, and is mainly used for communications and in ovens.

Microwave arcing: Sparks or flashes resulting from microwaves hitting metallic surfaces.

Millisecond (msec): One thousandth of a second Milli being Latin for 1000.

Modelling: Term used for computer manipulation to improve a figure or test an idea.

Modulator/demodulator (MODEM): The electronics of a circuit that converts an analogue input to a digital code and vice versa- most used in communications.

Moka (Mocha): An Italian type of coffee developed in the area of the town of Mocha.

Monitor: Computer or flat screen providing an image through the use of pixels.

Motor: A device powered by electricity having a stationary stator and a moving rotor connected together to move the rotor electromechanically.

Moving average: Computing an average value from a fixed set of data values, as more data is added and continuing the computation process of averaging more data.

Moore's Law: An empirical law termed by Gordon Law, the CEO of Intel in 1965 who observed that the number of transistors in a dense integrated circuit doubles about every two years.

Mouse: The hand-held pointing device that detects 2D motion and translates this to the movement of an arrow on a computer monitor display.

Multiplexer: A device that changes a parallel set of digital inputs to a single serial output, or vice-versa.

National Aeronautical and Space Administration (NASA): U.S. national space agency.

Nebula: An interstellar cloud of gas in the universe, which is not easily seen by the naked eye unless using an infra-red lens that responds to light in the IR range.

Neon: A chemical element (atomic number 10) that is a colourless inert gas.

Nespresso: Espresso type of coffee machine marketed by Nescafe.

Neural network: Software containing algorithms that can be taught to recognise patterns and then flag them as required. Because of the manner of the patterns and their connections, this is similar to how it is believed the synapses (neurotransmitters) of the brain operates.

Network: A group of interconnected devices.

Number series: A group of numbers that may represent a data set or combined may represent a single value.

Nyquist: Named after Harry Nyquist who developed the theory of under sampling a number series, to explain how poor sampling could result in a false or *aliased* representation of the series.

Ohm: A unit of electrical resistance, which according to Ohm's law, is the resistance when a current of 1 Ampere is subjected to a potential of 1 Volt across two points in a circuit.

On-field, Off-field: Activities within the field of play is on-field, and those outside of the field of play are off-field.

Organic LED (OLED): In an LED a film of organic compound emits light in response to an electric current through it. Considered to provide more colourful LED displays than without it.

Optical Character Recognition (OCR): A device that translates images of typed or handwritten text into machine coding.

Optical drive: A device that takes computer discs, and uses laser light for burning (writing) data to or reading from them – it takes computer discs such as CDs or DVDs.

Outlier: A point outside of the normal scatter of points on a graph and is not part of the trend of points.

Oxide: A chemical compound that contains at least one oxygen atom and one other element.

Oxidation: The process of a chemical reaction in which oxide atoms change state (aka *Redox*).

Oxygen: The chemical element 'O' with an atomic number of 8, which readily forms oxides with most other elements and compounds.

Ozone: A pale blue gas (seen in the sky) with a pungent smell having a chemical formula of O_3.

Parabola: A curve that is approximately U-shaped.

Parallel computing (Supercomputer): A very fast computer that is made from multiple computers stacked on each other, so that coding enters each computer in parallel and the speed of operation is therefore much faster than a single computer.

Pattern recognition: Automated recognition of patterns with applications in many areas of signal processing. May use cross-correlation as the basic mathematical method of recognition.

Percolator: A device that causes the movement of fluid through a group of solids, such as hot water passing through coffee beans.

Performance: How efficiently something responds to an input of data or effect.

Personal computer (PC): Small computers such as laptops that can be used by individuals for their personal applications.

pH (power of hydrogen): A chemistry scale used to specify the degree of acidity of a liquid solution.

Photoconductive: A material that becomes conductive due to absorption of electromagnetic energy, including light.

Photodiode: A semiconductor diode which when exposed to light, develops a voltage or changes electrical resistance.

Photoelasticity: The optical properties of a material change when deformed mechanically.

Photoresistor: An electrical component that changes resistance when light is shone on it (aka LDR and *photo-conductive cell*).

Picture Element (Pixel): An element of an array providing a single light point which is usually observed on a picture on a flat panel, monitor or screen.

Piezoelectric: A material that produces an electrical signal output when compressed, and vice-versa (used in phones to change sound to signal).

Piping: Lines on a diagram that appear equivalent to pipes, formed to show the direction of flow of data or fluid.

Piping & Instrumentation Diagram (P&ID): A diagram showing lines of fluid travel seen as pipes containing fluid flowing from one process to another.

Pitot tube: A tube used to measure the flow of a fluid velocity invented by Henri Pitot, and commonly seen on the exterior of aircraft to determine the speed when flying.

Pixel: See Picture Element.

Plasma: A hot gas cloud of ions and free electrons that can be influenced by an electromagnetic field.

Pod: A singular vessel or housing containing a liquid, or landing craft used in space travel.

Polarity: The state of having two opposite poles of an electric or magnetic circuit in which electrons travel from one end to the other.

Polarize: To cause light waves to separate into two entities that are distinctly separate.

Polynomial: An equation containing variables and coefficients that involves addition, subtraction, multiplication, or division.

Poppet: A valve, mushroom in shape, with a flat end that is lifted in and out of an opening by a rod.

Portafilter: A device used in a piston-driven espresso machine, which contains tamped ground coffee in its basket and is fixed in position by a handle.

Potentiometer: A variable resistor that either rotates or slides, having three terminals to act as a voltage divider since the voltage between the central wiper terminal and either end changes as it is moved or wiped (aka *rheostat*).

Power: The electrical energy provided to a circuit and in DC is represented by volt-amperes.
Process Flow Diagram (PRD): A diagram displaying the flow of a product through a process and often associated with P&IDs.
Primary: The first part of a device, such as the primary copper winding in a transformer which also contains a secondary winding to change voltage levels.
Prediction: The ability to use existing data to forecast the likely form or direction of future data.
Predictive analytics: The method of dissection and analysis of a set of numbers or figures to be allow a prediction of future numbers to be made.
Predictive maintenance: The ability to analyse equipment failures to allow a prediction to be made of when the equipment should be maintained in order to avoid the failures.
Pressure: The physical force exerted on or against an object in contact with it.
Pressure controller: A device that controls the pressure flowing in a line.
Preventative maintenance: The application of equipment maintenance in order to avoid future failures, used in conjunction with predictive maintenance.
Printer: A device that applies a method to print characters onto a medium such as paper.
Probability (P): The branch of mathematics that determines how likely an event will occur, and often is a number between 0 and 100% (e.g. P50 is the probability of 50% of an event occurring).
Probe: A projecting terminal sometimes used as an electrode in an electrochemical process.
Process: A series of mechanical or chemical actions to modify the state of one or more objects.
Program: When used with a computer, a group of computer lines or statements that manipulate data to produce a required output. When used with a process, a series of actions on objects to produce a required object at the output.
Pulse: A short-term electrical surge of energy or pressure in a liquid.
Pump: A device that moves fluid by a mechanical action.
Pyrolytic: The thermal decomposition of materials at high temperature, such as when heating an oven to remove food (*Pyro* Greek for heat, *lytic* Greek for separation).
Pyrometer: An instrument for measuring high temperature, especially in furnaces.
Pythagoras: Greek mathematician around 500 BC who develop the mathematical theory behind geometry.
Quadratic: An algebraic equation in which has three known numbers and x is unknown. Uses the square of x to develop the value for x (Latin 'quadratus' meaning *square*).
Quantum computer: A future computer that uses quantum theory in terms of 'superposition' and 'entanglement' to perform computations. Instead of typically 24-bit formats we have in present-day computers, it has the potential for many thousands of bits, and so its speed of operation will be almost immediate.
Quantum computing instantaneous (QCI): A title given to instantaneous operations and displays produced through computing with a quantum computer.
Quantum dot LED (QLED): Samsung developed very tiny dot pixels for large TV screens or monitors which are considered brighter than the normal pixel, and Samsung

called the method Quantum dot. However, it has little to do with the maths of quantum theory.

Quarks: A subatomic particle carrying an electrical charge, and are part of quantum mechanics having been experimentally proven to exist.

Qubit: A quantum bit (aka *qbit*), which like normal computer bits, is a unit of information but having two states.

Quick Response code (QR): A two-dimensional bar code (in x and y directions) that is used with electronic readers for identification.

R computer language: A computer language for statistical computing and graphics production, invented at Auckland University by **R**oss Ihaka and **R**obert Gentleman in 1991, it was developed from the S-language.

Radio Detection and Ranging (RADAR): A device that uses radio waves to determine range to an object, by transmitting waves out and receiving the waves reflected back from the object. When mounted on a turntable and rotating, it can provide the track and angle of a moving object (such as an aeroplane in the sky).

Radar gun: Similar to the radar, this hand-held device is used to measure speed of a moving object, transmitting radio waves and receiving their reflections to show how fast an object is travelling with respect to the observer.

Radiation: The transmission of waves which are generally electromagnetic, from a source (such as the sun).

Radiator: Usually a heat exchanger used to transmit heat to warm a building by running hot water through a system of pipes.

Radio: A device that is used for receiving and transmitting radio waves for communication purposes.

Radio Frequency IDentification (RFID): Using an electromagnetic field is a form of identification code is used for tracking objects such as parcels, in which a parcel bearing the tag passes through a field which is modified. The RFID is either active (it transmits the code) or passive (it modifies the field).

Raster: A scan along a horizontal line, followed by the next line down thereby building a rectangular pattern of parallel lines forming a graphical display (as used on screens and visual monitors).

Reaction: A chemical process of change in which one set of chemicals interact with another.

Read-After-Write (RAW): The ability to read magnetic data immediately after writing it in order to display what is written, as a continuous check that what is written is correct.

Reality technology (-ies): Devices which give the brain the impression that what is being seen is real, even though it is computer developed.

Receive: Accepting properties which have been transmitted from a source, usually some form of wave energy.

Recording: Taking data and formatting it into a specific form which can be written into a medium (such as a database) that allows the later retrieval of the data.

Recursive: Returning back to the start of a process, such as when taking output data and feeding it back to the input for comparison or in order to modify the input data.

Reflections: The received energy after energy (e.g. light) has been transmitted to an object and bounces back from the object.

Refracting (refracted): The process of waves travelling to an object and having their direction of travel changed slightly, but continuing on.

Refrigerator: A device that cools objects typically food.

Refrigerant gas: The gas that readily transmits heat when compressed, and removes heat (cools) when expanded or released.

Regression analysis: The mathematical method in statistics of analysing data which has already been recorded, in order to determine a trend to allow a prediction of future data to be made.

Remote: The ability to perform acts that can influence an operation, physically from some distance from the operation.

Residual: Often referred to as the error in a computational process, or the remanent pieces remaining from a process.

Resistance: The natural tendency of a material to resist the flow of electrical current through it.

Resolution: The level of detail offered by rastered digital images which often depends on the number of pixels on the display screen or monitor.

Retinal recognition: The transmission and reflection of light to and from the retina in the eye, in order to correlate the reflected image with that stored in the database. See also iris recognition.

Repeater: Some form of booster or amplification that is placed in a line to increase the signal (such as a pressure pump in a line in which the fluid pressure is decreasing).

Reverse-cycle: The ability to change the operation to perform in the opposite direction, such as changing a heater to a cooler as used in a reverse cycle air-conditioner.

Rinse-aid: An additive in dishwashing, which causes the dishes to have a surface that is water repellent (hydrophobic) so that the drips do not adhere to the dishes and dry faster.

Robot (-ics): A device that has motors controlled by a minicomputer (CPU) which receives radio signals to make the CPU operate in a semi-automated manner.

Rods: The components (aka *photoreceptor cells*) on the retina that function in low light and assist with vision at night (where cones function in bright light) providing signals to the brain.

Roller: Tubular devices that can maintain a process such as supporting a conveyor belt.

Root Mean Squared (RMS): The mathematical process that is the square root of the mean of the squares of a set of numbers. When applied to AC currents, it is used as the equivalent value of the average DC current.

Rotor: A device or spindle that rotates and is mainly found in motors where its magnetic field interacts with stationary magnets (stator) to cause a force that makes the rotor turn.

Sacrificial anode: A metallic rod that is commonly used in water heaters and has a slight polarity difference with the heater body (aka *galvanic cathodic protection*), so that if water corrodes the heater's metal over time, instead the active metallic rod will corrode (but is easily replaced).

Sample rate (rate): The time period over which a sample of a quantity or value is taken.

Satellite: A device placed in space within the earth's orbit that receives and transmits radio waves acting as a communications platform.

Satellite Navigation (Sat-Nav): A series of satellites orbiting around the earth that are used to provide positional data to a computer or smartphone (see also GPS).

Scan: The ability to travel in a straight line and perform an action, such as transmitting a series of pulses or reading a series of magnetic points.

Science, Technology, Engineering, Maths (STEM): Trendy jargon used by education administrators to encourage the education of these topics in preparation for the coming era of automation.

Secondary: A copper winding normally found on a current or voltage transformer that provides the output using an input electromagnetic field current/voltage.

Seismometer: An instrument that uses a mass suspended on a spring housed within a permanent magnet, to give an electrical signal output when the magnet is physically vibrated by seismic pressure waves.

Semiconductor: A material, typically silicon, that has an impurity in it which causes it to respond to electrical signals in a controlled manner, and can be manufactured as transistors or integrated circuit chips.

Sensor: A device that naturally or is built to respond (with electrical signals) to changes in conditions around it.

Serial: Digital data that is continuously flowing along a single line.

Set-point: The value of a threshold level established to control an operation.

Shadowing: Another name for *simulating* an operation.

Shinkansen (bullet train): The Japanese train that runs very fast throughout Japan.

Simulator (-ion): A computer system that mimics a process and can provide all of the operational values of the process without the hardware- used for training operators and running in parallel with the operational process to ensure it is operating correctly.

Similarity index: A value that represents how a process is performing compared with how it performs when running perfectly- a measure of efficiency of an operation.

Skycam, Spider-cam: The remote camera system operated at sports venues, which is controlled by wires from the venue stands, so that the camera provides a view of the action from and above the sports field. So called due to the wire connections giving it the appearance of a spider in the air.

Slalom: To move from one side to another while travelling along a line.

Slope: The degree of tilt or gradient that an object or graph may exhibit.

Smartphone (cell): A battery-powered mobile device that has the primary function of a phone, but can also be used as a minicomputer using the internet to perform calculations using different *Application* (app) software seen as icons.

Smart TV: A TV that is connected to the internet, which not only provides normal television functions, but can also screen movies and other data retrieved from the internet.

Smart technology: Devices that use the internet to connect users quickly with applications serving different purposes.

Smart watch: An electronic timepiece that also contains minicomputing devices which monitor local conditions as well as use the internet to connect users to other applications.

Smell: A chemical reaction within the nostrils that provides an electrical signal to the brain, describing the flavour of a gas or other substance causing the reaction.

Smoke: Airborne particles and gases emitted when a material combusts (burns).

Snickometer: A name given by TV companies to the tiny pressure sensor transmitter that can be put on cricket stumps, which when vibrated (by movement of the wicket or air) provides a signal output to inform viewers of a vibration of the wicket or air. A *snick* is the term given when a ball briefly touches a cricket bat.

Soccer (football): The round-ball game played mainly using the foot with the objective of putting the ball into the opposition's goal net. Also referred to as *the world game.*

Solar: Power derived usually from photovoltaic panels as a result of radiant light absorption by the panels.

Solenoid: A device that has a linear movement as the result from a change in an electromagnetic field- usually found in vehicles to assist starting the engine.

Sound Navigation And Ranging (SONAR): A device built to transmit and receive pressure waves underwater, to detect objects and determine underwater distances to objects, using the sound velocity in water of 1500 m/s.

Software: A series of computer code representing instructions to operate the computer, in contrast to *hardware* from which the computer is built.

Sound: The vibration of particles that propagates through air, but also through liquid or solid. The sound vibrations cause the eardrum to vibrate, sending electrical pulses to the brain representing the sound wave characteristics (aka *frequency*).

Sphygmomanometer: A device used to measure blood pressure, containing a gauge, inflatable cuff, hand-held bladder and connection piping.

Spectrum: A band of frequencies, usually of electromagnetic or sound waves.

Speech Interpretation and Recognition Interface (Siri): A software package developed by Apple Inc that recognises voice patterns and is programmed to respond to the patterns with an action.

Speed Test: A smartphone app that transmits a signal to a server, which is then responded to by the server, and since there is no lengthy data set involved, the app provides a graphical representation of the transmission/reception and computation of transmission speeds.

Speedo: A device that displays the velocity a vehicle is travelling.

Split-system: Air-conditioning plant in which the cold/hot air is output from a unit at one location within a building, and the heat exchanger is located in another building- connected to each other using pipes containing refrigerant.

Spreadsheet: Software that provides a graphical form of a numerical table, in which data can be inserted in some boxes, and the table automatically computes output in other boxes. The table is printed out on paper in the form of a spread of data in a table on a sheet of paper.

Spring: A metal compressed into a continuous circular shape that allows pressure to be applied at both ends to compress or expand it- used to suppress motion.
Spin: The act of rotation of a cylinder, device or other particle.
Stacker: The form of adding a number of devices or levels on top of one another to make the sum of them taller.
Standard Deviation (SD, σ): A measure in statistics of the variation or dispersion of a set of values, with a low SD being values close to the mean while a high value means high deviation- can be an indication of the accuracy of a data set.
Stator: The fixed part of a motor containing fixed electromagnets that set up the motor's electromagnetic force field (see also *stator*).
Statistics (Stats): Mathematical discipline that provides a method to collect, sort and analyse data.
Steam oven: A device containing a volume of space in which heat and steam may be applied for cooking food.
Steam wand: A pen-type of device used in making coffee, for passing steam into the coffee to make it froth.
Step-down: The act of a fixed reduction in current or voltage with the input to a transformer being reduced to give a step-down in output value at its output.
Stethoscope: A medical acoustic instrument used for listening to sounds made internally by the human body.
Strain: The distance an object may stretch or reduce in length due to the application of a pressure.
Stress: The pressure applied to a body to cause it to undergo strain.
Stumps: The three rod-like vertical lengths of wood that have two short horizontal lengths of rod-like wooden *bails* resting across their tops, form the *wicket* in cricket. When the bails are knocked off their perch on the stumps, the batter (using the bat to shield the ball from hitting the stumps) is dismissed ('out').
Subject Matter Expert (SME): Persons who have a deep knowledge of a topic, area or discipline sufficient to be considered as of the greatest knowledge.
Sunlight: The part of the electromagnetic spectrum of frequencies (wavelengths) that can be seen with the eye.
Supervisory Control And Data Acquisition (SCADA): A device that has software which acquires, processes and interprets incoming data, in order to automatically control a continuous industrial process.
Surface Acoustic Wave (SAW): A pressure wave or vibration that travels along the top few microns or millimetres of a material's surface, with properties that can be monitored and used.
Symbol: Any characteristic that represents some meaningful figure or fact to someone.
Synchronisation: When a process which is regular in time is linked with another process, thereby providing data or other attributes to the other process, and vice-versa.
Synthetic Aperture Radar (SAR): A form of radar used by a satellite to scan the surface as the satellite travels over it, in order to produce 2D or 3D map of the surface.

Systolic: The blood pressure of the heartbeat when the heart muscle contracts to pump blood from its chambers into the arteries (opposite being *diastolic*).

Taste: The chemical reaction when the tongue's taste buds (so called because they have the appearance of budding flowers) contact a material, with the reaction causing an electrical signal to travel to the brain, describing the attributes of the material.

Terabyte (Tb): A unit representing 10^{12} bytes (actually 1,000,000,000,000 bytes).

Techie: A person who knows a lot or thinks they know a lot about technology.

Television (TV): A graphical display of data (*vision*) as a result of a transmission of the data over a distance from another location.

Teflon (Polytetrafluoroethylene – PTFE): A tough synthetic resin mainly used for coating non-stick cooking utensils, seals and bearings.

Tennis: The game of striking a ball using a racquet in which the striking surface is constructed of a stretched string matrix, in order to make the ball travel over a net and land within a defined area.

Thermal: Heat or temperature.

Thermistor: An electrical resistor in which the value of resistance depends on the surrounding temperature.

Thermocouple: An electrical device consisting of two conductors having different properties, forming a junction that produces a voltage depending on the temperature.

Thermomix: A type of kitchen appliance made by Vorwerk of Germany used for cooking and mixing food, consisting of a heating element, a motor, a weight scale and a microprocessor that is connected to the internet.

Thermostat: A device that regulates temperature by performing programmed actions.

Tidal: The rise and fall in water height as a result of the sun and moon's gravitational interaction (*pull*) and the rotation of the earth.

Time series: A group of code or numbers, which can be moved through a process in a set period of time.

Time (temporal) sampling: An amount of samples taken of values over a period of time.

Timing: A period in time when data is synchronised together.

Toroid: A device composed of a circular ferrous ring, having a copper coil wrapped around parts of it, to allow magnetic field interaction, often used as an electrical transformer or inductor.

Top loader: A washing machine that has clothes loaded into it from the top (see also front loader).

Track: The pathway taken by any object which is travelling; an electrical conducting strip running along a printed circuit board; or a strip of magnetised dots which run along a length of magnetic tape.

Trajectory: The direction of any object as it travels along a path or track.

Transfer function: The formula or gain applied to a signal that is passing through an amplifier or feedback loop of a closed circuit.

Transformer: An electrical device that takes an input signal and changes it to give a different signal at its output.

Transistors: Semiconductor device used in electronics that is used to amplify or switch signals and power.
Transmit: To pass an electrical signal from one point to another, via radio or TV broadcast.
Transmission: A line or stream of signals being transmitted.
Triangulation: The geometrical process of determining a position using three known points (*tri-* being Greek and Latin for three).
Tumble dryer: An appliance used to dry clothes by rotating a drum, thereby allowing the clothes within it to tumble through air and receive maximum exposure to air.
Tune: A change in frequencies of sound or signal that allows a process to operate more efficiently.
Turbine (turbo): A device that produces continuous power in which a rotor fitted with vanes revolves by the force of a fast-moving fluid.
Ultra-High Definition (UHD): The highest resolution digital TV monitor that has a typical pixel arrangement of 3840 pixels across × 2160 vertically.
Ultrasound: The sound frequency range above the highest audible (hearing) limit of 20 KHz and up to GigaHertz (GHz), and is used to detect objects or fractures in materials.
Ultraviolet (UV): The electromagnetic frequencies in the range of 8×10^{14} to 3×10^{16} Hz, which are between visible light and X-rays, and are used to remove microbiological contamination from water as well as detecting oil in materials such as rock (which glows).
Umpire: The term used for the referee (judge or top official) in cricket, baseball, tennis and other games (a shortened term from French *noumpere* meaning 'no-equal' or 'no-side', which was a requirement when English cricketers played French cricketers in the 1800s).
Universal Product Code (UPC): A bar code used for tracking items, written as a series of 2D wide and narrow lines representing numbers that a computer code reader uses to compare with its database.
Universal Serial Bus (USB): A computer industry standard for connecting peripheral devices such as a memory stick (aka 'thumb-drive', which is a single integrated circuit with a connection) at a computer input.
Up-load: The act of passing data from one point to another (see also 'download'), such as passing data from a computer to a USB memory stick.
Vacuum: An appliance that uses suction to remove dirt and other bits (evacuating them) from surfaces; also a space devoid of matter or lack of gaseous pressure.
Valve: A device that is mechanical or electrical, which regulates or controls the flow of fluid or current by opening various passageways (derived from Latin 'valva' – the moving part of a door).
Vapour: A substance in its gas phase at a temperature lower than critical point (above which it changes phase into a liquid).
Variable: In an equation, this is a function which can vary, for example in an equation with a term 2x, the variable is number 2 with x as the fixed quantity.

Variance: In statistics, the variance is the squared deviation of a variable from its mean value, which is how far a set of numbers is spread out from their average trend.

Velocity: The speed at which something is travelling.

Vector: The direction and magnitude of a line, e.g. along a trajectory that an object has travelled.

Ventilation grill: A grid that allows air to travel through to provide air or exhaust moisture.

Vibrate (-ion): The mechanical motion of oscillating around an equilibrium point.

Video-Assisted Referee (VAR): The use of a person to referee or umpire a game using a video display of the game, which provides vision from all angles and which may give the VAR referee a better positional view of an infringement than can be seen by the on-field official.

Virtual reality (VR): A device which when worn over the eyes, gives the wearer a vision in 3D which has the appearance of reality, but in fact is not real but virtually real (see also augmented reality AR)

Virtual screen: A screen seen in 3D wearing VR glasses that can display data, a keyboard or other controllable objects.

Voice: The audible sound made when air passes through the larynx (aka *voice box*) vibrating the vocal cords to produce vibrating air and hence, sound out of the mouth.

Volume: Area × height of an object or space; also the amplitude of an electrical signal or sound.

Vortex: The flow of a fluid in a circular manner, such as in water whirlpools.

VR head-set: The device that can be put on the head covering the eyes containing two screens which are computer-controlled to display images in 3D.

Wafer: A thin layer that can be stacked together to form a sandwich or composite matrix of layers.

Washing machine: An appliance that automatically flushes water through clothing in order to wash it and remove dirt and other greasy particles in the closing.

Watson: The IBM computer programmed to have a large database, so that anyone asking it a technical question can be provided with the technical answer from its database.

Wavelength: The distance from peak to peak or trough to trough along a steady wave.

Webinar: A seminar presented by video link using the internet.

Wheatstone bridge: An electrical circuit developed by Charles Wheatstone in 1843, which is used to measure an unknown electrical value by using three known resistors and determining the fourth resistance value (referred to as a *bridge* because it balances values from two sides as current flows through it).

Wiggle trace: The ink line drawn over a time, as a varying voltage moves a pen in a variable manner along a direction at right angles to and along that of a moving paper.

Wind: The rapid change of air pressure as it passes by objects.

Wireless Fidelity (*Wifi*): An electronic method to transmit and receive radio frequency signals over short distances, developed by the Commonwealth Scientific & Industrial Research Organisation (CSIRO), the government research body of Australia.

Women's Tennis Association (WTA): The tennis association regulating women's tennis around the world.

Word: A group of computer words that are recognised by the computer as having a value e.g. '101' in binary represents 5.

Word pattern: The format or layout of a series of computer words or symbols which can be recognised by other computers.

X-Rays: Electromagnetic waves that radiate over the frequency band of 10^{16} to 10^{20} Hz being positioned between UV and Gamma rays. Used in radiography to see images of bones, CT scanning and angiography to produce images of blood vessels.

Zirconium (Zr): The dense chemical element (atomic number 40) is a strong metal similar to titanium and is used among other attributes as an alloying agent for its strong resistance to corrosion.

Index

Italicised and **bold** pages refer to figures and tables, respectively

G

H

I

K

L

Q

R

W

X

Z

Printed in the United States
by Baker & Taylor Publisher Services